NEW FRONTIERS IN NANOCHEMISTRY

Concepts, Theories, and Trends

Volume 3
Sustainable Nanochemistry

NEW FRONTIERS IN NANOCHEMISTRY

Concepts, Theories, and Trends

Volume 3
Sustainable Nanochemistry

Edited by

Mihai V. Putz, PhD, Dr.-Habil., MBA

Professor, Faculty of Chemistry, Biology, Geography,
Laboratory of Computational and Structural Physical Chemistry for
Nanosciences and QSAR, West University of Timișoara, Romania

PI-1, Laboratory of Renewable Energies – Photovoltaics,
National Research and Development Institute for Electrochemistry
and Condensed Matter (INCEM), Timișoara, Romania

APPLE ACADEMIC PRESS

Apple Academic Press Inc.
4164 Lakeshore Road
Burlington ON L7L 1A4
Canada

Apple Academic Press Inc.
1265 Goldenrod Circle NE
Palm Bay, Florida 32905
USA

First issued in paperback 2021

Exclusive worldwide distribution by CRC Press, a member of Taylor & Francis Group

No claim to original U.S. Government works

New Frontiers in Nanochemistry: Concepts, Theories, and Trends, Volume 3: Sustainable Nanochemistry
ISBN 13: 978-1-77463-176-8 (pbk)
ISBN 13: 978-1-77188-779-3 (hbk)

New Frontiers in Nanochemistry, 3-volume set
International Standard Book Number-13: 978-1-77188-780-9 (Hardcover)
International Standard Book Number-13: 978-0-36707-782-2 (eBook)

Library and Archives Canada Cataloguing in Publication

Title: New frontiers in nanochemistry : concepts, theories, and trends / edited by Mihai V. Putz, PhD, Dr.-Habil., MBA.

Names: Putz, Mihai V., editor.

Description: Includes bibliographical references and indexes. | Contents: Volume 1. Structural nanochemistry -- Volume 2. Topological nanochemistry -- Volume 3. Sustainable nanochemistry.

Identifiers: Canadiana (print) 20190106808 | Canadiana (ebook) 20190106883 | ISBN 9781771887779 (v. 1 ; hardcover) | ISBN 9781771887786 (v. 2 ; hardcover) | ISBN 9781771887793 (v. 3 ; hardcover) | ISBN 9780429022937 (v. 1 ; ebook) | ISBN 9780429022944 (v. 2 ; ebook) | ISBN 9780429022951 (v. 3 ; ebook)

Subjects: LCSH: Nanostructured materials. | LCSH: Nanochemistry.

Classification: LCC TA418.9.N35 N49 2019 | DDC 541/.2—dc23

Library of Congress Cataloging-in-Publication Data

CIP data on file with US Library of Congress

About the Editor

Mihai V. Putz, PhD, Dr.-Habil., MBA

Professor, Laboratory of Computational & Structural Physical Chemistry for Nanosciences & QSAR, West University of Timișoara; Laboratory of Renewable Energies-Photovoltaics, National Research & Development Institute for Electrochemistry & Condensed Matter (INCEM), Romania

Mihai V. Putz, PhD, MBA, Dr.-Habil, is a laureate in physics (1997), with a post-graduation degree in spectroscopy (1999), and a PhD degree in chemistry (2002). He bears with many postdoctoral stages: in chemistry (2002–2003) and in physics (2004, 2010, 2011) at the University of Calabria, Italy, and Free University of Berlin, Germany, respectively. He is currently a Full Professor of theoretical and computational physical chemistry at his alma mater, West University of Timisoara, Romania.

He has made valuable contributions in computational, quantum, and physical chemistry through seminal works that appeared in many international journals. He is an Editor-in-Chief of the *International Journal of Chemical Modeling* and *New Frontiers in Chemistry*. In addition, he has authored and edited many books. He is a member of many professional societies and has received several national and international awards from the Romanian National Authority of Scientific Research (2008), the German Academic Exchange Service DAAD (2000, 2004, 2011), and the Center of International Cooperation of Free University Berlin (2010). He is the leader of the Laboratory of Computational and Structural Physical Chemistry for Nanosciences and QSAR at the Biology-Chemistry Department of West University of Timisoara, Romania, where he conducts research into the fundamental and applicative fields of quantum physical chemistry and QSAR. He is also a PI-1 at the Laboratory of Renewable Energies – Photovoltaics, National Research and Development Institute for Electrochemistry and Condensed Matter (INCEM), Timișoara, Romania. Among his numerous awards, in 2010 Mihai V. Putz was declared, through a national competition, the Best Researcher of Romania, while in 2013 he was recognized among the first Dr.-Habil. in Chemistry in Romania. In 2013, he was appointed

Scientific Director of newly founded Laboratory of Structural and Computational Physical Chemistry for Nanosciences and QSAR at his alma mater university. In 2014, he was recognized by the Romanian Ministry of Research as Principal Investigator of the first degree at the National Institute for Electrochemistry and Condensed Matter (INCEMC), Timisoara, and was also granted full membership in the International Academy of Mathematical Chemistry.

Recently, Mihai V. Putz expanded his interest to strategic management in general and to nanosciences and nanotechnology strategic management in particular. In this context, between 2015–2017, he attended and finished as the promotion leader of the MBA in Strategic Management of Organizations – The Development of the Business Space specialization program at the West University of Timişoara, the Faculty of Economics and Business Administration. While from 2016, he was engaged in the doctoral school of the same faculty, advancing new models of strategic management in the new economy based on frontier scientific inclusive ecological knowledge.

Contents

Contributors

Sorana D. Bolboacă
Iuliu Haţieganu University of Medicine and Pharmacy Cluj-Napoca

Bogdan Bumbăcilă
Laboratory of Computational and Structural Physical Chemistry for Nanosciences and QSAR,
Biology-Chemistry Department, Faculty of Chemistry, Biology, Geography at West University of
Timişoara, Pestalozzi Street No. 16, Timişoara, RO–300115, Romania

Sergiu A. Chicu
Institute of Chemistry Timisoara of the Romanian Academy, Mihai Viteazul Boulevard 24, RO–300223
Timişoara, Romania. Permanent address: Siegstr. 4, 50859 Koeln, Germany

Nicoleta A. Dudaş
Laboratory of Computational and Structural Physical Chemistry for Nanosciences and QSAR,
Biology-Chemistry Department, West University of Timişoara, Pestalozzi Str. No. 16,
Timişoara 300115, Romania

Corina Duda-Seiman
Laboratory of Computational and Structural Physical Chemistry for Nanosciences and QSAR,
Biology-Chemistry Department, Faculty of Chemistry, Biology, Geography at West University of
Timişoara, Pestalozzi Street No. 16, Timişoara, RO–300115, Romania

Daniel Duda-Seiman
Department of Medical Ambulatory, and Medical Emergencies, University of Medicine and Pharmacy
"Victor Babes," Avenue C. D. Loga No. 49, RO–300020 Timisoara, Romania

Harish Dureja
Department of Pharmaceutical Sciences, M.D. University, Rohtak–124001, India

Rohit Dutt
School of Medical and Allied Sciences, GD Goenka University, Gurugram–122103, India

Mirela I. Iorga
National Institute for Research and Development in Electrochemistry and Condensed Matter,
Department of Applied Electrochemistry, 300569 Timisoara, Romania

Adriana Isvoran
Department of Biology-Chemistry, West University of Timisoara, 16 Pestalozzi, Timisoara (Romania) |
Advanced Environmental Research Laboratory, 4 Oituz, Timisoara, Romania,
E-mail: adriana.isvoran@e-wut.com

Lorentz Jäntschi
Technical University of Cluj-Napoca, Romania

Naveen Khatri
Faculty of Pharmaceutical Sciences, Pt. B.D. Sharma University of Health Sciences, Rohtak–124001,
India

A. K. Madan
Faculty of Pharmaceutical Sciences, Pt. B.D. Sharma University of Health Sciences, Rohtak–124001,
India, Tel.: +91 98963 46211, E-mail: madan_ak@yahoo.com

Sergio Madurga
Department of Materials Science & Physical Chemistry and Research Institute of Theoretical and Computational Chemistry (IQTCUB) of Barcelona University, C/ Martí i Franquès, 1, 08028-Barcelona, Spain, E-mail: s.madurga@ub.edu

Francesc Mas
Department of Materials Science & Physical Chemistry and Research Institute of Theoretical and Computational Chemistry (IQTCUB) of Barcelona University, C/ Martí i Franquès, 1, 08028-Barcelona, Spain

Ana-Matea Mikecin
Rudjer Boskovic Institute, Division of Molecular Medicine, Bijenicka 54, 10000 Zagreb, Croatia, E-mail: Ana-Matea.Mikecin@irb.hr

Grdisa Mira
Rudjer Boskovic Institute, 10000 Zagreb, Croatia, E-mail: grdisa@irb.hr

Marius C. Mirica
National Institute for Research and Development in Electrochemistry and Condensed Matter, Department of Applied Electrochemistry, 300569 Timisoara, Romania

Isabel Pastor
Small Biophysics Lab of Condensed Matter Department, Barcelona University, C/ Martí i Franquès, 1, 08028-Barcelona (Spain) and Carlos III Health Institute, Madrid, Spain

Laura Pitulice
Department of Biology-Chemistry, West University of Timisoara, Pestalozzi Street No. 16, 44, Timisoara, RO–300115, Romania | Advanced Environmental Research Laboratory, 4 Oituz, Timisoara, Romania

Ana-Maria Putz
Institute of Chemistry Timisoara of the Romanian Academy, 24 Mihai Viteazul Bld., Timisoara 300223, Romania

Mihai V. Putz
Laboratory of Computational and Structural Physical Chemistry for Nanosciences and QSAR, Biology-Chemistry Department, Faculty of Chemistry, Biology, Geography at West University of Timişoara, Pestalozzi Street No. 16, Timişoara, RO–300115, Romania | Laboratory of Renewable Energies-Photovoltaics, R&D National Institute for Electrochemistry and Condensed Matter, Dr. A. Paunescu Podeanu Str. No. 16, 144, RO–300569 Timişoara, Romania, Tel.: +40-256-592-638, Fax: +40-256-592-620, E-mails: mv_putz@yahoo.com, mihai.putz@e-uvt.ro

James Garnett Sawyer II
241 Lexington Avenue, Buffalo, NY 14222, USA

Marina A. Tudoran
Laboratory of Structural and Computational Physical Chemistry for Nanosciences and QSAR, Biology–Chemistry Department, West University of Timisoara, Pestalozzi Street No. 44, Timisoara, RO–300115, Romania | Laboratory of Renewable Energies-Photovoltaics, R&D National Institute for Electrochemistry and Condensed Matter, Dr. A. Paunescu Podeanu Str. No. 144, Timisoara, RO–300569, Romania

Lemi Türker
Department of Chemistry, Middle East Technical University, 06800, Ankara, Turkey

Serhat Variş
Department of Chemistry, Middle East Technical University, 06800, Ankara, Turkey

Eudald Vilaseca
Department of Materials Science & Physical Chemistry and Research Institute of
Theoretical and Computational Chemistry (IQTCUB) of Barcelona University,
C/ Martí i Franquès, 1, 08028-Barcelona, Spain, E-mail: eudald.vilaseca@ub.edu

Marla Jeanne Wagner
407 Norwood Avenue, Buffalo, NY 14222, USA

Abbreviations

(Q)SAR	(quantitative) structure-activity relationship
3D	three-dimensional
3MP5O	3-methyl-pyrazole-5-one
6D	six-dimensional
A.D.	Anno Domini (after the birth of Jesus)
ACA	anticancer activity
ACEI	angiotensin-converting enzyme inhibitor
AIDS	acquired immune deficiency syndrome
AL	allene-like
AMO	atomic mass one
ANCSI	National Authority for Scientific Research and Innovation
ART	anti-retroviral therapy
ASA	accessible solvent area
ASA	accessible surface area
B.C.	before Christ (before the birth of Jesus)
B.C.E	before the common era
B.P	before present
BAS	bioactive substrate
BESs	bioelectrochemical systems
BH	bulk heterojunction
BOS	balance of system
BPDE	benzo[a]pyrene-7,8-diol-9,10-epoxide
C.E.	common era
CA	cancer
CA	carbonic anhydrase
cART	combination anti-retroviral therapy
CB-1	cannabinoid receptor-1
CB-2	cannabinoid receptor-2
CBRI	Chester Beatty Research Institute
CC	cell cycle
CDK	cyclin-dependent kinases
CDKI	cyclin-dependent kinases inhibitors
CHORI	Children's Hospital Oakland Research Institute
CKI	CDK inhibitors

CLRNT	colorant
CNTs	carbon nanotubes
COG	clusters of orthologous groups
COX–2	cycloxygenase–2
CS	contact surface
CSC	cancer stem cell
CTAB	cetyl trimethyl ammonium bromide
CVC	critical vesicle concentration
DCCMAM	dual coordinate cuboctahedron model of atomic mass
DCIJP	direct ceramic ink-jet printing
DDS	drug delivery systems
DHFR	dihydrofolate reductase
DMPA	2,2-dimethoxy–2-phenylacetophenone
DNA	deoxyribonucleic acid
DPP-4	dipeptidyl-peptidase IV
DSSC	dye-sensitized solar cells
DT	decision tree
EA	ellagic acid
EAC	Ehrlich Ascites carcinoma
ED	electron diffraction
EM	electromagnetic
EQE	external quantum efficiency
ESC	embryonic stem cells
ESCK	elastase substrate catalyzed kinetics
FG	frustrated geometry
FG2PV	fullerene-based double-graphene photovoltaics
GAMESS	general atomic and molecular electronic structure system
GARFT	glycinamide ribonucleotide formyltransferase
GIP	glucose-dependent insulinotropic peptide
GLN	amino acid glutamine
GLP-1	glucagon-like peptide–1
GPCRs	G-protein-coupled receptors
GPLS	genetic partial least squares
GR	general relativity
GROMACS	Groningen machine for chemical simulations
GSS	genotypic sensitivity scores
HAART	highly active antiretroviral therapy
HIV	human immunodeficiency virus
HPM	high-performance multi-crystalline
hSC	human stem cell

IADD	*in silico*-aided drug design
ICR	Institute of Cancer Research
INN	immune neural network
iPSC	induced pluripotent stem cell
IVR	intramolecular vibration redistribution
KOG	eukaryotic orthologous groups
LCD	liquid crystal display
LFER	linear free energy relationship
LILT	low-intensity-low-illumination
LTR	long terminal repeats
MAA	moving average analysis
MC	macromolecular crowding
mc-Si	multi-crystalline silicon
MD	molecular dynamic
MDs	molecular descriptors
MF	modified Fenton
MFD	molecular fragment dynamics
MNDO	modified neglect of diatomic overlap
MRI	magnetic resonance imaging
MS	molecular surface
MSD	minimal steric difference
MTD	minimal topological difference method
NDAs	new drug applications
NG	nobel gases
NIR	near infra-red
NOCT	nominal operating cell temperature
NRTI	nucleoside-like reverse transcriptase inhibitor
OFML	optofluidic maskless lithography
OH	octahedral
OPLS-AA	all-atom optimized potentials for liquid simulations
P3HT	poly(3-hexylthiophene)
PAH	polycyclic aromatic hydrocarbon
PbS QDs	lead sulfide quantum dots
PCE	power conversion efficiency
PE	polyethylene
PEG	polyethylene glycol
PEG-DA	polyethylene glycol diacrylate
PETT	phenethylthiazolethiourea
PEVA	poly(ethylene-vinyl acetate)
PIs	protease inhibitors

PPAR	peroxisome proliferator-activated receptors
PPARG	PPAR gamma
PPRE	peroxisome proliferator hormone responsive elements
PTE	periodic table of the elements
PV	photovoltaic
PV/T	photovoltaic-thermal
PVC	polyvinyl chloride
PVP	polyvinyl pyrrolidone
QCARs	quantitative cationic–activity relationships
QDSC	quantum dot solar cell
QFPRs/QFARs	quantitative features—property/activity relationships
QG	quantum gravity
QSAR	quantitative structure-activity relationships
QSPR/QSAR	quantitative structure-property/activity relationships
QSSA	quasi-steady-state approximation
QWs	quantum wells
RAL	raltegravir
RAL-GLU	RAL glucuronide
RAM	radar absorbing materials
REACH	registration, evaluation, and authorization of chemicals
RF	random forest
RNA	ribonucleic acid
RXR	retinoid receptor X
SC	stem cell
SFL	stop flow lithography
SGLT2	sodium/glucose cotransporter 2
Si QD	silicon quantum dot
SMIfp	SMILES fingerprint
SMOSC	small molecule organic solar cell
SP	square-planar
SPR	surface plasmon resonance
SPS	solar power satellite
STP	standard temperature and pressure
SVM	support vector machine
SWNTs	single-wall carbon nanotubes
TB	resistant tuberculosis
TBT	tributyltin
TF	term frequency
TF-IDF	term frequency-inverse document frequency
TH	tetrahedral

TIs	topological indices
TOF	turn over frequency
TS	thymidylate synthetase
TWI	terminal Wiener index
UGT1A1	UDP-glucuronosyltransferase 1A1
UV	ultraviolet
vdWS	van der Waals surface

General Preface to Volumes 1–3

The nanosciences, just born on the dawn of the 21st century, widely require dictionaries, encyclopedias, handbooks to fulfill its consecration, and development in academia, research, and industry.

Therefore, this present editorial endeavor springs from the continuous demand on the international market of scientific publications for having a condensed yet explicative dictionary of the basic and advanced up-to-date concepts in nanochemistry. It is viewed as a combination (complementing and overlapping) on various notions, concepts, tools, and algorithms from physical-, quantum-, theoretical-, mathematical-, and even biological-chemistry. The definitions given in the integrated volumes are accompanied by essential examples, applications, open issues, and even historical notes on the persons/subjects, with relevant literature and scholarly contributions.

The current mission is to prepare a premiere referential work for graduate students, PhDs and post-docs, researchers in academia, and industry dealing with nanosciences and nanotechnology. From the book format perspective, the volumes are imagined as practical and attractive as possible with about 130 essential terms as coming from about 60 active scholars from all parts of the globe, explaining each entry with a minimum of five pages (viewed as scientific letters/essays or short course about it), containing definitions, short characterizations, uses, usefulness, limitations, etc., and references – while spanning more than 1,600 pages in the present edition. This effort resulted in this unique and up-to-date *New Frontiers in Nanochemistry: Concepts, Principles, and Trends*, a must for any respected university and the individual updated library! It will also support the future more than necessary decade-editions with new additions in both entries and contributors worldwide!

On the other side, the broad expertise of the editor – in the non-limitative fields of quantum physical-chemical theory, nanosciences and quantum roots of chemistry, computational and theoretical chemistry, quantum modeling of chemical bonding, atomic periodicity and scales of chemical indices, molecular reactivity by chemical indices, electronegativity theory of atoms in molecule, density functionals of electronegativity, conceptual density functional theory, graphene and topological defects of graphenic ribbons, quantitative structure-activity/property relationships (QSAR/QSPR), effector-receptor complex interaction and quantum/logistic enzyme kinetics, and many other related scientific branches – assures that both the educated and generally interested reader in science and technology will be benefited from the book in many ways. They may be listed as:

- an introduction to general nanochemistry;
- an introduction to nanosciences;
- a resource for fast clarification of the basic and modern concepts in multidisciplinary chemistry;
- inspiration for a new application and transdisciplinary connection;
- an advocate for unity in natural phenomena at nano-scales, merging between mathematical, physics and biology towards nanochemistry;
- a reference for both academia (for lessons, essays, exams) as well in R&D industrial sectors in planning and projecting new materials, with aimed properties in both structure and reactivity;
- a compact yet explicated collection of updated scientific knowledge for general, university, and personal libraries; and
- a self-contained book for personal, academic, and technological instruction and research.

Accordingly, the specific aims of the present book are:

- to be a concise and updated source of information on the nanochemistry fields;
- to comprehensively cover nanosystems: from quantum theory to atoms, to molecules, to complexes and chemical materials;
- to be a necessary resource for every-day use by students, academics, and researchers;
- to present informative and innovative contents alike, presented in a systematical and alphabetically manner;
- to present not only definitions but consistent explicative essays on nanochemistry in a short essay or paper/chapter format; and
- to be written by leading and active experts and contributors in nanochemistry worldwide.

All in all, the book is aimed to give supporting material for relevant multi- and trans-disciplinary courses and disciplines specific to nanosciences and nanotechnology in all the major universities in Europe, the Americas, and Asia-Pacific, among which nanochemistry is the core for the privileged position in between physics (elementary properties of quantum particles) and biology (manifested properties of bodies by the environment/forces/ substances influences). They may be non-exclusively listed as:

- *Nanomaterials: Chemistry and Physics*
- *Introduction to Nanosciences*
- *Bottom-Up Technology: Nanochemistry*
- *Nanoengineering: Chemistry and Physics*
- *Sustainable Nanosystems*
- *Ecotoxicology*
- *Environmental Chemistry*
- *Quantum Chemistry*
- *Structural Chemistry*
- *Physical Chemistry–Chemical Physics*

With the belief that, at least partly, this present editorial endeavor succeeds in the above mission and purposes. The editor of this publishing event heartily thank all the contributors for their dedication, inspiration, insight, and generosity, along with the truly friendly and constructive Publisher, Apple Academic Press (jointly with CRC Press at Taylor & Francis), and especially to the President, Mr. Ashish Kumar, and the rest of the AAP team.

Special Preface to Volume 3: Sustainable Nanochemistry

The present final volume of *New Frontiers in Nanochemistry: Concepts, Theories, and Trends* addresses the *necessity of "resolving" the knowledge limitations with a social and economic impact, which are the main problems of humankind in the second decennium of the 21st century*, namely, as the document "Horizon 2020" states, specifically:

- The "Safe, Ecological and Efficient Energy" problem;
- The "Longer and Healthier Life" problem.

For this reason, sustainable nanochemistry is referred to as those contributions resolving the research problems that succeed in *repositioning, from the fundamental perspective, the matter structure and interaction* (photons, electrons, atoms, molecules, biomolecules); however, the approach should give a "salt" rather than merely a "pulse" of quality in life comfort and in exploring the resources of nature for a sustainable future.

For instance, one can describe the chemical–biological interaction and toxicity by the topological and algebraic models as given through the Spectral-SAR dictionary voice, for instance, among other related dictionary entries. Having as the final objective the designing of the new drugs with specific action, from active substances or pharmacophore and generic substances synthesized as a result of some topological projections, various algorithms were developed for better understanding and controlling the mechanism of binding action of ligand: the chemical substance, toxicant, respectively the "target" structure, meaning the structure that is chosen to be structural optimized by the allosteric interaction mode, binds with the receptor (the biologic organism sites, at the cellular level, which can be biomolecules of enzymic type, metabolic activators or an inhibitor. Toxicity is this way characterized by the type of bonding mechanism identified, the innovative algorithms correlating the ligand-receptor interaction, and the substrate-enzyme affinity, by reformulating the problem of quantitative structure-activity (biological), QSAR, for instance by the algebraic approach with Spectral-SAR variant.

Ultimately, one may consider the "semi-molecules" with simple conjugate bonds broken in such manner that can be able to form molecular chains with primary and/or secondary branches, more adapted to the one similar with "lock-and-key" bond mechanism in accord with the Fisherian principle of the drug; this way it was made the essential step in bringing from virtual a new molecule considered only topo-computational "decomposed" (*SMILES— Simplified Molecular-Input Line-Entry System type*) on the level of "real" conceptual-interaction mechanism and bonding by lipo-cellular transduction under this fragmentary form (see the SMILES entry in this volume).

Other promising QSAR and 3D-QSAR with high prediction capacity are focused on toxicological potential (high in anti-HIV composition and for any other processes of cellular apoptosis in different degenerative diseases, as in arteriosclerosis, Alzheimer type, etc.), contributing to the so-called functional medicine by the proposed pharmacotoxicology and pharmacody-namics, conceptual-computational but also with synthesis perspectives of a pharmaceutics laboratory.

In the same spirit, the other included voices of this third and final volume of the multi-volume package dedicated to *New Frontiers in Nanochemistry: Concepts, Theories, and Trends* unfolds *the topological* (so applicative) themes relevant to it, accordingly with the specifically sustainable nano-research (see the first paragraph). They all came from eminent international scientists and scholars and span the A-to-V sustainable nanochemistry (toxicology and renewable energy) entries such as: anti-HIV and anti-carcinogenic agents; drug development; computational bio-, eco-, pharmacology; ecotoxicology; (logistic) enzyme kinetics; "digitalized" structure-activity relationships; sterility and chemical modeling; solar cells and photovoltaic phenomena; molecular crowding and molecular (van der Waals) interaction; among many more.

This third volume of the 3-volume *New Frontiers in Nanochemistry: Concepts, Theories, and Trends* contains about 34 chapters from contributors coming from three continents (Europe, America, and Asia), and from six countries (USA, India, Turkey, Spain, Germany, and Romania) in multiple first-rate explicative dictionary voices. Let's hope they will be heard worldwide and will positively (in an ecolo-progressist manner) influence the 21st-century macro-destiny of the Earth from the nano-sustainable perspective!

Fugit irreparabile tempus!

Heartily yours,
—**Mihai V. Putz**
(Timisoara, Romania)

CHAPTER 1

Anti-HIV Agents

BOGDAN BUMBĂCILĂ[1] and PUTZ[1,2]

[1]Laboratory of Computational and Structural Physical Chemistry for Nanosciences and QSAR, Biology-Chemistry Department, Faculty of Chemistry, Biology, Geography at West University of Timişoara, Pestalozzi Street No. 16, Timişoara, RO–300115, Romania

[2]Laboratory of Renewable Energies-Photovoltaics, R&D National Institute for Electrochemistry and Condensed Matter, Dr. A. Paunescu Podeanu Str. No. 144, RO–300569 Timişoara, Romania, Tel.: +40-256-592-638, Fax: +40-256-592-620, E-mail: mv_putz@yahoo.com; mihai.putz@e-uvt.ro

1.1 DEFINITION

Human immunodeficiency virus infection (HIV) and acquired immune deficiency syndrome (AIDS) are conditions caused by the HIV, a virus which interferes with the immune system of the host causing in time opportunistic infections and development of cancers which rarely happen to people with competent immune systems. The late symptoms of the infections are severe and are referred together as "syndrome" – AIDS (Moore et al., 1999).

There is no cure for the infection or vaccine for its prophylaxis. Still, antiretroviral pharmacological treatment can slow the course of the disease and may lead to a near-normal life expectancy. New research is indicating that treatment should be recommended as soon as the diagnosis is made. Without a treatment, the average survival time after infection is 11 years (Deeks, 2013; WHO 2015).

The classes of therapeutical agents are:

- Entry inhibitors (fusion inhibitors) – EI.
- Reverse-transcriptase inhibitors (nucleoside analogs – NRTI, non-nucleoside analogs – NNRTI, and nucleotide analogs – NtRTI).

- Integrase inhibitors – INSTIs.
- Protease inhibitors – PI.
- Maturation inhibitors – MI.

These drug classes are usually used in combination, and their combinations are called antiretroviral therapy (ART), combination antiretroviral therapy (cART) or highly active antiretroviral therapy (HAART). Today, typical combinations include two NRTIs and one NNRTI/PI/INSTI (Moore et al., 1999; U.S. Department of Health and Human Services, 2015).

1.2 ENTRY INHIBITORS

Entry inhibitors, also known as fusion inhibitors, are a class of antiretroviral drugs. They interfere with the binding, fusion, and entry of an HIV virion to a human cell.

There are several proteins involved with the HIV virion entry process. They are:

- CD4 – a protein receptor found on the surface of the helper *T* Cells of the human immune system.
- gp120 – an HIV protein that binds to the CD4 receptor.
- CXCR4 and CCR5 – chemokine co-receptors found on the surface of the Helper *T* Cells and macrophages.
- gp41 – an HIV protein that penetrates the host cell's membrane.

HIV enters into a cell by following a few steps:

- HIV gp120 is binding to the CD4 receptor.
- HIV gp120 suffers a conformational change, which increases its affinity for a co-receptor and exposes HIV gp41.
- HIV gp120 binds to a co-receptor – either CCR5 or CXCR4.
- HIV gp41 penetrates the HIV lipid membrane and the T-cell membrane.
- The viral core – the capsid is entering into the cell.

1.2.1 CCR5 CO-RECEPTOR ANTAGONISTS

HIV enters in the cell by attaching its lipid membrane gp120 (glycoprotein 120) to the CD4 receptor on the host. A conformational change in gp120 occurs, therefore, allowing it to bind also to co-receptor CCR5 expressed on the host cell. The co-attachment is triggering the expression of gp41, which creates a bridge between the viral envelope and the cell membrane. The process is called fusion, because a gap in the host cell membrane is formed, allowing the viral nucleocapsid to enter into the cell. By blocking the CCR5 co-receptor, the expression of the gp41 is not possible so that the viral envelope will not establish the contact with the membrane.

1. Maraviroc

Maraviroc is the only one approved today for therapy.

2. Aplaviroc

Aplaviroc is proved to be toxic for the liver.

3. *Vicriviroc*

Vicriviroc is proved to be not quite effective in the clinical trials, so it was abandoned.

4. *Cenicriviroc*

Cenicriviroc is currently in clinical studies.

1.2.2 FUSION INHIBITORS

1.2.2.1 ENFUVIRTIDE

It is extremely expensive, unstable in aqueous solution, and it is administered as a subcutaneous injection twice daily.

Because the patient has to prepare the solution with the lyophilized powder before its administration and because of its costs, enfuvirtide is maintained as an alternative when other anti-HIV agents failed. It is an oligopeptide, and its formula is:

Acetyl-Tyr-Thr-Ser-Leu-Ile-His-Ser-Leu-Ile-Glu-Glu-Ser-Gln-
Asn-Gln-Gln-Glu-Lys-Asn-Glu-Gln-Glu-Leu-Leu-Glu-Leu-
Asp-Lys-Trp-Ala-Ser-Leu-Trp-Asn-Trp-Phe-NH$_2$

It binds to gp41 and changes its conformation; thus, it prevents the formation of the entry pore in the host cell membrane with the entry of the viral nucleocapsid. (Lalezari et al., 2003)

1.2.3 GP120 ANTAGONISTS

They are blocking the gp120; thus, they prevent the attachment of this viral protein with the CD4 receptor on the host cell membrane.

1. *Fostemsavir*

Because gp120 is a viral protein and very well conserved, this drug offers promises to those patients who developed resistance to other therapies, because it is very unlikely to promote independent of CD4-binding viral particles.

1.3 REVERSE TRANSCRIPTASE INHIBITORS

1.3.1 NUCLEOSIDE ANALOGS

*   The first class of antiretroviral drugs introduced in therapy.
*   In order to manifest their effect, their molecules have to be incorporated into the viral DNA; they are activated, therefore, in the cell, by the addition with the help of cellular kinase enzymes of three phosphate groups to their deoxyribose part, to form triphosphates.

- The molecules compete with natural occurring nucleosides to prevent their incorporation into the viral DNA, and thus the complete reverse transcription is prevented.
- They practically inhibit the RNA-dependant AND-polymerase.

1.   *Zidovudine/Azidothymidine*

- Thymidine analog.
- Slows replication of HIV but does not stop it entirely.
- It is mostly used to prevent HIV transmission, such as from mother to child during birth or after a needle stick injury.
- For treatment, it is always used combined with other drugs because HIV can become resistant to it.

2.   *Didanosine*
- Adenosine analog.

3. *Zalcitabine*

- Pyrimidine/deoxycytidine analog.
- Less potent than other NRTIs.
- It has a very low half-life, so it has to be frequently administered–3 times a day.

4. *Stavudine*

- Thymidine analog;
- It has long-term and irreversible side-effects, so because of them today is less frequently used.

5. *Lamivudine*
   - Cytidine analog.

6. *Abacavir*
   - Guanosine analog.

7. *Emtricitabine*
   - Cytidine analog.

8. *Elvucitabine*

9. *Entecavir*
   - Guanosine analog.

10. *Apricitabine*

11. *Amdoxovir*

- Currently in clinical studies.

12. *Festinavir*

- Not yet into clinical studies; it is believed to have an improved safety (Weinberg, 2012).

1.3.2   NUCLEOTIDE ANALOGS

- Phosphorylated nucleoside analogs.

1. Tenofovir Disoproxil

- It prevents the formation of the 5′ to 3′ phosphodiester linkage during the DNA chain elongation reverse-transcripted because the molecule lacks an –OH group on the 3′ carbon.
- Together with emtricitabine, it is now available as TRUVADA, a combination which was demonstrated in 2015 as 100% efficient as pre-exposure prophylaxis to HIV.

1.3.3 NON-NUCLEOSIDE REVERSE-TRANSCRIPTASE INHIBITORS

- Molecules that inhibit directly, non-competitively, the reverse transcriptase enzyme; by binding to the enzyme, they produce a conformational change of its three-dimensional structure.

1. Efavirenz

- The molecule binds to a distinct site, other than the active site, known as the NNRTI pocket.
- It is not active on HIV-2 reverse transcriptase, which has a different structure of the pocket than HIV-1 reverse transcriptase (Ren et al., 2002).

2. *Nevirapine*

- Clinical studies proved that prophylaxis of mother to child transmission during birth with single-dose nevirapine in addition to zidovudine is more effective than zidovudine alone (Lallemant et al., 2004).

3. *Delavirdine*

- Currently rarely used because of its short half-life (thus the high frequency of administration).

4. *Atevirdine*
   - Currently in clinical studies.

5. *Etravirine*
   - Apparently, it does not produce HIV resistance.

6. *Rilpivirine*

- Higher potency, longer half-life, and reduced side-effects than all the other NNRTIs.

1.4 INTEGRASE INHIBITORS

Integrase inhibitors or integrase strand transfer inhibitors are drugs that are blocking the integrase activity, an enzyme that inserts the viral DNA obtained from the activity of reverse transcriptase into the DNA of the host cell.

1.4.1 RALTEGRAVIR

Initially approved for patients resistant to other HAART drugs, it was shown that the treatment with this drug led to undetectable viral loads, sooner than those taking similarly potent NNRTIs or PIs. Today, the studies are directed to the ability of this molecule to eradicate the virus from the latent reservoirs. It was also proved that raltegravir has anti-viral activity on herpes viruses (Savarino, 2006).

1.4.2 ELVITEGRAVIR

- It shares the molecular core with the fluoroquinolone antibiotics.

1.4.3 DOLUTEGRAVIR

- Commercialized under the brand name of Tivicay.
- It is an antiretroviral medicine used along other drugs in the anti-HIV therapy.
- It is also used in post-exposure prophylaxis for preventing HIV installation after the potential exposure.

1.4.4 BI 224436

- Currently in clinical studies.
- It binds to another site of the integrase enzyme; thus, it has a slightly different mechanism of action than raltegravir, dolutegravir, and elvitegravir which are binding to the catalytic site of the enzyme (Levin, 2011).

1.4.5 MK-2048

- Currently in clinical studies developed by Merck & Co.
- It appears to be four times more potent that raltegravir – it inhibits the integrase enzyme four times longer in time than raltegravir.
- It is also investigated as an option of pre-exposure prophylaxis (Mascolini, 2009).

1.4.6   CARBOTEGRAVIR (GSK744)

- It was found that if packed as nanoparticles it has a half-life of 21 to 50 days so, in this formulation, it can be administered as once in three months.
- Currently in clinical studies (Borrell, 2014).

1.5   PROTEASE INHIBITORS

Protein inhibitors are inhibiting the viral replication by blocking the HIV proteases, enzymes that are important for the proteolysis of protein precursors that are necessary for the production of infectious mature viral particles. Because these drugs are highly specific, there is a risk, as in antibiotics, of the development of drug-resistant mutated viral particles. To reduce this risk, it is common to use several drugs together with different mechanisms of action and thus targeting different steps in the HIV replication cycle (Rang, 2007).

1.5.1   SAQUINAVIR

- The first drug in this class approved for therapy.

1.5.2   RITONAVIR

- Today, it is used as a booster for other protease inhibitors.

1.5.3   INDINAVIR

- It interacts with food and beverages, so it has many restrictions in this direction.

1.5.4   NELFINAVIR

1.5.5 AMPRENAVIR AND FOSAMPRENAVIR (PHOSPHATE-ESTER PRO-DRUG)

- Initially used as an antihypertensive agent.

1.5.6 LOPINAVIR

- It is also effective against HPV.

1.5.7   ATAZANAVIR

1.5.8   TIPRANAVIR

- The resistance to this molecule requires too many mutations to take place, so it is an alternative for the resistance to other protease inhibitors (Doyon et al., 2005).

1.5.9 DARUNAVIR

- Its structure allows the molecule to create the highest number of hydrogen bonds with the protease's active site than other protease inhibitors (Ghosh et al., 2007).

1.6 PORTMANTEAU INHIBITORS

Portmanteau inhibitors are molecules designed to have reverse transcriptase and integrase inhibitory activities or, more recently, integrase, and entry (CCR5 blocking) inhibitory activities.

The molecules are lab-designed, having dual-cores, corresponding to the one needed for the reverse transcriptase/entry blocking activity and to that one of the integrase inhibitory activity.

Trends in HIV/AIDS management include the development of formulations of two, three, or more drugs, combined in a single tablet to facilitate the patient's adherence to the treatment. Still, physic-chemical interactions between drugs or incompatibilities between drugs and the excipients make these formulations hard to achieve.

The newest strategy in the designing process of HIV medication is the creation of a single molecule which targets the virus at different levels, thus improving the patient's compliance and reducing the mechanisms of drug resistance.

In 2011, H.S. Bodiwala synthesized portmanteau inhibitors of HIV-1 integrase and CCR5 co-receptor.

R =

8   -CH$_3$

9   -CH(CH$_3$)$_2$

10 -CH$_2$CH(CH$_3$)$_2$

11

12

13

14

15

16

17

18

19

20

21

5,6,7 - intermediates, no proven activity

KEYWORDS

- **acquired immune deficiency syndrome**
- **anti-retroviral therapy**
- **combination antiretroviral therapy**
- **highly active antiretroviral therapy**
- **human immunodeficiency virus**

REFERENCES

Bodiwala, H. S., Sabde, S., Gupta, P., Mukherjee, R., Kumar, R., Garg, P., Bhutani, K. K., Mitra, D., & Singh, I. P., (2011). Design and synthesis of caffeoyl-anilides as portmanteau inhibitors of HIV-1 integrase and CCR5. *Bioorganic & Medicinal Chemistry, 19*(3), 1256–1263.

Borrell, B., (2014). Long-acting shot prevents infection with HIV analog–Periodic injection keeps monkeys virus-free and could confer as long as three months of protection in humans. *Nature, 4* March (doi:10.1038/nature.2014.14819).

Deeks, S. G., (2013). The end of AIDS: HIV infection as a chronic disease. *The Lancet, 382*(9903), 1525–1533.

Doyon, L., Tremblay, S., Bourgon, L., Wardrop, E., & Cordingley, M., (2005). Selection and characterization of HIV-1 showing reduced susceptibility to the non-peptidic protease inhibitor tipranavir. *Antiviral Research, 68*(1), 27–35.

Ghosh, A. K., Dawson, Z. L., & Mitsuya, H., (2007). Darunavir, a conceptually new HIV-1 protease inhibitor for the treatment of drug-resistant HIV. *Bioorganic Medicinal Chemistry, 15*(24), 7576–7580.

Guidelines for the Use of Antiretroviral Agents in HIV-1-Infected Adults and Adolescents, 2015. https://aidsinfo.nih.gov/contentfiles/lvguidelines/adultandadolescentgl.pdf-US Department of Health and Human Services.

Lalezari, J. P., Eron, J. J., Carlson, M., Cohen, C., Dejesus, E., Arduino, R. C., Gallant, J. E., & Volberding, P., (2003). A phase II clinical study of the long-term safety and antiviral activity of enfuvirtide-based antiretroviral therapy. *AIDS, 17*(5), 691–698.

Lallemant, M., Jourdain, G., Le Coeur, S., Mary, J. Y., Ngo-Giang-Huong, N., Koetsawang, S., Kanshana, S., McIntosh, K., & Thaineua, V., (2004). Single-dose perinatal nevirapine plus standard zidovudine to prevent mother-to-child transmission of HIV-1 in Thailand. *The New England Journal of Medicine, 351*, 217–228.

Levin, J., (2011). *BI 224436, a Non-Catalytic Site Integrase Inhibitor, is a Potent Inhibitor of the Replication of Treatment-Naïve and Raltegravir-Resistant Clinical Isolates of HIV-1.* Conference Reports for NATAP. 51th ICAAC Chicago, IL; September 17-20 (http://www.natap.org/2011/ICAAC/ICAAC_34.htm).

Mascolini, M., (2009). Merck Offers Unique Perspective on Second-Generation Integrase Inhibitor. *10th International Workshop on Clinical Pharmacology of HIV Therapy*. Amsterdam April 15-17 (http://www.natap.org/2009/PK/PK_10.htm).

Moore, R. D., & Chaisson, R. E., (1999). Natural history of HIV infection in the era of combination antiretroviral therapy. *AIDS, 13*(14), 1933–1942.

Rang, H. P., Dale, M. M., Ritter, J. M., & Flower, R. J., (2007). *Rang and Dale's Pharmacology* (6th edn.). Churchill Livingstone Elsevier, Philadelphia (eBook ISBN: 9780702040740; pp. 844; https://www.elsevier.com/books/rang-and-dales-pharmacology-e-book/rang/978-0-7020-4074-0).

Ren, J., Bird, L. E., Chamberlain, P. P., Stewart-Jones, G. B., Stuart, D. I., & Stammers, D. K., (2002). Structure of HIV-2 reverse transcriptase at 2.35: A resolution and the mechanism of resistance to non-nucleoside inhibitors. *Proceedings of the National Academy of Sciences of the United States of America, 99*(22), 14410–14415.

Savarino, A., (2006). A historical sketch of the discovery and development of HIV-1 integrase inhibitors. *Expert Opinion on Investigational Drugs, 15*(12), 1507–1522.

Steigbigel, R. T., Cooper, D. A., & Kumar, P. N., (2008). Raltegravir with optimized background therapy for resistant HIV-1 infection. *The New England Journal of Medicine, 359*(4), 339–354.

Wainberg, M. A., (2012). The need for development of new HIV-1 reverse transcriptase and integrase inhibitors in the aftermath of antiviral drug resistance. *Scientifica, 1*–28.

World Health Organization. http://www.who.int/hiv/pub/guidelines/en/–Guidelines HIV (2015).

CHAPTER 2

Biopolymers

LORENTZ JÄNTSCHI[1] and SORANA D. BOLBOACĂ[2]

[1]*Technical University of Cluj-Napoca, Romania*

[2]*Iuliu Haţieganu University of Medicine and Pharmacy Cluj-Napoca, Romania*

2.1 DEFINITION

In its general acceptation, biopolymers are polymers produced by living organisms. The two main characteristics encounter here are: (i) to be a polymer, e.g., to have repeating (monomer) units; and (ii) to be produced by living organisms, e.g., to be produced from a DNA (deoxyribonucleic acid) encoded information.

2.2 HISTORICAL ORIGIN(S)

It was not concluded if RNA (Joyce, 2002) or DNA (Vreeland et al., 2000, 2002) acted first to replicate for life. The convenient supposition is that the prebiotic world was made by a mixture of small organic molecules (such as short-chain fatty acids and amino acids) that produced relatively short peptides (Andras and Andras, 2005). The emergence of proteins (as polypeptides) possibly it brought to the emergence of the encapsulated reproduction of sequences of proteins later turned into the advanced function of it by developing memories of replication. The biological memories, as we know, are made from nucleic acids (Walker, 1972).

The biopolymers groups together are the polymers of biological origin (see Figure 2.1).

In the natural context, the trophic chain is responsible for the synthesis of BioPoly from simpler ones to complex ones (see Figure 2.2).

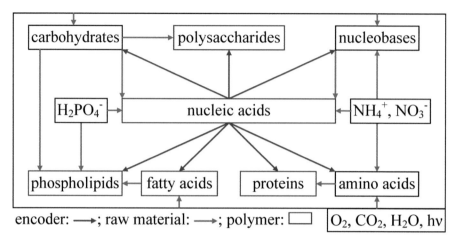

FIGURE 2.1   Biopolymers and their biosynthetic routes.

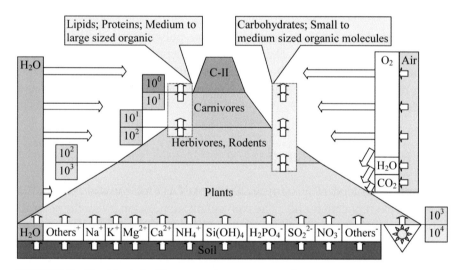

FIGURE 2.2   Trophic chain, biomass, and energy conversion.

It was identified about 100 amino-acids till 1966, and about 240 to 1979 (Fowden et al., 1979); most having a rather restricted distribution, and many of them appear to be products of interaction of one or other of 21 classical amino acids (for lists of 21(A), 16(B), and 13(C), see Vickery, 1972). Other groupings include 22 being genetically encoded (Srinivasan et al., 2002), which are usually contained in the human amniotic fluid (Levy and Montag, 1969), from which and 21 are usually found in proteins (Vickery and Schmidt, 1931).

2.3 NANO-SCIENTIFIC DEVELOPMENT(S)

Cytosine (C), guanine (G), adenine (A), and thymine (T) are used for building of the biological memories in DNA (deoxyribonucleic acid) and Cytosine (C), guanine (G), adenine (A), and uracil (U) – a demethylated form of thymine (T) are used for building of the biological memories in the RNA (ribonucleic acid) being paired when the nucleic acids have double strands (see Figure 2.3). Cytosine (C), thymine (T), and uracil (U) are pyrimidine $(C_4H_4N_2)$ derivatives.

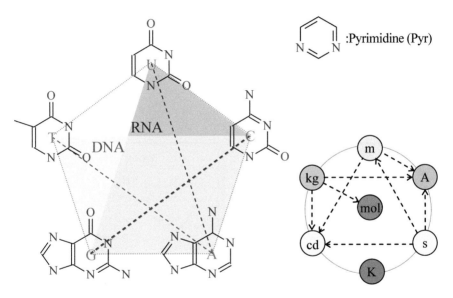

FIGURE 2.3 Nucleobases (left) and fundamental physical units (right) and their linkages.

The pairing is made via two (in the case of A and T/U) or three (in the case of *C* and G) hydrogen bonds (see Figure 2.4).

The presence of the double bonds (see Figure 2.3) produces an almost planar arrangement for the pair (see Figure 2.5) and the Glu-Pho strain arranges in a 3D form of a helix (see Figure 2.6).

The helix of the DNA (see Figure 2.6; for simplicity, only the phosphorus atoms were depicted to scale) was constructed from crystallographic experimental data providing the arrangement of a DNA decamer (Qiu et al., 1997).

A series of math was used in order to determine the equation(s) of the helix(es). Thus, firstly relative to its direction of the propagation, the helix can be rotated and therefore three angular variables must be used to parameterize this rotation (see Equation 1):

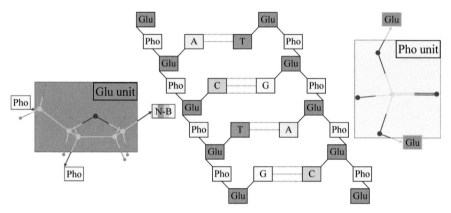

FIGURE 2.4 Pairing of the nucleobases in DNA.

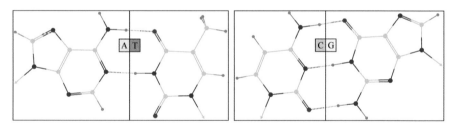

FIGURE 2.5 Bridging of the bases in DNA.

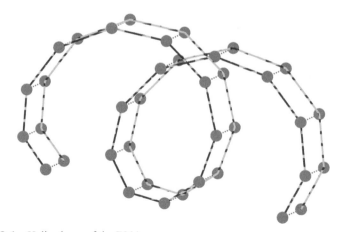

FIGURE 2.6 Helix shape of the DNA.

$x'_i = \cos(c_1) \cdot x_i - \sin(c_1) \cdot z_i$ $y'_i = \sin(c_0) \cdot \sin(c_1) \cdot x_i + \cos(c_0) \cdot y_i + \sin(c_0) \cdot \cos(c_1) \cdot z_i$ $z'_i = \cos(c_0) \cdot \sin(c_1) \cdot x_i - \sin(c_0) \cdot y_i + \cos(c_0) \cdot \cos(c_1) \cdot z_i$ $x''_i = \cos(c_2) \cdot x'_i + \sin(c_2) \cdot y'_i$ $y''_i = -\sin(c_2) \cdot x'_i + \cos(c_2) \cdot y'_i$ $z''_i = z'_i$	**Equation 1.** Equations to be used to align the helix relative to a arbitrary z' axis

Then, the z" coordinates must be aligned on a line (first constraint in Equation 2) and the x" and y" coordinates relatively to a shift (c_4 and c_5) should be on a circle (the last two constraints in Equation 2):

$z''_i \sim c_3 + c_6 \cdot t_i$ $x''_i \sim c_4 + c_7 \cdot \sin(2 \cdot \pi \cdot c_8 \cdot t_i)$ $y''_i \sim c_5 + c_7 \cdot \cos(2 \cdot \pi \cdot c_8 \cdot t_i)$	**Equation 2.** Constraints to be used to obtain the helix equations

The helix parameters (c_6, c_7, and c_8), the translation shifts (c_3, c_4, and c_5) and the rotation ones (c_0, c_1, and c_2) can be determined by minimizing the residuals corresponding to constraints from Equation 2).

The DNA has a double helix shape (see Figure 2.4). Because of this, the other series of eight parameters (from c_{0+9} to c_{8+9}) must be used to parameterize the second helix. Fortunately, the two helixes are linked (via hydrogen bonds, see Figures 2.3 and 2.4) which allows the defining of a series of identities (see Equation 3), and thus only the translation shifts (c_3, c_4, and c_5) are actually supplementary variables to be identified.

$c_{0+9} = -c_0$ $c_{1+9} = c_1 - \pi$ $c_{2+9} = c_2 + \pi$	$c_{6+9} = c_6$ $c_{7+9} = c_7$ $c_{8+9} = c_8$	**Equation 3.** Identities linking the helixes equations of the DNA

By using these equations (Equations 1–3) all variables (from c_0 to c_{8+9}) were identified for the data given in (Qiu et al., 1997), as given in Equation 4, when explained variance was 503.97 Å^2, unexplained variance was 24.71 Å^2, and the probability of a wrong model is $p_F = 7.2 \times 10^{-6}$.

$c_0 = -1.74438 \times 10^{-2}$	$c_1 = 4.33624 \times 10^{-1}$	$c_2 = -9.84365 \times 10^{-1}$	**Equation 4.**
$c_9 = -c_0$	$c_{10} = c_1 - \pi$	$c_{11} = c_2 + \pi$	Identified parameters for the used data to depict Figure 2.6
$c_3 = -1.75024 \times 10^{1}$	$c_4 = -1.09980 \times 10^{1}$	$c_5 = -1.27403 \times 10^{1}$	
$c_{12} = -2.28491 \times 10^{1}$	$c_{13} = -7.15844 \times 10^{0}$	$c_{14} = -1.49852 \times 10^{1}$	
$c_6 = 2.94075 \times 10^{0}$	$c_7 = 9.33651 \times 10^{0}$	$c_8 = -1.02345 \times 10^{-1}$	
$c_{6+9} = c_6$	$c_{7+9} = c_7$	$c_{8+9} = c_8$	

A number of usually 20, sometimes 22 (see Table 2.1) amino-acids are decoded from the nucleic acids by the genetic code when groups of three bases (a 'codon') encode the amino-acid to be built (Crick et al., 1961).

TABLE 2.1   Decoding of the Amino Acids from the Nucleic Acids

1	3\2	U	C	A	G
U	U	UUU → Phe	UCU → Ser	UAU → Tyr	UGU → Cys
	C	UUC → Phe	UCC → Ser	UAC → Tyr	UGC → Cys
	A	UUA → Leu	UCA → Ser	UAA → Xxx or Stop	UGA → Zzz or Stop
	G	UUG → Leu	UCG → Ser	UAG → Yyy or Stop	UGG → Trp
C	U	CUU → Leu	CCU → Pro	CAU → His	CGU → Arg
	C	CUC → Leu	CCC → Pro	CAC → His	CGC → Arg
	A	CUA → Leu	CCA → Pro	CAA → Gln	CGA → Arg
	G	CUG → Leu	CCG → Pro	CAG → Gln	CGG → Arg
A	U	AUU → Ile	ACU → Thr	AAU → Asn	AGU → Ser
	C	AUC → Ile	ACC → Thr	AAC → Asn	AGC → Ser
	A	AUA → Ile	ACA → Thr	AAA → Lys	AGA → Arg
	G	AUG → Met or Start	ACG → Thr	AAG → Lys	AGG → Arg
G	U	GUU → Val	GCU → Ala	GAU → Asp	GGU → Gly
	C	GUC → Val	GCC → Ala	GAC → Asp	GGC → Gly
	A	GUA → Val	GCA → Ala	GAA → Glu	GGA → Gly
	G	GUG → Val	GCG → Ala	GAG → Glu	GGG → Gly

Remark: generics Xxx, Yyy, and Zzz (and Stops) are differently decoded by different ribosomes.

Hydrophobic interactions in cellular processes support the idea that the thermodynamic potential arising from an incompatibility of organic compounds with water is likely to be the driving force of evolution (Black, 1973). Thus, the hydrophobicity is responsible for a multitude of biological facts. The pK_a of Cytosine is 12.2 (primary one) and 4.5 (secondary one), for Guanine is 12.3 (primary), 9.2 (secondary), and 3.3 (amide), for Adenine is 9.80 (primary), 4.15 (secondary), for Thymine is 9.70 and for Uracil is 10.0.

A list of 32 amino acids is given in Table 2.2. For further details, including the hydrophobicity scales, see Bolboacă and Jäntschi (2008).

TABLE 2.2 List of 32 Amino Acids

Name/3l/mol/use/no	2D structural formula	Remarks
Histidine His $C_6H_9N_3O_2$ E/1		Responsible for histamine biosynthesis
Isoleucine Ile $C_6H_{13}NO_2$ E/2		Almost exclusively in proteins and enzymes
Leucine Leu $C_6H_{13}NO_2$ E/3		Almost exclusively in proteins and enzymes Supplementary Ref. (Braconnot, 1820)
Lysine Lys $C_6H_{14}N_2O_2$ E/4		Positive charge on the aliphatic side chain
Methionine Met $C_5H_{11}NO_2S$ E/5		Initiate protein synthesis
Phenylalanine Phe $C_9H_{11}NO_2$ E/6		Aromatic amino acid from proteins
Threonine The $C_4H_9NO_3$ E/7		Involved in porphyrin metabolism
Tryptophan Trp $C_{11}H_{12}N_2O_2$ E/8		Must be obtained from the diet (1.1%)

TABLE 2.2   *(Continued)*

Name/3l/mol/use/no	2D structural formula	Remarks
Valine Val $C_5H_{11}NO_2$ E/9		Hold proteins together
Arginine Arg $C_6H_{14}N_4O_2$ C/10		Present in active sites of enzymes
Cysteine Cys $C_3H_7NO_2S$ C/11		Present in active sites and protein tertiary structure Supplementary Ref. (Baumann, 1884)
Glutamine Gln $C_5H_{10}N_2O_3$ C/12		Easily cross the blood–brain barrier
Pyrrolysine Pyl $C_{12}H_{21}N_3O_3$ C/13		Biosynthesis of proteins in some methanogenic archaea and bacterium
Ornithine Orn $C_5H_{12}N_2O_2$ C/14		Role in the urea cycle
Proline Pro $C_5H_9NO_2$ C/15		Role in synthesis of collagen
Selenocysteine Sec $C_3H_7NO_2Se$ C/16		Building block of selenoproteins

TABLE 2.2 *(Continued)*

Name/3l/mol/use/no	2D structural formula	Remarks
Serine Ser $C_3H_7NO_3$ C/17		Present in active site of serine proteases
Taurine Tau $C_2H_7NO_3S$ C/18		Involved in bile acid biochemistry Is not an amino-acid by the definition of amino-acids Supplementary Ref. (Tiedemann & Gmelin, 1827)
Tyrosine Tyr $C_9H_{11}NO_3$ C/19		Used to build neurotransmitters and hormones
Alanine Ala $C_3H_7NO_2$ N/20		Used in the biosynthesis of proteins
Asparagine Asn $C_4H_8N_2O_3$ N/21		Used at the active sites of enzymes Supplementary Ref. (Vauquelin & Robiquet, 1806)
Aspartic acid Asp $C_4H_7NO_4$ N/22		Intermediate in the citric acid cycle Supplementary Ref. (Henry & Plimmer, 1912)
Glutamic acid Glu $C_5H_9NO_4$ N/23		Found on the surface of proteins
Glycine Gly $C_2H_5NO_2$ N/24		Acts as a neurotransmitter antagonist Supplementary Ref. (Braconnot, 1820)

TABLE 2.2 *(Continued)*

Name/3l/mol/use/no	2D structural formula	Remarks
2-Aminoisobutyric acid Aib $C_4H_9NO_2$ O/25		Found in some antibiotics of fungal origin Supplementary Ref. (Clarke & Bean, 1931)
Citrulline Cit $C_6H_{13}N_3O_3$ O/26		Works to detoxify and eliminate unwanted ammonia Supplementary Ref. (Wada, 1930)
Dehydroalanine Dha $C_3H_5NO_2$ O/27		Found in peptides of microbial origin Supplementary Ref. (Gavaret et al., 1980)
γ-Aminobutyric acid Gaba $C_4H_9NO_2$ O/28		Is inhibitory neurotransmitter in the mammalian central nervous system Supplementary Ref. (Fricke & Parts, 1938)
Homocysteine Hcy $C_4H_9NO_2S$ O/29		Biosynthesized from methionine Supplementary Ref. (Butz & du Vigneaud, 1932)
Hydroxyproline Hyp $C_5H_9NO_3$ O/30		Used in structural proteins like collagen Supplementary Ref. (Henry & Plimmer, 1912)
Lanthionine Lth $C_6H_{12}N_2O_4S$ O/31		Found in bacterial cell walls Supplementary Ref. (Horn et al., 1941)
Cystine – $C_6H_{12}N_2O_4S_2$ –/32		It is a site of redox reactions and a mechanical linkage allowing proteins to retain their 3D structure Supplementary Ref. (Wollaston, 1810)

Legends: 3l: three letters acronym (JCBN, 1983); mol: molecular formula; E: essential amino acids (McCollum, 1931); C: essential amino acids only in certain cases (Reeds, 2000; Fürst and Stehle, 2004), N: nonessential amino-acids but occurs in living organisms (Fürst and Stehle, 1952); O: other amino acids.

2.4 NANO-CHEMICAL APPLICATION(S)

A chain of the nucleic acids (see Figures 2.1 and 2.4) alternates the Pho and the Glu units, being from this point of view a copolymer of these monomers.

Proteins (with more than 50 amino acids) and peptides (with less than 51 amino acids) have a primary structure linearly chaining the amino acids (mmdb_id 106025, pdb_id 1JXW from NCBI-protein databank depicted in Figure 2.7), a secondary structure stabilized by hydrogen bonds, a tertiary structure stabilized by a hydrophobic core (but also through salt bridges, hydrogen bonds, and disulfide bonds) and a quaternary structure when several protein molecules becomes subunits of a single protein complex.

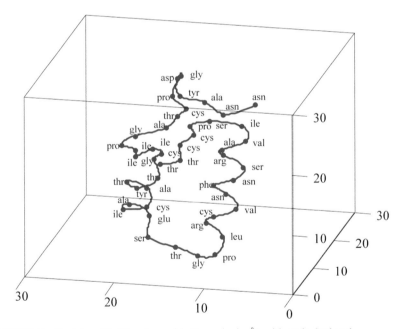

FIGURE 2.7 Protein with 47 amino acids (to scale, in Å) with ends depicted.

Polysaccharides are polymers of carbohydrates in which monosaccharide units are bound from linear to highly branched, through glycosidic linkages (see Figure 2.8). Hydrolysis gives the constituent monosaccharides or oligosaccharides.

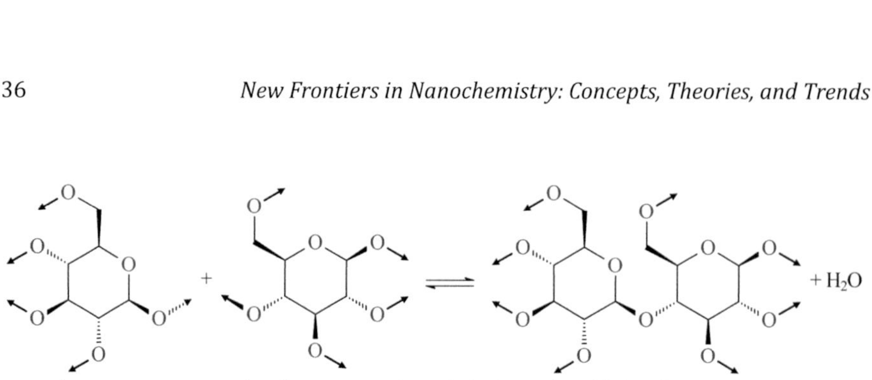

FIGURE 2.8 Formation of the polysaccharides.

With few exceptions (such as is deoxyribose, $H-(C=O)-(CH_2)-(CHOH)_3-H$, $C_3O_4H_{10}$) monosaccharides have the molecular formula $(CH_2O)_n$, where n ranges from 2 (only one with $n = 2$, diose, $H-(C=O)-(CH_2)-OH$, $C_2O_2H_4$) to usually 7 ($n = 3$ triose, $n = 4$ tetrose, $n = 5$ pentose, $n = 6$ hexose, $n = 7$ heptose). The molecular structure of a monosaccharide can be written as $H(CHOH)_x(C=O)(CHOH)_y H$, where $x + y + 1 = n$ to have $(CH_2O)_n$ as the molecular formula. The most important monosaccharide, glucose (depicted as a monomeric unit in Figure 2.8), is a hexose. Examples of heptoses include the ketoses mannoheptulose and sedoheptulose. Monosaccharides with eight or more carbons are rarely observed as they are quite unstable. In the next table are given the monosaccharides for n from 3 to 6 (see Table 2.3). From $n = 5$, the monosaccharides are stable also in their cyclic tautomeric form (see Figure 2.9), the lactol being prevalent in nature against aldose, while smaller ones (e.g., $n = 3$ and $n = 4$) may cyclize by dimerization (when resulted cyclic monosaccharides have $n = 6$, and $n = 8$, respectively, see Figure 2.10).

TABLE 2.3 Monosaccharides from Trioses to Hexoses

n =	Aldoses		Ketoses
3	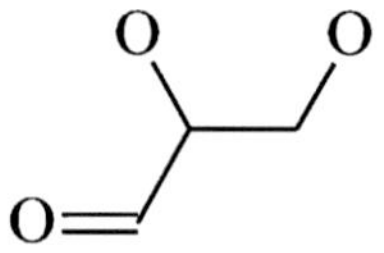 D-glyceraldehyde		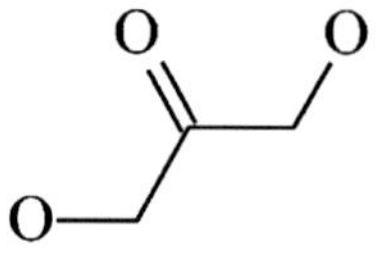 D-dihydroxyacetone
4	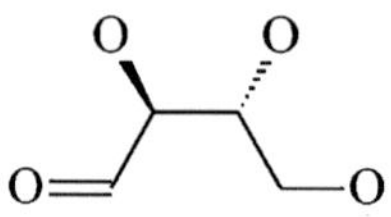 D-erythrose	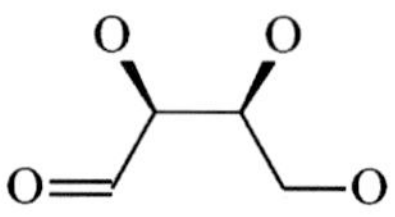 D-threose	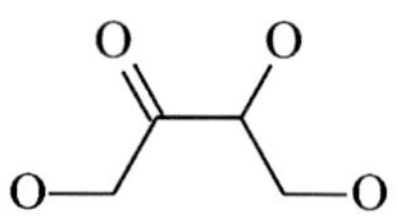 D-erythrulose

TABLE 2.3 *(Continued)*

n =	Aldoses		Ketoses

5

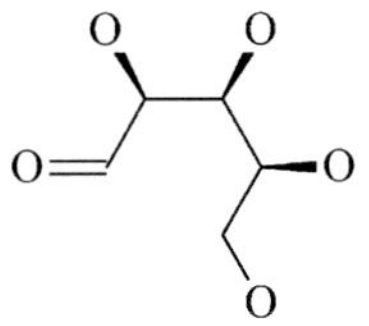

D-ribose

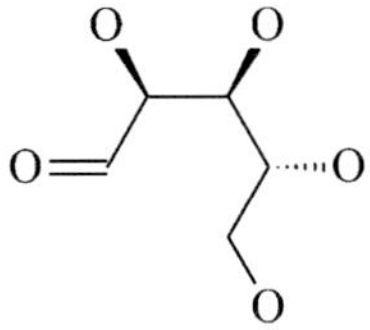

D-arabinose

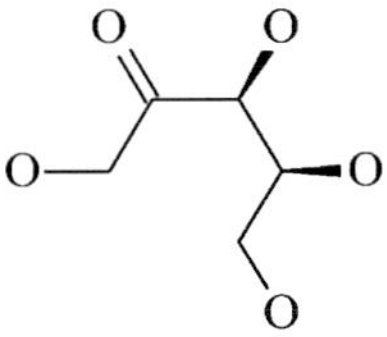

D-xylulose

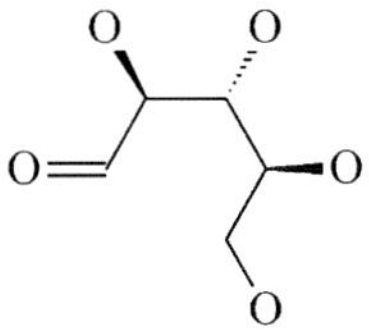

D-xylose

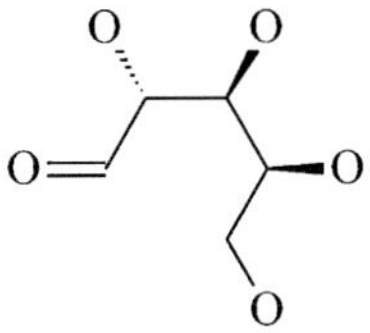

D-lyxose

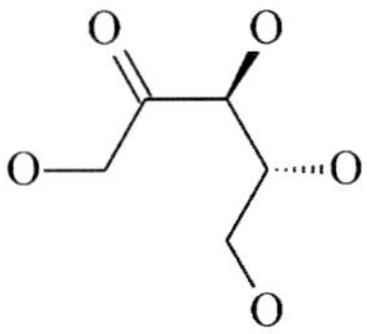

D-ribulose

6

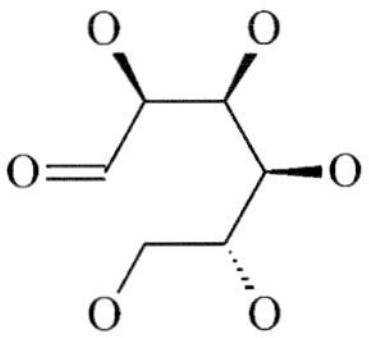

D-talose

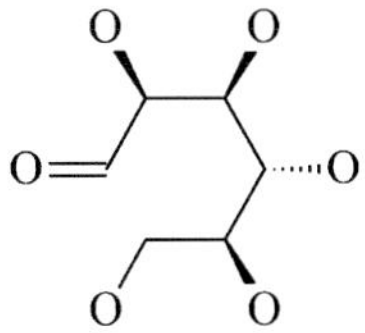

D-gulose

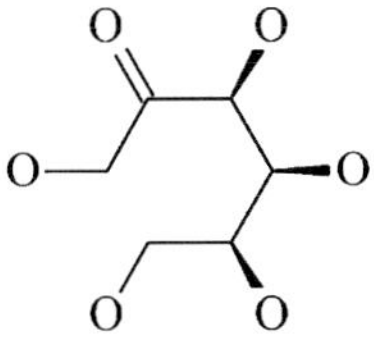

D-psicose

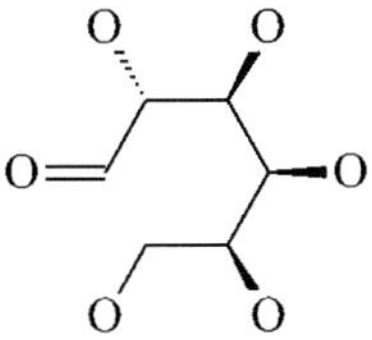

D-altrose

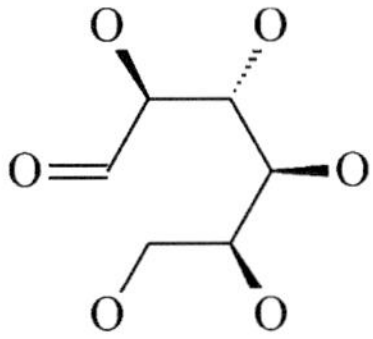

D-glucose

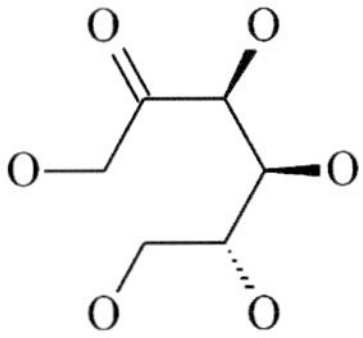

D-tagatose

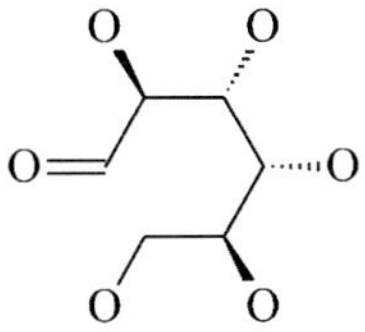

D-galactose

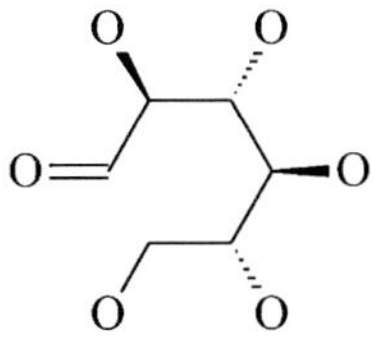

D-idose

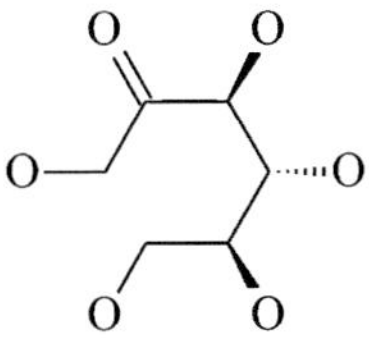

D-sorbose

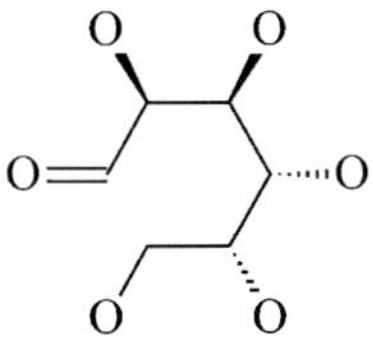

D-mannose

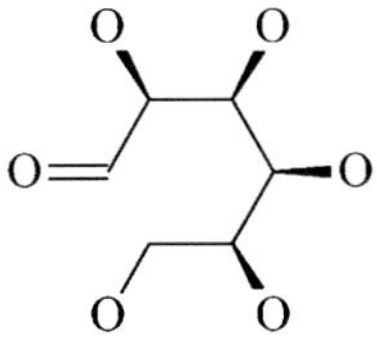

D-allose

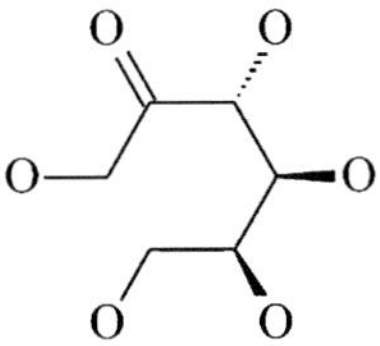

D-fructose

TABLE 2.3   *(Continued)*

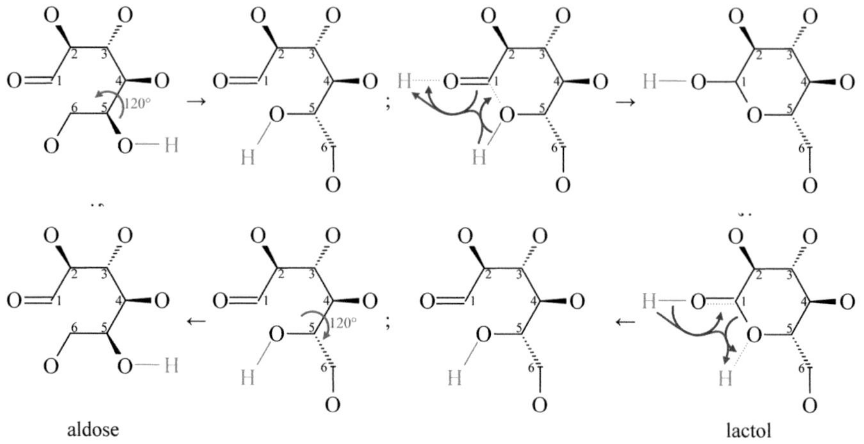

aldose   lactol

FIGURE 2.9   Transition states in tautomerization of a monosaccharide (e.g., D-glucose here).

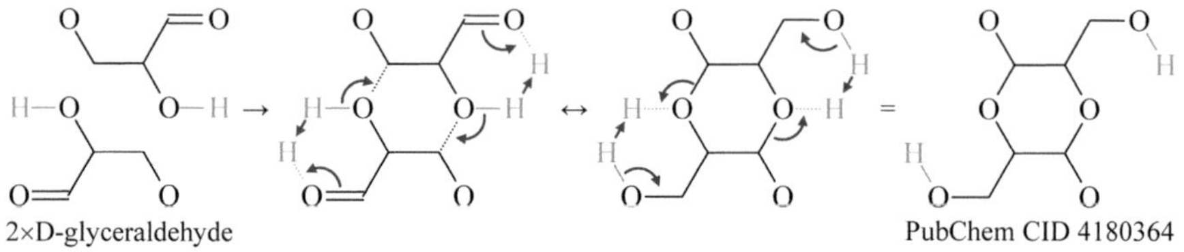

2×D-glyceraldehyde   PubChem CID 4180364

FIGURE 2.10   Dimerization and tautomerization D-glyceraldehyde.

Dimers as two units of monosaccharides are the first level of polymerization in carbohydrates (monosaccharides, disaccharides, oligosaccharides, and polysaccharides). Table 2.4 gives some common encountered representatives (the carbons of the chain are conventionally numbered from 1 to n, starting from the end which is closest to >C=O group).

In Figure 2.11, a polysaccharide made from 14 units of glucose is given. The glucose units are connected together through positions 1–4 (1 to 4, see Figure 2.9 or Table 2.4) in a helix geometrical arrangement.

It should be noted that in a polymeric arrangement of the monosaccharides, the geometrical arrangement of the oxygen atoms relative to the plane of the glycosidic cycle determines the geometry of the polymer. By taking as an example the glucose molecule, the topology determines four different arrangements of two glucose units, while the geometrical arrangement increases the number of isomers to 12 (see Table 2.4).

TABLE 2.4 Disaccharides Condensation from Units of Monosaccharides (e.g., D-Glucose)

Monosaccharide 1		Monosaccharide 2		Disaccharides	Remarks
	+		$\rightarrow$ $+$ H_2O		2×Glucose in alpha-alpha 1-1 linkage. α,α-Trehalose
					2×Glucose in alpha-beta 1-1 linkage. α,β-Trehalose
					2×Glucose in beta-beta 1-1 linkage. β,β-Trehalose

TABLE 2.4 (Continued)

Monosaccharide 1	Monosaccharide 2	Disaccharides	Remarks
			2×Glucose in alpha 1-2 linkage. α-Kojibiose
			2×Glucose in alpha 1-2 linkage. β-Kojibiose
			2×Glucose in alpha 1-2 linkage. α-Sophorose

TABLE 2.4 *(Continued)*

Monosaccharide 1	Monosaccharide 2	Disaccharides	Remarks
	+ $\xrightarrow{}$ + H$_2$O		2×Glucose in beta (α,β) 1-3 linkage. Laminarabiose
			2×Glucose in alpha (α,β) 1-3 linkage. Nigerose
	+ $\xrightarrow{}$ + H$_2$O		2×Glucose in alpha 1-4 linkage. D-Maltobiose Obtained from hydrolysis of malt & starch

TABLE 2.4 *(Continued)*

Monosaccharide 1	Monosaccharide 2	Disaccharides	Remarks
			2×Glucose in beta 1-4 linkage. D-Cellobiose Obtained from hydrolysis of cellulose
			2×Glucose in alpha (α,β) 1-6 linkage. Isomaltose
			2×Glucose in beta (α,β) 1-6 linkage. Gentiobiose

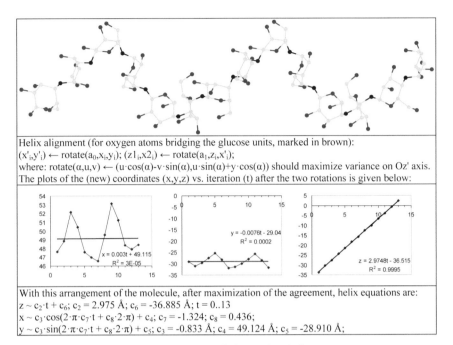

Helix alignment (for oxygen atoms bridging the glucose units, marked in brown):
$(x'_i, y'_i) \leftarrow$ rotate(a_0, x_i, y_i); $(z1_i, x2_i) \leftarrow$ rotate(a_1, z_i, x'_i);
where: rotate$(\alpha, u, v) \leftarrow (u \cdot \cos(\alpha) - v \cdot \sin(\alpha), u \cdot \sin(\alpha) + y \cdot \cos(\alpha))$ should maximize variance on Oz' axis.
The plots of the (new) coordinates (x, y, z) vs. iteration (t) after the two rotations is given below:

$x = 0.003t + 49.115$
$R^2 = 3E{-}05$

$y = -0.0076t - 29.04$
$R^2 = 0.0002$

$z = 2.9748t - 36.515$
$R^2 = 0.9995$

With this arrangement of the molecule, after maximization of the agreement, helix equations are:
$z \sim c_2 \cdot t + c_6$; $c_2 = 2.975$ Å; $c_6 = -36.885$ Å; $t = 0..13$
$x \sim c_3 \cdot \cos(2 \cdot \pi \cdot c_7 \cdot t + c_8 \cdot 2 \cdot \pi) + c_4$; $c_7 = -1.324$; $c_8 = 0.436$;
$y \sim c_3 \cdot \sin(2 \cdot \pi \cdot c_7 \cdot t + c_8 \cdot 2 \cdot \pi) + c_5$; $c_3 = -0.833$ Å; $c_4 = 49.124$ Å; $c_5 = -28.910$ Å;

FIGURE 2.11   Polysaccharide of 14 units of glucose in a helix arrangement.

By taking even a simpler case, with only one glucose unit, there are five different positions susceptible to provide different geometrical arrangements, but not all of them equivalent, because the 5th is connected with a carbon. Calculating for four positions, $2^4 = 16$ gives each geometrical isomer twice (8 isomers by keeping position 5 fixed), and letting now the position 5 to be placed on one or another side of the cycle formal plane, and it is arrived back to 16 geometrical isomers (see Table 2.5). From these ones, only half of them (8, the "D-" type ones) are listed in Table 2.3.

It should be noted that in a polymeric arrangement of the monosaccharides, the geometrical arrangement of the oxygen atoms relative to the plane of the glycosidic cycle determines the geometry of the polymer. By taking as an example the glucose molecule, the topology determines four different arrangements of two glucose units, while the geometrical arrangement increases the number of isomers to twelve (see Table 2.4).

Polymers of monosaccharides are therefore many. The most important ones are:

• Glycogen – a branched polysaccharide of glucose serving as the main energy storage in animals and fungi (Tebb, 1898).

TABLE 2.5   Enumerating the Isomers of Glucose

From 1 to 4 can be "a" or "b", but being "a" or being "b" is irrelevant (e.g. is the same with the one having everywhere "b" in place of "a" and vice-versa).	All 16:	Only 8:	Again 16:
	aaaa, aaab, aaba, aabb, abaa, abab, abba, abbb, baaa, baab, baba, babb, bbaa, bbab, bbba, bbbb.	aaaa, aaab, aaba, aabb, abaa, abab, abba, abbb.	aaaac, aaabc, aabac, aabbc, abaac, ababc, abbac, abbbc, aaaad, aaabd, aabad, aabbd, abaad, ababd, abbad, abbbd.

- Cellulose – a linear chain polysaccharide of glucose with the molecular formula $(C_6(H_2O)_5)_n$ from several hundred to many thousands of repeated units (Updegraff, 1969) is structural component of the primary cell wall of green plants, many forms of algae and the oomycetes, being the most abundant biopolymer on Earth (Klemm et al., 2005).
- Starch – a part linear (amylose) part branched (amylopectin) polysaccharide of glucose serving as the main energy storage in most green plants (Payen, 1839).

Peptidoglycan (also known as murein) is a copolymer of N-acetylglucosamine and N-acetylmuramic acid as alternating monosaccharides (see Figure 2.12). To the N-acetylmuramic acid a chain of three to five amino acids is attached (conjugation not depicted in Figure 2.1). Taking a mesh-like layer from outside of the plasma membrane of most bacteria, it forms a part of the cell wall (see Figure 2.13). Peptidoglycan is about 90% of the dry weight of Gram-positive bacteria and only 10% of Gram-negative ones, serving thus as the primary determinant of the characterization of bacteria type (Gram, 1884).

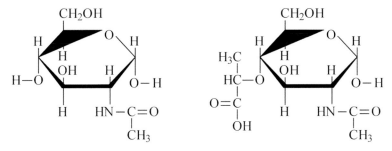

FIGURE 2.12   N-acetylglucosamine (left) and N-acetylmuramic acid (right).

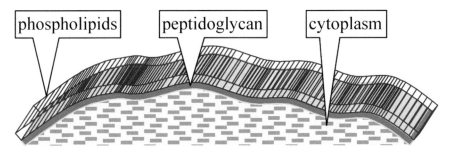

FIGURE 2.13   Two biopolymers (phospholipids and peptidoglycan) creates the cells wall.

The structure of a phospholipid molecule generally consists of a (hydrophilic) phosphate head and hydrophobic fatty acid (R-COO) tails (see Figure 2.14).

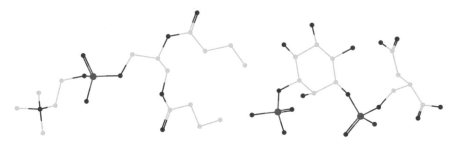

FIGURE 2.14   Dibutanoyllecithin (left) and Phosphatidylinositol 3-phosphate (right).

2.5   MULTI-/TRANS-DISCIPLINARY CONNECTION(S)

A series of biopolymers based bio-composites are synthesized by the living organisms and are currently used as is or with minimal chemical treatment, the following list including only the most common ones:

- Latex, as stable water-based emulsion of 30–45% (see Jacob et al., 1993) isoprene polymers with minor impurities of other organic compounds is found mostly in *Magnoliopsida* class of land plants (see Metcalfe, 1967). Rubbers, as its thermal treatment with sulfur (or vulcanization), were patented in 1839 (Hayward, 1838).
- Guncotton, obtained nitrating cellulose through exposure to nitric acid were discovered in 1833 (see Braconnot, 1833), later plasticized with camphor as celluloid (see Parkes, 1855).
- Casein which is about 80% of the proteins in cow milk simply treated with caustic-alkali solution (Spitteler, 1899), and later plasticized with formaldehyde as Galalith (Siegfeld, 1904).
- Polylactide is derived from corn starch, tapioca roots, chips, starch or sugarcane, the polymerization of the lactide being favored by the presence of potassium carbonate (Carothers et al., 1932).
- Wool contains up to 90% keratin (Cardamone et al., 2005), a member of scleroprotein group shaped like rods or wires. Silk contains two main proteins, sericin, and fibroin (Hakimi et al., 2007).

2.6 OPEN ISSUES

Sun passes the sky from east to west, and it is possibly responsible for the selectivity of the bio-synthesizers among chiral molecules. Thus, the D-isomer (D-glucose), also known as dextrose, occurs widely in nature, but the L-isomer (L-glucose) does not.

The appearance of the biological organisms is at the conjunction of a series of coincidences from at least are the earth's mass, its position in the biologically suitable region and possibly to the presence of the moon, involved in tides, a natural way of producing cyclic movements of a chemical equilibrium.

The worlds of prebiotic chemistry and primitive biology lie on opposite sides of the defining moment for life, when Darwinian evolution first began to operate (Figure 2.1). Before that time, chemical processes may have led to a substantial level of complexity. Depending on the nature of the prebiotic environment, available building blocks may have included amino acids, hydroxy acids, sugars, purines, pyrimidines, and fatty acids. These could have combined to form polymers of largely random sequence and mixed stereochemistry (handedness). Some of the polymers may have had special properties, such as adherence to a particular mineral surface, unusual resistance to degradation, or the propensity to form supramolecular aggregates. Eventually every polymer, no matter how stable, would have succumbed to degradation.

KEYWORDS

- **DNA**
- **nucleic acids**
- **polysaccharide**
- **proteins**
- **RNA**

REFERENCES AND FURTHER READING

Andras, P., & Andras, C., (2005). The origins of life–the 'protein interaction world' hypothesis: Protein interactions were the first form of self-reproducing life and nucleic acids evolved later as memory molecules. *Medical Hypotheses, 64*(4), 678–688.

Baumann, E., (1884). About cystin and cystein (In German). *Z Phys Chem, 8*(4), 299–305.

Black, S., (1973). A theory on the origin of life. *Advances in Enzymology and Related Areas of Molecular Biology, 3*, 193–234.

Bolboacă, S. D., & Jäntschi, L., (2008). Modeling analysis of amino acids hydrophobicity. *MATCH-Communications in Mathematical and in Computer Chemistry, 60*(3), 1021–1032.

Braconnot, H. M., (1820). On the conversion of animal matters into new substances by means of sulfuric acid (In French). *Ann. Chim. Phys. Ser., 2*(13), 113–125.

Braconnot, H., (1833). From the transformation of several vegetable substances into a new principle (In French). *Ann Chem Phys, 52*, 290–294.

Butz, L. W., & Du Vigneaud, V., (1932). The formation of a homolog of cysteine by the decomposition of methionine with sulfuric acid. *Journal of Biological Chemistry, 99*(1), 135–142.

Cardamone, J. M., Nuñez, A., Garcia, R. A., & Aldema-Ramos, M., (2009). Characterizing wool keratin. *Research Letters in Materials Science*, 147175, p. 5.

Carothers, W. H., Dorough, G. L., & Van Natta, F. J., (1932). Studies of polymerization and ring formation A. The reversible polymerization of six-membered cyclic esters. *The Journal of the American Chemical Society, 54*(2), 761–772.

Clarke, H. T., & Bean, J. H., (1931). α-Aminoisobutyric acid. *Organic Syntheses, 11*, 4.

Crick, F. H. C., Barnett, L., Brenner, S., & Watts-Tobin, R. J., (1961). General nature of the genetic code for proteins. *Nature, 192*(4809), 1227–1232.

Fowden, L., Lea, J. P., & Bell, E. A., (1979). The nonprotein amino acids of plants. *Advances in Enzymology and Related Areas of Molecular Biology, 50*, 117–175.

Fricke, H., & Parts, A. G., (1938). The dielectric absorption and dielectric constant of solutions of aliphatic amino acids. *Journal of Physical Chemistry, 42*(1), 1171–1185.

Fürst, P., & Stehle, P., (1952). Nonessential amino acids and nitrogen utilization. *Nutrition Reviews, 10*(1), 10–11.

Fürst, P., & Stehle, P., (2004). What are the essential elements needed for the determination of amino acid requirements in humans? *J. Nutr., 134*(6), 1558–1565.

Gavaret, J. M., Nunez, J., & Cahnman, H. J., (1980). Formation of dehydroalanine residues during thyroid hormone synthesis in thyroglobulin. *The Journal of Biological Chemistry, 255*(11), 5281–5285.

Gram, C. H., (1884). About the isolated coloring of schizomycetes in cut and dry preparations (In German). *Forts Med., 2*, 185–189.

Hakimi, O., Knight, D. P., Vollrath, F., & Vadgama, P., (2007). Spider and mulberry silkworm silks as compatible biomaterials. *Composites Part B: Engineering, 38*(3), 324–337.

Hayward, N., (1838). *Improvement in the Mode of Preparing Caoutchouc with Sulfur for the Manufacture of Various Articles*. Patent US000001090 from Feb. 24, 1839.

Henry, R., & Plimmer, A., (1912). *The Chemical Constitution of the Proteins: Analysis*. London: Longmans, p. 208.

Horn, M. J., Jones, B. D., & Ringel, S. J., (1941). Isolation of a new sulfur-containing amino acid (lanthionine) from sodium carbonate-treated wool. *Journal of Biological Chemistry, 138*(1), 141–149.

Jacob, J. L., Dauzac, J., & Prevôt, J. C., (1993). The composition of natural latex from *Hevea brasiliensis. Clin. Rev. Allergy, 11*, 325–337.

JCBN IUPAC-IUBMB, (1983). Nomenclature and symbolism for amino acids and peptides. *Pure Appl. Chem., 56*(5), 595–624.

Joyce, G. F., (2002). The antiquity of RNA-based evolution. *Nature, 418*, 214–221.

Klemm, D., Heublein, B., Fink, H. P., & Bohn, A., (2005). Cellulose: Fascinating biopolymer and sustainable raw material. *Angewandte Chemie International Edition, 44*(22), 3358–3393.

Levy, H. L., & Montag, P. P., (1969). Free amino acids in human amniotic fluid. A quantitative study by ion-exchange chromatography. *Pediatric Research, 3*(2), 113–120.

McCollum, E. V., (1931). Relationship between diet and dental caries. *Journal of Dental Research, 11*(4), 553–571.

Metcalfe, C. R., (1967). Distribution of latex in the plant kingdom. *Economic Botany, 21*(2), 115–127.

Parkes, A., (1855). *Certain Preparations of oils for, and Solutions Used When Waterproofing, and for the Manufacture of Various Articles by the Use of Such Compounds*. Patent UK2359 from Oct. 22.

Payen, M., (1839). Elementary composition of the starch of various plants, of its most concrete portions, of those most easily disintegrated, of the products of its dissolution, and the atomic weight of starch and dextrin. *J. Franklin Inst., 27*(2), 123–124.

Qiu, H., Dewan, J. C., & Seeman, N. C., (1997). A DNA decamer with a sticky end: The crystal structure of d-CGACGATCGT. *Journal of Molecular Biology, 267*(4), 881–898.

Reeds, P. J., (2000). Dispensable and indispensable amino acids for humans. *J. Nutr., 130*(7), 1835S–1840S.

Siegfeld, M., (1904). Galalith. *Z Angew Chem* ('galalith' is in English) *17*(48), 1816–1818.

Spitteler, A., (1899). *Manufacture of Transparent Products from Impure Paranucleoproteids*. Patent US672541 A from Apr. 23, 1901.

Srinivasan, G., James, C. M., & Krzycki, J. A., (2002). Pyrrolysine encoded by UAG in Archaea: Charging of a UAG-decoding specialized tRNA. *Science, 296*(5572), 1459–1462.

Tebb, M. C., (1898). Hydrolysis of glycogen. *The Journal of Physiology, 22*(5), 423–432.

Tiedemann, F., & Gmelin, L., (1827). Some new elements of the bile of the oxen (In German). *Ann Phys., 85*(2), 326–337.

Updegraff, D. M., (1969). Semimicro determination of cellulose in biological materials. *Analytical Biochemistry, 32*(3), 420–424.

Vauquelin, L. N., & Robiquet, P. J., (1806). The discovery of a new plant principle in *Asparagus sativus*. *Annales de Chimie, 57*, 88–93.

Vickery, H. B., (1972). The history of the discovery of the amino acids. II. A review of amino acids described since 1931 as components of native proteins. *Advances in Protein Chemistry, 26*, 81–171.

Vickery, H. B., & Schmidt, C. L. A., (1931). The history of the discovery of the amino acids. *Chemical Reviews, 9*(2), 169–318.

Vreeland, R. H., Rosenzweig, W. D., & Powers, D. W., (2000). Isolation of a 250 million-year-old halotolerant bacterium from a primary salt crystal. *Nature, 407*, 897–900.

Vreeland, R. H., Straight, S., Krammes, J., Dougherty, K., Rosenzweig, W. D., & Kamekura, M., (2002). *Halosimplex carlsbadense* gen. nov., sp. nov., a unique *Halophilic archaeon*, with three 16S rRNA genes, that grows only in defined medium with glycerol and acetate or pyruvate. *Extremophiles, 6*, 445–452.

Wada, M., (1930). About citrulline, a new amino acid in the press juice of the watermelon, *Citrullus vulgaris* subsp (In German). *Biochem Z., 224*(S), 420–429.

Walker, I., (1972). Biological memory. *Acta Biotheoretica, 21*(3), 203–235.

Wollaston, W. H., (1810). On cystic oxide, a new species of urinary calculus. *Philosophical Transactions of the Royal Society of London, 100*, 223–230.

CHAPTER 3

Cancer and Anticancer Activity

GRDISA MIRA[1] and ANA-MATEA MIKECIN[2]

[1]Rudjer Boskovic Institute, 10,000 Zagreb, Croatia, E-mail: grdisa@irb.hr

[2]Rudjer Boskovic Institute, Division of Molecular Medicine,
Bijenicka 54, 10,000 Zagreb, Croatia, E-mail: Ana-Matea.Mikecin@irb.hr

3.1 DEFINITION

Cancer (CA) is a disease which involves cell transformation, deregulation of apoptosis, proliferation, invasion, angiogenesis, and metastasis. Anticancer activity (ACA) is a term used for the action of different compounds against or tending to prevent CA. The developing knowledge about CA biology suggests that anticancer treatment with high doses of the anticancer drug is not always appropriate, because it may cause different side effects. Thus, the scientists have found an alternative approach to anticancer treatment in synergistic organic compounds which already exist in many therapeutic plants and oils. Natural peptides have many advantages in comparison with synthetic anticancer compounds. They are mainly toxicologically safe, have a wide spectrum of therapeutic action, exhibit less side effects as compared to synthetic drugs and are better absorbed in the intestinal tract. The compounds with ACA may influence CA through various mechanisms such as antioxidant, antimutagenic, and antiproliferative, enhancement of immune function and surveillance, enzyme induction and enhancing detoxification, as well as modulation of multidrug resistance.

3.2 HISTORICAL ORIGIN(S)

CA, also known as a malignant tumor or malignant neoplasm, is a group of diseases involving abnormal cell growth with the potential to invade or spread to other parts of the body. It begins when cells in a part of the body start to grow out of control. Not all tumors are cancerous; benign tumors do not spread to other parts of the body. There are over 100 different known CAs that affect humans.

The oldest description of human CA was discovered in an Egyptian papyrus (between 3000–1500 BC). It referred to breast tumors. Some of the earliest evidence of CA is found among fossilized bone tumors dating back to the Bronze Age (1900–1600 BC) and human mummies dating back 2400 years ago. The origin of word 'cancer' is credited to Hippocrates (460–370 BC), who first recognizes the difference between benign and malignant tumors. His writings describe the CAs of many body sites. At that time there was no treatment for the disease, only cauterization, a method to destroy tissue with a hot instrument called "the fire drill."

In 1713, Italian doctor Bernardino Ramazzini reported a high incidence of breast CA in nuns and 1775 Percival Pott described CA of the scrotum as an occupational CA in chimney sweeps. In 1761, the first link between tobacco and CA was described, although the research on that issue started many years later (in the 1950s and early 1960). The first autopsy postmortem was performed in 1761 by Giovanni Morgagni.

In the 19th century, scientific oncology was born, using a modern microscope in studying diseased tissue. Rudolf Virchow provided the scientific basis for the pathological study of CA. This method not only allowed a better understanding of the damage CA had done, but also aided the development of CA surgery. The pathologist could also tell the surgeon whether the operation had completely removed the CA. At the same time, Stephen Paget devised a theory about metastatic tumor cells.

Development of drugs for the treatment of CA has a long history. Many scientists in the latest 90th put an effort on searching effective anticancer drug, either in natural products or newly synthesized compounds.

3.3   NANO-SCIENTIFIC DEVELOPMENT(S)

CA is the consequence of uncontrolled proliferation following the loss of function of tumor suppressor genes and the activation of oncogenes. CA cells have the ability to grow in various tissue environments (Knudson, 2001). Tumor suppressor genes, which control the proliferation signals, are usually inactivated in CA cells. CA can be caused by chemical cancerogenesis, ionizing radiation, viral or bacterial infection, and hereditary factors.

Early research on the cause of CA was summarized in 1958 by Haddow (1958). The first chemical carcinogen was identified in 1928–29, and the carcinogenic substance was identified in 1933. The author concluded that might be a connection between chemical structures of carcinogens with the genetic material, as the source of the chemical mechanism of action. Later, in 1964, the ongoing research into the causes of CA was summarized (Brookes and Lawley, 1964). They referred to the competing hypotheses that carcinogens reacted mainly with proteins versus mainly with DNA. Their hypothesis was proved by microsome test (McCann et al., 1975). The results have shown that 90% of known carcinogens caused DNA damage or mutations. Some of cause of CA could be avoid such as tobacco, alcohol, diet (especially meat and fat), food additives, occupational exposures (including aromatic amines, benzene, heavy metals, vinyl chloride), pollution, industrial products, medicines, and medical procedures, UV light from the sun, exposure to medical x-rays, and infection. Many of them are DNA damaging agents. Recently, the role of DNA damage and reduced expression of DNA repair genes are summarized in an article by Bernstein et al., (2013). Characterization of the potential mechanisms of carcinogenesis in regard to the types of genetic and epigenetic changes that are associated with CA development is an important part of searching for potential treatment and target for therapy. Many conventional therapies cause serious side effects, and the researchers look after a drug for better CA treatment with a milder side effect. Many chemical compounds, biologically active peptides were synthesized and tested for potential ACA (Grdisa et al., 1995, 1998; Dogan-Koruznjak et al., 2003; Zambrowicz et al., 2013). Recently, days the ACA moved from the drugs who kill tumor cells towards to the drugs with acting on molecular targets, such as proteins for regulation of cell cycle, proteins involved in regulation of apoptosis, angiogenesis, etc. (Vermeulen et al., 2003; Bergers and Benjamin, 2003; Biroccio et al., 2003; Schulze-Berkamen and Krammer, 2004) (Figure 3.1).

3.4 NANO-CHEMICAL APPLICATION(S)

CA causes many human deaths. Current therapies include surgery (i.e., removal of the solid tumor), chemotherapy (administration of anticancer drugs) and radiotherapy (treatment with X-rays) and these treatments are not satisfactory. Thus, the aim of each therapeutic strategy is to influence tumor cells with less toxicity for normal cells. Over the past few decades great advances with anticancer drugs, especial with specific drugs, have been made in patient care and treatment. Sometimes it is very difficult to deliver specific drugs (biologically active peptides and proteins) into the cells. In the area of drug delivery, a big contribution has been made with nanoparticles (Connor et al., 2005; Podsiadlo et al., 2007; Ghosh et al., 2008; Lehner et al., 2012; Shutova and Lvov, 2012). But progress in drug delivery systems for specific clinical applications remains a challenging mission. The future evolution of nanomaterials has to be directed to the development of stimuli responsiveness and multi-functional nanosystems that permit the optimization of diagnosis and therapy.

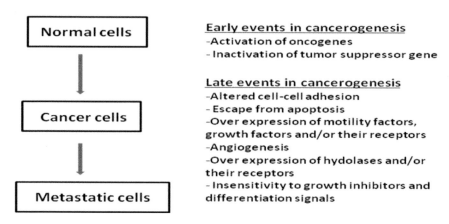

FIGURE 3.1   Stages in cancerogenesis with cellular and genetic events involved in the formation and evolution of cancer.

3.5   MULTI-/TRANS-DISCIPLINARY CONNECTION(S)

CA is a biomedically complex group of diseases involving cell transformation, deregulation of apoptosis, proliferation, angiogenesis, and metastasis. Thus, to resolve so complex disease, many approaches are necessary for the different field of natural science (biology, biochemistry, molecular biology). However, the treatment of such disease requires specific drugs and methods, available through the combination of knowledge in organic chemistry, physics, and mathematics.

3.6   OPEN ISSUES

So far there are no known therapies that eliminate CA. The most important challenge is the understanding of the accurate mechanisms involved in the development of CAs. Specific targeted therapy of tumors is a promising approach. It is well known that CA cells proliferate without control and avoid apoptosis. The angiogenesis is enhanced, and the expression of survival proteins is higher. All these events are the challenge for developing efficient CA therapies. Recent insight into CA cell biology has led to the pharmaceutical development of novel targeted anticancer drugs. Nanotechnology could help in the efficient delivery of specifically targeted drugs into the cells as well as improved their action (increasing of bioavailability).

KEYWORDS

- **anticancer activity**
- **cancer**
- **treatment of cancer**

REFERENCES AND FURTHER READING

Bergers, G., & Benjamin, L. E., (2003). Tumorigenesis and the angiogenic switch. *Nature Reviews Cancer, 3*, 401–410.

Bernstein, C., Prasad, A. R., Nfonsam, V., & Bernstein, H., (2013). DNA damage, DNA repair, and cancer (Chapter 16). In: Chen, C., (ed.), *New Research Direction in DNA Repair* (pp. 413–465). CC BY 3.0 license. doi: 10.5772/53919, ISBN 978–953–51–1114–6.

Biroccio, A., Leonetti, C., & Zupi, G., (2003). The future of antisense therapy: Combination with anticancer treatment. *Oncogene, 22*, 6579–6588.

Brookes, P., & Lawley, P. D., (1964). Evidence for the binding of polynuclear aromatic hydrocarbons to the nucleic acids of mouse skin: Relation between carcinogenic power of hydrocarbons and their binding to deoxyribonucleic acid. *Nature, 202*, 781–784. doi: 10.1038/202781a0.

Connor, E. E., Mwamuka, J., Gole, A., Murphy, C. J., & Wyatt, M. D., (2005). Gold nanoparticles are taken up by human cells but do not cause acute cytotoxicity. *Small, 1*, 325–327.

Dogan-Koruznjak, J., Grdisa, M., Slade, N., Zamola, B., Pavelic, K., & Karminski-Zamola, G., (2003). Novel derivatives of benzo[*b*]thieno[2,3-*c*]quinolones: Synthesis, photochemical synthesis, and antitumor evaluation. *Journal of Medicinal Chemistry, 46*, 4516–4524.

Ghosh, P., Han, G., De, M., Kim, C. K., & Rotello, V. M., (2008). Gold nanoparticles in delivery applications. *Advanced Drug Delivery Reviews, 60*, 1307–1315.

Grdisa, M., Kralj, M., Eckert-Maksic, M., Maksic, Z. B., & Pavelic, K., (1994). 6-amino-6-deoxy ascorbic acid induces apoptosis in human tumor cells. *Journal of Cancer Research and Clinical Oncology, 121*, 98–102.

Grdisa, M., Lopotar, N., & Pavelic, K., (1998). Effect of a 17-member azalide on tumor cell growth. *Chemotherapy, 44*, 331–336.

Haddow, A., (1958). Chemical cancerogens and their modes of action. *British Medical Bulletin, 14*, 79–92. PMID 13536366.

Lehner, R., Wang, X., Wolf, M., & Hunziker, P., (2012). Designing switchable nanosystems for medical application. *Journal of Controlled Release, 161*, 307–316.

McCann, J., Choi, E., Yamasaki, E., & Ames, B. N., (1975). Detection of cancerogens as mutagens in the Salmonella/microsome test: Assay of 300 chemicals. *Proceedings of the National Academy of Science USA, 72*, 5135–5139. doi: 10.1073/pnas.72.12.5135.

Podsiadlo, P., Sinani, V. A., Bahng, J. H., Kam, N. W., Lee, J., & Kotov, N. A., (2007). Gold nanoparticles enhance the antileukemia action of a 6-mercaptopurine chemotherapeutic agent. *Langmuir, 24*, 568–574.

Schulze-Berkamen, H., & Krammer, P. H., (2004). Apoptosis in cancer–implication for therapy. *Seminar Oncology, 31*, 90–119.

Shutova, T. G., & Lvov, Y. M., (2012). Encapsulation of natural polyphenols with antioxidant properties in polyelectrolyte capsules and nanoparticles. In: Diederich, M., & Noworyta, K., (eds.), *Natural Compounds as Inducers of Cell Death* (Vol. 1, pp. 215–235). Springer Netherlands.

Vermulen, K., Van Bockstaele, D. R., & Berneman, Z. N., (2003). The cell cycle: A review of regulation, deregulation, and therapeutic targets in cancer. *Cell Proliferation, 36*, 131–149. doi: 10.1007/978-94-007-4575-9-9.

Zambrowicz, A., Timmer, M., Polanowski, A., Lubec, G., & Trziszka, T., (2013). Manufacturing of peptides exhibiting biological activity. *Amino Acids, 44*, 315–320.

CHAPTER 4

Cancer/Anti-Cancer Chemotherapy: Pharmacological Management

BOGDAN BUMBĂCILĂ,[1] CORINA DUDA-SEIMAN,[1]
DANIEL DUDA-SEIMAN,[2] and MIHAI V. PUTZ[1,3]

[1]*Laboratory of Computational and Structural Physical Chemistry for Nanosciences and QSAR, Biology-Chemistry Department, Faculty of Chemistry, Biology, Geography at West University of Timișoara, Pestalozzi Street No. 16, Timișoara, RO–300115, Romania*

[2]*Department of Medical Ambulatory, and Medical Emergencies, University of Medicine and Pharmacy "Victor Babes," Avenue C. D. Loga No. 49, RO–300020 Timisoara, Romania*

[3]*Laboratory of Renewable Energies-Photovoltaics, R&D National Institute for Electrochemistry and Condensed Matter, Dr. A. Paunescu Podeanu Str. No. 144, RO–300569 Timișoara, Romania, Tel.: +40-256-592-638, Fax: +40-256-592-620, E-mail: mv_putz@yahoo.com; mihai.putz@e-uvt.ro*

4.1 DEFINITION

Cancer is a group of conditions implying abnormal cell growth in a large variety of tissues, cells that possess the ability to invade other tissues (organs). Today, there are described more than 100 types of cancer in humans (WHO, Fact sheet N°297, Updated February, 2015).

Cancer is treated today usually by combining different methods: surgery, pharmacological therapy (cytotoxic chemotherapy), hormonal therapy, targeted therapy (immunotherapy), and radiation therapy. The choice depends on the characteristics of the tumor/cancer like its localization, the characterization of the tumor, the stage of the disease but by far most important it is personalized on the patient and its status.

The treatment of cancer has evolved over thousands of years. Surgical methods were already performed in ancient Egypt; hormonal therapy and radiation therapy were proposed in the nineteenth century. It is true that chemotherapy was introduced only in the twentieth century and at the end of the twentieth century, researchers developed targeted therapies and immuno-therapies (Sudhakar, 2009).

Chemotherapy refers to anti-cancer drug treatments. These drugs have a cytotoxic effect on cancer cells. Anti-cancer drugs interfere with the cell division in various ways, but unfortunately, they are not specific, being active also on other rapidly dividing cell types. Though the medication is not specific only for diseased cells or for different types of cancer, it possesses certain specificity because cancer cells cannot repair their interfered DNA, for example, but the normal cells can. Today, the chemotherapy is usually combinational: there are used at once several drugs, with different mechanisms of action to inhibit the mitosis in different ways. This way, the chances that the cancer cells develop resistance to the treatment are reduced, the efficacy of the treatment is improved, and the secondary effects of the treatment are also reduced, because the drug doses of the regimens can be lowered (Corrie et al., 2008).

4.2   HISTORICAL ORIGINS

History of anti-cancer chemotherapy begins only in the twentieth century, somewhere around 1940. During the First World War, a toxic gas, yperite (mustard gas), bis(2-chloroethyl) sulfide was found to be extremely cyto-toxic on the lymphoid and myeloid cell lines. Starting from this compound, in the late 1930s, three related compounds were developed, named nitrogen mustards. They were found to be effective in the treatment of lymphoma, by scientists from Yale School of Medicine, United States. The first approved chemotherapy drug was mustine (Goodman et al., 1946).

- bis(2-chloroethyl)ethylamine

- bis(2-chloroethyl)methylamine (mustine)

- tris(2-chloroethyl)amine

Shortly after the Second World War, Sydney Farber, MD from Harvard Medical School discovered that folic acid stimulates the proliferation of cancer cells observing this in children with acute lymphoblastic leukemia. He had the idea to introduce therapies with molecules that possess similar chemical structures to that of folic acid but which block the folate-requiring enzymes when binding to them. The first two anti-folates were aminopterin and amethopterin (or methotrexate). His study proved in 1948 that these molecules are inducing remissions of the disease in children with leukemia (Wright et al., 1951).

In 1965, there have already been discovered more than 10 anticancer drugs. Three American physicians proposed that anticancer therapy should be conducted with combinations of drugs, with different mechanisms of action. They named this pharmacological strategy combination chemotherapy or polychemotherapy.

Since then, numerous substances were introduced and marketed for anticancer therapy. Also, new strategies for the treatment were developed– including pharmacological approach before and after a surgical method, radiation therapy, targeted therapies using monoclonal antibodies which migrate directly to the tumor site after the administration and directly and specifically inhibits the metabolic pathway through which the abnormal division takes place.

4.3 NANO-CHEMICAL IMPLICATIONS

4.3.1 *DRUG CLASSES*

4.3.1.1 *ALKYLATING AGENTS*

They attach an alkyl group to the DNA molecule of the diseased cell which will cause a breakage of the strands and eventually, the death of the cell. These drugs have an effect during every phase of its life cycle.

There are three accepted mechanisms of action:

1. Attachment of alkyl groups to the DNA *G* bases in the nucleotides. This modification of the DNA causes its disruption by the DNA-repairing enzymes which are trying to replace the alkylated bases. The DNA transcription to RNA is also interrupted.
2. Formation of crossed linkages between 2 *G* bases from different strands of the same DNA molecule or even between different DNA molecules. These bridges prevent the DNA replication (new synthesis) and also its transcription to RNA.
3. Induction of mispairing of the nucleotides. For example, an alkylated guanine (G) base could pair with time (T), which normally could not happen because *G* pairs with *C* and *T* with A. If the altered pairing is not corrected, a mutation can be born, and the new DNA molecule cannot fulfill its normal (biological active) 3D conformation. Therefore, it cannot replicate (Siddik, 2005) (Table 4.1).

4.3.1.2 *ANTIMETABOLITES*

These drugs are chemicals that inhibit the use of a natural metabolite. Their chemical structures resemble one of the metabolites, and their mechanism of action consists in their ability to block the cell growth and division by blocking enzymes that the natural metabolite is stimulating (Peters et al., 2000).

Antimetabolites are interfering with DNA synthesis; therefore, cell division and cellular growth and because cancer cells have a dividing potency higher than normal cells, the cancer tissues' growth is affected more.

Antimetabolites act as purines or pyrimidines, important components of DNA. They block the incorporation of these substances in the *S* phase of the

TABLE 4.1 Alkylating Agents

Pharmaco-logical class	Pharmacological Subclass	Drug	Mechanism of action	Other observations
1. Alkylating agents	1.1. Nitrogen mustards	1.1.1. Mechlorethamine, mustine, chlormethine	• It binds to the N7 nitrogen atom of the DNA base guanine, crosslinking its two strands, therefore it stops the transcription process, therefore the duplication of the cell (Takimoto et al., 2008)	• It is the first used alkylating agent
		1.1.2. Cyclophosphamide	• It interferes with the DNA replication and the cancer cells mitosis by forming G-G DNA cross-linkages (Takimoto et al., 2008)	• It is, in fact, a pro-drug of 4-hydroxy cyclophosphamide & aldophosphamide (tautomers), converted to it by the liver cytochrome P450 enzymes (Huttunen et al., 2011)
		1.1.3. Ifosfamide	• It binds to the N7 nitrogen atom of the DNA base guanine establishing intra- and inter- DNA strand cross-linkage (drugbank.ca)	• Requires activation by CYP450 microsomal liver enzymes to active metabolites in order to exert its cytotoxic effects (drugbank.ca)
		1.1.4. Melphalan	• DNA alkylation	
		1.1.5. Chlorambucil	• DNA alkylation	
		1.1.6. Uramustine	• DNA alkylation	• Hybrid compounds with uramustine and distamycin A rests in the molecule were synthesized, having an increased cytotoxic activity (Baraldi et al., 2002)
		1.1.7. Bendamustine	• The alkylating agent causing intra-strand and inter-strand cross-links between DNA bases (Tageja et al. 2010)	
		1.1.8. Trophosphamide	• DNA alkylation	

TABLE 4.1   *(Continued)*

Pharmacological class	Pharmacological Subclass	Drug	Mechanism of action	Other observations
	1.2. Nitrosoureas	1.2.1. Carmustine	• DNA alkylation	• It is used together with an alkyl guanine transferase inhibitor (O(6)-benzylguanine). This drug increases the efficacy if carmustine by inhibiting the DNA repairing by the formation of interstrand cross-linkage between N1 of guanine and N3 of cytosine (Qian et al., 2013)
		1.2.2. Lomustine	• DNA alkylation	• It can be successfully associated with monoclonal antibodies for improving outcomes in recurrent cancers (clinicaltrials.gov)
		1.2.3. Semustine	• DNA alkylation	• Apparently, it is potentially carcinogenic (cancer.org)
		1.2.4. Ethylnitrosourea	• DNA alkylation	• It is one of the most potent mutagenic substances, so it has its use restricted to genetic laboratories (Cordes, 2005)
		1.2.5. Streptozocin	• DNA alkylation	
		1.2.6. Fotemustine	• DNA alkylation	
		1.2.7. Nimustine	• DNA alkylation	
		1.2.8. Ranimustine	• DNA alkylation	
	1.3. Alkyl sulphonates	1.3.1. Busulfan	It forms DNA intrastrand links between adenine and guanine or between guanine and guanine (Iwamoto et al., 2004)	
		1.3.2. Treosulfan		• It has less severe toxicity and side effects than busulfan (Slatter et al., 2011)

TABLE 4.1 *(Continued)*

Pharmacological class	Pharmacological Subclass	Drug	Mechanism of action	Other observations
		1.3.3. Mannosulfan		
1.4. Triazenes		1.4.1. Dacarbazine	• The molecule is methylating guanine at the O(6) and N(7) positions (Gerulath et al., 1972)	
		1.4.2. Temozolomide	• The molecule is methylating guanine at the O(6) and N(7) positions; • Some tumor cells are able to repair the alkylation of DNA; thus, they are lowering the therapeutic activity of temozolomide by expressing a protein—O(6)-alkylguanine DNA alkyltransferase (AGT) encoded by a gene named MGMT; • Sometimes, when temozolomide is really necessary, MGMT gene is switched off using genetical engineering techniques (Jacinto et al., 2007, Hegi et al., 2005)	• It is, in fact, a prodrug of dacarbazine (Newlands et al., 1997)
1.5. Non-classical alkylating agents		1.5.1. Procarbazine	• The mechanism of action is not fully understood; • Metabolism yields another active metabolite and hydrogen peroxide—which results in the breaking of DNA strands (Pratt et al., 1994)	

TABLE 4.1 *(Continued)*

Pharmaco-logical class	Pharmacological Subclass	Drug	Mechanism of action	Other observations
		1.5.2. Altretamine	• It is presumed to be an alkylating agent; its catabolism produces formaldehyde (Damia et al., 1995)	
		1.5.3. ThioTEPA	• It is an alkylating agent which generates other alkylating metabolites (Schellens et al., 2005)	
	1.6. Alkylating-like agents	1.6.1. Cisplatin	• An aqua-ligand formed immediately after its administration binds to Guanine in DNA; DNA cannot repair, and mechanisms of cellular apoptosis are therefore activated (Trzaska, 2005)	• It does not have alkyl rests in its molecule thus it is not a true alkylating agent
		1.6.2. Carboplatin	• It forms interstrand crosslinks which last more than those formed by cisplatin after its chemical activation (Anthony, 2002)	
		1.6.3. Nedaplatin		• It has been developed for reducing cisplatin's toxicity
		1.6.4. Oxaliplatin	• It is the only platinum compound in its series of analogs to form both intra- and interstrand cross-links (Graham et al., 2004)	

TABLE 4.1 *(Continued)*

Pharmaco-logical class	Pharmacological Subclass	Drug	Mechanism of action	Other observations
		1.6.5. Satraplatin	• Latest studies show that comparing to other platinum-based anti-cancer drugs, and it does not develop resistance (it produces DNA crosslinks that are not recognized by the repairing proteins, so the DNA remains damaged) (Choy et al., 2008)	• It is a prodrug
		1.6.6. Dicycloplatin	Under study	• Currently in clinical studies; • It appears to be more soluble and more stable in solutions than other platinum-based drugs (Apps et al., 2015)
		1.6.7. Pyriplatin	Under study	
		1.6.8. Phenanthriplatin	Under study	• Studies show it has increased selectivity for cancer cells • It appears it enters the cell transported by a biological molecule in its intact form (Apps et al., 2015)
		1.6.9. Triplatin tetranitrate	Under study	

cell cycle (the phase of the semiconservative replication of DNA), stopping the cell's division (Takimoto et al., 2008).

The main classes of antimetabolites are:

- Antimetabolites base analogs;
- Dihydrofolate reductase inhibitors (antifolates); and
- Thymidylate synthetase inhibitors;

1. Antimetabolites Base Analogs

Antimetabolites base analogs (altered nucleic bases) are two types. They are purine analogs and pyrimidine analogs.

2. Dihydrofolate Reductase Inhibitors

They are also called antifolates because they block the effects of folic acid (vitamin B9). Folic acid's primary action in the organism is to function as a cofactor for methyltransferase enzymes involved in thymidine, purine, and sulfur-containing amino acids synthesis. The majority of antifolates inhibit these enzymes when they participate as mischievous cofactors (as they have resemblances in their chemical structures with the one of folic acid). Therefore they stop the cell division, nucleic acids synthesis, and repair and protein synthesis (Takimoto, 1996). Some of the molecules in this class can selectively inhibit folate's actions in microbial organisms and not in the human organism that is parasitized by these (Gangjee et al., 2007). For example, trimetrexate can be given to AIDS suffering patients with *Pneumocystis carinii* infection (Allegra et al., 1987).

These molecules act in the *S* phase of the cell cycle, acting specifically during nucleic acids' synthesis, so they have a powerful effect on the cells which are rapidly dividing (e.g., cancer cells but unfortunately also the gastrointestinal tract mucous cells, myeloid precursor cells–so being explained their secondary effects). Of course, the cancer cell can develop resistance to these agents, too. The mechanism through which the resistance is developed is somehow feedback. In response to a decreased tetrahydrofolate, the cell begins to produce more enzyme dihydrofolate reductase (DHFR). This enzyme converts dihydrofolate to tetrahydrofolate, a methyl group carrier necessary for purines synthesis. Because the antifolates are competitive inhibitors of DHFR, its increased amounts in the cell can overcome the drug inhibition. But recently, other mechanisms of resistance were discovered, like impaired transport via the reduced folate carrier. (Gorlick et al., 1996)

3. Thymidylate Synthetase (TS) Inhibitors

Thymidylate synthetase (TS) is an enzyme that catalyzes the biotransformation of deoxyuridine monophosphate to deoxythymidine monophosphate. Thymidine is one of the nucleotides present in DNA structure.

If the catalytic enzyme is inhibited, deoxyuridine monophosphate will rise, and deoxythymidine monophosphate will not be sufficient for DNA synthesis, and the structure of the nucleic acid will not be completed during the *S* phase of the cell cycle (Skeel, 2003) (Table 4.2).

4.3.1.3   ANTI MICROTUBULE AGENTS

These drugs are, in fact, plant-derived chemical agents that block cell division by preventing microtubule function.

Microtubules are protein components of the cytoskeleton, found throughout the cytoplasm of the cell. Their parts (α and β-tubulins) permit their rapid assembly and disassembly. They play a very important role during cell division. Their primary function is to connect with the chromosomes, help them split and then move the new chromosomes to the daughter cells (Chabner et al., 2005).

4.3.1.3.1   Vinca alkaloids

Vinca alkaloids are now synthetically produced. They prevent the formation of the microtubules.

4.3.1.3.2   Taxanes

Taxanes are preventing the disassembly of microtubules; therefore, the cells exposed to taxanes cannot complete the mitosis. These drugs can also slow the blood vessels growth (the angiogenesis process) which is an essential characteristic which is needed to slow the tumor growth and also the metastasis.

4.3.1.3.3   Podophyllum alkaloids

Podophyllum alkaloids are shown in Tables 4.3–4.5.

TABLE 4.2 Antimetabolites

Pharmacological Class	Pharmacological Subclass	Drug	Mechanism of action	Other observations
2. Antimetabolites	2.1. Antimetabolite base analogs	2.1.1. Azathioprine	• Its active metabolite methyl-thioinosine monophosphate is an inhibitor of the purine synthesis which blocks an enzyme involved in a particular step of the synthesis process, but other mechanisms consisting in the inducement of cell apoptosis and DNA interfering were also described (Cara et al., 2004)	• Purine analog • It is the main immunosuppressive cytotoxic drug • It especially prevents the clonal expansion of lymphocytes in the induction phase of the immune response • It affects both the cellular and the humoral immunity (Elion, 1989; Maltzman et al., 2003)
		2.1.2. Mercaptopurine	• It competes with the natural purines: hypoxanthine and guanine for an enzyme involved in the generation of purine nucleotides through the purine salvage pathway and is itself converted to two metabolites which inhibit another enzyme, involved in de novo purine synthesis (Sahasranaman et al., 2008)	
		2.1.3. Tioguanine	• It inhibits through its metabolites enzymes implied in the synthesis of guanine, and it is incorporated in the RNA molecule leading to a form which cannot be translated in the ribosomes (Oncea et al., 2008)	

TABLE 4.2 *(Continued)*

Pharmacological Class	Pharmacological Subclass	Drug	Mechanism of action	Other observations
		2.1.4. Fludarabine	• It inhibits DNA synthesis by blocking the DNA polymerase and the ribonucleotide reductase (Rai et al., 2000)	
		2.1.5. Pentostatin	• Its resemblance with adenosine inhibits the enzyme adenosine deaminase, involved in the purine metabolism (Sauter et al., 2008)	
		2.1.6. Cladribine	• Purine analog, it inhibits the adenosine deaminase	
		2.1.7. Nelarabine	• Purine analog	
		2.1.8. Clofarabine	• Purine analog	
		2.1.9. Cytarabine	• It inhibits DNA and RnA polymerases and the nucleotide reductase, important enzymes for DNA synthesis (Chhikara et al., 2010)	• This molecule has its sugar component altered
		2.1.10. 6-azauracil	• Pyrimidine analog	
			• It inhibits an enzyme involved in the RNA synthesis process (Timar et al., 1969)	
		2.1.11. Gemcitabine	• Cytidine "faulty" analog	
	2.2. Dihydrofolate reductase inhibitors	2.2.1. Methotrexate	• Competitively inhibits dihydrofolate reductase; thus, it inhibits DNA synthesis through thymidine synthesis (Herfarth, 2012)	

TABLE 4.2 *(Continued)*

Pharmacological Class	Pharmacological Subclass	Drug	Mechanism of action	Other observations
		2.2.2. Edatrexate	Under study	• It may overcome tumor resistance to methotrexate, which loses its activity after it is polyglutamated (www.cancer.gov)
		2.2.3. Trimetrexate	Under study	
		2.2.4. Piritrexim	Under study	
		2.2.5. Pemetrexed	Under study	• Also named multitarget antifolate • This molecule inhibits the three important enzymes involved in purine and pyrimidine synthesis: dihydrofolate reductase (DHFR), thymidylate synthetase (TS) and glycinamide ribonucleotide form-yltransferase (GARFT) (McLeod et al., 2000)
	2.3. Thymidylate synthetase inhibitors	2.3.1. Raltitrexed	Under study	
		2.3.2. 5-Fluorouracil	Under study	
		2.3.3. Pemetrexed	Under study	
		2.3.4. Nolatrexed	Under study	
		2.3.5. Capecitabine	Under study	• Five fluorouracil prodrug
		2.3.6. Floxuridine	Under study	

TABLE 4.3 Antimicrotubule Agents

Pharmacological Class	Pharmacological Subclass	Drug	Mechanism of action	Other observations
3. Antimicrotubule agents	3.1. Vinca alkaloids	3.1.1. Vinblastine	• At low concentrations, it suppresses microtubule dynamics, and at higher concentrations, it reduces microtubule polymer mass (Ravina, 2011)	
		3.1.2. Vincristine	• It binds to a structural protein (tubulin), interfering its polymerization process, stopping the cell from separating its chromosomes during metaphase (Ravina, 2011)	
		3.1.3. Vindesine	Under study	
		3.1.4. Vinorelbine	Under study	• The first semisynthetic vinca alkaloid
		3.1.5. Vinflunine	Under study	
		3.1.6. Nocodazole	Under study	• It IS NOT a vinca alkaloid, but it has the same mechanism of action • Apparently, it decreases the oncogenic potential of cancer cells (Li, et al., 2013)

TABLE 4.3 (Continued)

Pharmacological Class	Pharmacological Subclass	Drug	Mechanism of action	Other observations
	3.2. Taxanes	3.2.1. Paclitaxel	• It stabilizes the microtubule polymer and protects it from disassembly thus inducing cell apoptosis (Bharadwaj et al., 2004)	
		3.2.2. Docetaxel	• It acts as a microtubule disassembly inhibitor, and it also inhibits s protein: apoptosis-blocking bcl–2 oncoproteins thus generating apoptosis via two mechanisms (Lyseng-Williamson et al., 2005)	
	3.3. Podophyllum alkaloides	3.3.1. Podophyllotoxin	Under study	• It is also inhibiting the microtubule formation, and it is used to produce two other drugs with different mechanisms of action: etoposide and teniposide

TABLE 4.4 Topoisomerase Inhibitors

Pharmacological Class	Pharmacological Subclass	Drug	Mechanism of action	Other observations
4. Topoisomerase inhibitors	4.1. Topoisomerase I inhibitors	4.1.1. Irinotecan	Under study	• it is active through its metabolite–7-ethyl-10-hydroxy-camptothecin
		4.1.2. Topotecan	• It fixates the topoisomerase cleavage complex to the DNA strand where the complex acts; the DNA replication cannot continue (Pommier, 2006)	
	4.2. Topoisomerase II "poisons."	4.2.1. Etoposide	• It forms a complex with DNA and the topoisomerase II enzyme, preventing the relegation of the 2 DNA strands, thus promoting their breakage (Willmore et al., 2004)	• It is a semisynthetic derivative of podophyllotoxin
		4.2.2. Doxorubicin	• It stabilizes the topoisomerase II cleavage complex after it has broken the double-stranded structure of DNA; thus, the structure of the double helix is maintained, and the DNA replication process is arrested	
			• Its structure, when metabolized, leads to free radicals, probably enhancing its cytotoxic properties (Tacar et al., 2013)	
		4.2.3. Mitoxantrone	• It intercalates DNA to cause intra and inter-strand crosslinking	
			• Inhibits topoisomerase II (Nitiss, 2009)	
		4.2.4. Pixantrone	Under study	
		4.2.5. Losoxantrone	Under study	

TABLE 4.4 (Continued)

Pharmacological Class	Pharmacological Subclass	Drug	Mechanism of action	Other observations
		4.2.6. Amsacrine	• DNA intercalator; • Topoisomerase II inhibitor	
		4.2.7. Amonafide	• Interferes with topoisomerase II binding to DNA and induces chromatin disorganization (Allen et al., 2011)	
		4.2.8. Teniposide	• It causes dose-dependent single and double-strand breaks in DNA (Cragg, 2005)	• It is a semisynthetic derivative of podophyllotoxin;
	4.3. Topoisomerase II catalytic inhibitors	4.3.1. Aclarubicin	• It acts as a topoisomerase I cleavage complex stabilizer but it also binds directly to DNA, preventing its access by topoisomerase II cleavage complex thus blocking the catalytic access of the enzyme complex (Hajji et al., 2005)	

TABLE 4.5 Cytotoxic Antibiotics

Pharmaco-logical Class	Pharmacological Subclass	Drug	Mechanism of action	Other observations
5. Cytotoxic antibiotics	5.1. Anthracyclines	5.1.1. Daunorubicin	• It intercalates the DNA molecule, and it inhibits the progression of the topoisomerase II complex which relaxes DNA supercoils (Momparler et al., 1976)	
		5.1.2. Doxorubicin	• It intercalates the DNA, stabilizing the topoisomerase II complex; • Like all the anthracyclines, it induces the eviction of histones from chromatin; thus, important mechanisms of DNA repairing are affected (Pang et al., 2013)	
		5.1.3. Idarubicin	• Similar to daunorubicin	
		5.1.4. Epirubicin	• DNA intercalator	
			• Topoisomerase II poison: it blocks the enzyme after the cleavage of the DNA strands (Bonfante et al., 1980)	
		5.1.5. Nemorubicin	• It has a different mechanism of action than the other anthracyclines, manifested probably through the activity of a metabolite but it is not yet explained;	
		5.1.6. Sabarubicin	Under study	• Much less cardiotoxic than its analogs (doxorubicin and daunorubicin) (Zucchi et al., 2000)

TABLE 4.5 *(Continued)*

Pharmaco-logical Class	Pharmacological Subclass	Drug	Mechanism of action	Other observations
		5.1.7. Valrubicin	Under study	
		5.1.8. Pirarubicin	Under study	
		5.1.9. Aclarubicin	Under study	
	5.2. Other antibiotics	5.2.1. Bleomycin A2	• It induces DNA strand breaks probably through the generation of free radicals • It inhibits the incorporation of thymidine into DNA (Takimoto et al., 2008)	
		5.2.2. Mitomycin C	• DNA crosslinker, it blocks the DNA replication (Tomasz, 1995)	
		5.2.3. Actinomycin D	• It binds to DNA at the transcription initiation complex and prevents the elongation of RNA chain by the RNA polymerase (Sobell, 1985)	

4.3.1.4 TOPOISOMERASE INHIBITORS

These molecules affect the activity of two enzymes: topoisomerase I and II. During DNA replication and transcription processes, the newly formed DNA strain/RNA strain forms supercoils because of the double-helix shape of the molecule. These two enzymes are producing local cuts at the phosphate rests into one of the strands (topoisomerase I) or in both of them (topoisomerase II) to release the tension. Of course, if the enzymes would not perform this activity, the torsion would eventually stop the ability of the RNA and DNA polymerases to continue to attach nucleotides down the newly formed DNA strand.

4.3.1.4.1 Topoisomerase I Inhibitors

Topoisomerase I inhibitors are obtained semisynthetically from *Camptotheca acuminata* – an ornamental tree.

4.3.1.4.2 Topoisomerase II Inhibitors

1. **Topoisomerase II "poisons"**: they are targeting the DNA-enzyme complex, forming covalent complexes with the enzyme; some of them promote an increased cleavage, and others prevent the relegation of the strands.
2. **Topoisomerase II "catalytic inhibitors"**: that block the enzyme's activity.

4.3.1.5 Cytotoxic Antibiotics

These molecules have many mechanisms of action, but overall they interrupt cell division. These mechanisms include:

- intercalation between the DNA strands thus preventing RNA synthesis;
- generation of highly active free radicals that damage intra-and inter-cellular components;
- DNA alkylation;
- binding to cellular membranes and altering ion transports;
- inhibition of topoisomerase II (Offermanns et al., 2008, Nitiss, 2009);

- anthracyclines; and
- other antibiotics.

4.4   TRANS-DISCIPLINARY CONNECTIONS

The great need for therapeutic agents in cancer therapy always led to extensive research in this domain. In the last 20 years, the steps for making a new drug available for therapy always include QSAR studies of the molecules before reaching into preclinical and clinical assessments. These QSAR/QSPR studies usually begin from choosing a series of analog molecules and describing a Hansch equation for each of the members of the series, taking into account characteristics like the cytotoxic potency but also the potential toxicity because it is known that all the chemical structures that possess anticancer activity are also inducing side effects because of their low specificity on the rapid dividing cell lines.

All computational approaches are followed up by in vitro studies, using malignant cell lines to verify the biological activity of the candidate molecules that possess the best QSAR profiles. The molecules that prove to be efficient will be included in clinical trials which are showing theirs in vivo efficacy/toxicity.

Starting from data obtained in the QSAR studies, novel molecules can be conceived because with every study the molecular libraries and databases are enriched: possible biological activities or contributions to the whole biological activity are described for different structural cores and functional groups. Most often, chemical syntheses of novel molecules are part of the big process of drug development, especially if the goal is to create compounds with multiple mechanisms of action, thus possible higher anticancer activities.

For example, Ketabforoosh, and his colleagues (2013) proved that six heterocyclic esters of caffeic acid had cytotoxic activity, comparable to doxorubicin's activity against HeLa, SK-OV-3, and HT-29 cancer cell lines. Starting from data found in the literature that catechol ring is essential for cytotoxicity and that anticancer effect is due to antioxidant mechanism of the catechol ring and methylation of hydroxyl groups present on the ring reduces the cytotoxic activity and that conjugated double bond on the carbon chain on the ring is also required for inhibition of cell growth, the researchers synthesized three series of six esteric derivatives. The QSAR studies were conducted using molecular descriptors such as hydrophobicity, topological indices, electronic parameters, and steric factors in order to find the effects of different structural properties on the biological activity of each compound. The values were calculated using Hyperchem soft. QSAR analysis proved that the cytotoxic activity influenced by the molecular shape of these compounds and incorporation of arylsulfonyl groups diminishes the activity. In vivo assessment of the activities using cancer cell lines revealed that the compounds are more active against HeLa cancer cell line compared to SK-OV-3 and HT-29 (Ketabforoosha et al., 2013).

An extensive study, conducted by Speck-Planche and his colleagues (2012) tried to develop a multi-target (mt) algorithm for the virtual screening and rational in silico discovery of anti-breast cancer agents against thirteen cell lines. Several hundred compounds, picked from a molecular database were tested, and the models developed correctly classified (by comparison with the in vitro assessments) as active or inactive more than 90% of the compounds.

4.5   OPEN ISSUES

In the drug discovery process, an evaluation of a great number of molecules has to be performed. It is estimated that a large amount (50%) of all known molecules are not yet described by pharmacological, ecological, mutagenic, and toxicological characteristics at all, and only 15% of them have complete profiles.

It is very difficult to find new drugs, and it was estimated that a new anti-cancer drug is found in 200,000 molecules. Of course, some of the chemical properties of the molecules (as the central heterocyclic nucleus, proper functional groups or certain atoms in key positions) needed for their anticancer activity are known, but theoretically, a very large series of analogs can be synthesized having all this information. The synthesis of all these molecules is impossible because of the production costs. So the best way to synthesize the best compound is to predict its activity and toxicity by mathematical or computer-assisted methods (Zhu, 2010).

Laboratory testing today for predicting pharmaco-toxicological profiles consists of three methods: in vivo, in vitro, and in silico. The in vivo methods, using laboratory animals often fail to predict well some very important characteristics like the mutagenic or the chronic toxic effects of the potential drugs because the human genome is different by those of the rodents that are usually tested on. A large number of the candidate analog molecules is also an impediment not only for synthesizing but also for testing all of them – the costs would be enormous, and the number of the animals would be too large. The in vitro methods (using artificial membranes or culture cells or systems that resemble living tissues) could spare the number of the animals used but the costs and the time spent for testing would still be too large (Lam, 1997).

Combinatorial chemistry today can build libraries of hypothetical molecules which could become real and in silico methods (using software which is characterizing chemical structures by parameters already described in the literature) allow the scientists to predict pharmacological or toxicological characteristics of the substances. So if a molecule could prove useful for therapy further research would be made (synthesis, preclinical, and clinical testing) saving important costs and time (Tyagi et al., 2013).

Other methods are used too, for finding new drugs and we will mention some of them: isolation and purification of natural products that are described by folk medicine or that are accidentally discovered as potentially efficient, testing metabolites or molecules that have precise structure modifications from some of the already in use drugs, direct organic synthesis of new molecules after the chemical structure of their target and the binding effect are well known or at least intuited (Tang, 2014).

KEYWORDS

- **alkylating agents**
- **anti-metabolites**
- **anti-microtubule agents**
- **cytotoxic antibiotics**
- **topoisomerase inhibitors**

REFERENCES AND FURTHER READINGS

Allegra, C. J., Chabner, B. A., Tuazon, C. U., Ogata-Arakaki, D., Baird, B., Drake, J. C., et al., (1987). Trimetrexate for the treatment of *Pneumocystis carinii* pneumonia in patients with acquired immunodeficiency syndrome. *The New England Journal of Medicine, 117,* 978–985.

Allen, S. L., & Lundberg, A. S., (2011). Amonafide: A potential role in treating acute myeloid leukemia. *Expert Opinion on Investigational Drugs, 20*(7), 995–1003.

Anthony, P. C., (2002). *Pharmacology Secrets, Chapter "Alkylating Agents, and Related Drugs,"* p. 276. Hanley & Belfus Inc. Medical Publishers, Philadelphia, USA.

Apps, M. G., Choi, E. H. Y., & Wheate, N. J., (2015). The state-of-play and future of platinum drugs. *Endocrine-Related Cancer, 22*(4), 219–233.

Baraldi, P. G., Romagnoli, R., Guadix, A. E., Pineda de las I. M. J., Gallo, M. A., Espinosa, A., Martinez, A., Bingham, J. P., & Hartley, J. A., (2002). Design, synthesis, and biological activity of hybrid compounds between uramustine and DNA minor groove binder distamycin A. *Journal of Medicinal Chemistry, 45*(17), 3630–3638.

Bevacizumab and Lomustine for Recurrent GBM–https://clinicaltrials.gov/ct2/show/NCT 01290939 (accessed on 24 February 2019).

Bharadwaj, R., & Yu, H., (2004). The spindle checkpoint, aneuploidy, and cancer. *Oncogene, 23*(11), 2016–2027.

Bonfante, V., Bonadonna, G., Villani, F., & Martini, A., (1980). Preliminary clinical experience with 4-epidoxorubicin in advanced human neoplasia. *Recent Results in Cancer Research 74,* 192–199.

Cancer–http://www.who.int/mediacentre/factsheets/fs297/en/ (accessed on 24 February 2019).

Cara, C. J., Pena, A. S., Sans, M., Rodrigo, L., Guerrero-Esteo, M., Hinojosa, J., García-Paredes, J., & Guijarro, L. G., (2004). Reviewing the mechanism of action of thiopurine drugs: Towards a new paradigm in clinical practice, medical science monitor. *International Medical Journal of Experimental and Clinical Research, 10*(11), 247–254.

Chabner, B., & Longo, D. L., (2005). *Cancer Chemotherapy and Biotherapy: Principles and Practice* (4th edn.). Philadelphia: Lippincott Williams & Wilkins.

Chhikara, B. S., & Parang, K., (2010). Development of cytarabine prodrugs and delivery systems for leukemia treatment. *Expert Opinion on Drug Delivery, 7*(12), 1399–1414.

Choy, H., Park, C., & Yao, M., (2008). Current status and future prospects for satraplatin, an oral platinum analog. *Clinical Cancer Research, 14,* p. 1633.

Cordes, P. S., (2005). N-ethyl-N-nitrosourea mutagenesis: Boarding the mouse mutant express. *Microbiology and Molecular Biology Reviews, 69*(3), 426–439.

Corrie, P. G., & Pippa, G., (2008). Cytotoxic chemotherapy: Clinical aspects. *Medicine, 36*(1), 24–28.

Cragg, G. M., & Newman, D. J., (2005). Plants as a source of anti-cancer agents. *Journal of Ethnopharmacology, 100*(1/2), 72–79.

Damia, G., & D'Incalci, M., (1995). Clinical pharmacokinetics of altretamine. *Clinical Pharmacokinetics, 28*(6), 439–448.

Elion, G., (1989). The purine path to chemotherapy. *Science, 244*(4900), 41–47.

Gangjee, A., Jain, H. D., & Kurup, S., (2007). Recent advances in classical and non-classical antifolates as antitumor and anti-opportunistic infection agents: Part I. *Anticancer Agents Medicinal Chemistry, 7*(5), 524–542.

Gerulath, A. H., & Loo, T. L., (1972). Mechanism of action of 5-(3,3-dimethyl–1-triazeno) imidazole–4-carboxamide in mammalian cells in culture. *Biochemical Pharmacology, 21*(17), 2335–2343.

Goodman, L. S., Wintrobe, M. M., Dameshek, W., Goodman, M. J., Gilman, A., & McLennan, M. T., (1946). Nitrogen mustard therapy. Use of methyl-bis(beta-chloroethyl) amine hydrochloride and tris(beta-chloroethyl)amine hydrochloride for Hodgkin's disease, lymphosarcoma, leukemia, and certain allied and miscellaneous disorders. *Journal of the American Medical Association, 132*(3), 126–132.

Gorlick, R., Goker, E., Trippett, T., Waltham, M., Banerjee, D., & Bertino, J. R., (1996). Intrinsic and acquired resistance to methotrexate in acute leukemia. *New England Journal of Medicine, 335,* 1041–1048.

Graham, J., Mushin, M., & Kirkpatrick, P., (2004). Oxaliplatin. *Nature Reviews Drug Discovery, 3*(1), 11–20.

Hajji, N., Mateos, S., Pastor, N., Dominguez, I., & Cortes, F., (2005). Induction of genotoxic and cytotoxic damage by aclarubicin, a dual topoisomerase inhibitor. *Mutation Research, 583*, 26–35.

Hegi, M. E., Diserens, A. C., Gorlia, T., Hamou, M. F., De Tribolet, N., Weller, M., et al., (2005). MGMT gene silencing and benefit from temozolomide in glioblastoma. *New England Journal of Medicine, 352*(10), 997–1003.

Herfarth, H. H., Long, M. D., & Isaacs, K. L., (2012). Methotrexate: Underused and ignored? *Digestive Diseases, 30*(3), 112–118.

Huttunen, K. M., Raunio, H., & Rautio, J., (2011). Prodrugs-from serendipity to rational design. *Pharmacological Reviews, 63*(3), 750–771.

Ifosfamide–http://www.drugbank.ca/drugs/DB01181 (accessed on 24 February 2019).

Iwamoto, T., Hiraku, Y., Oikawa, S., Mizutani, H., Kojima, M., & Kawanishi, S., (2004). DNA intrastrand cross-link at the 5'-GA–3' sequence formed by busulfan and its role in the cytotoxic effect. *Cancer Science, 95*(5), 454–458.

Jacinto, F. V., & Esteller, M., (2007). MGMT hypermethylation: A prognostic foe, a predictive friend. *DNA Repair, 6*(8), 1155–1160.

Ketabforoosha, S. H. E., Aminib, M., Vosooghia, M., Shafieea, A., Azizic, E., & Kobarfardd, F., (2013). Synthesis, evaluation of anticancer activity and QSAR study of heterocyclic esters of caffeic acid. *Iranian Journal of Pharmaceutical Research, 12*(4), 705–719.

Known and Probable Human Carcinogens, http://www.cancer.org/cancer/cancercauses/other-carcinogens/generalinformationaboutcarcinogens/known-and-probable-human-carcinogens (accessed on 24 February 2019).

Lam, K. S., (1997). Application of combinatorial library methods in cancer research and drug discovery. *Anticancer Drug Design, 12*(3), 145–167.

Li, J., Chen, X., Ding, X., Cheng, Y., Zhao, B., Lai, Z. C., et al., (2013). LATS2 suppresses oncogenic Wnt signaling by disrupting β-catenin/BCL9 interaction. *Cell Reports, 5*(6), 1650–1663.

Lyseng-Williamson, K. A., & Fenton, C., (2005). Docetaxel: A review of its use in metastatic breast cancer. *Drugs, 65*(17), 2513–2531.

Maltzman, J. S., & Koretzky, G. A., (2003). Azathioprine: Old drug, new actions. *Journal of Clinical Investigation, 111*(8), 1122–1124.

Mazzini, S., Scaglioni, L., Mondelli, R., Caruso, M., & Riccardi-Sirtori, F., (2012). The interaction of nemorubicin metabolite PNU–159682 with DNA fragments d(CGTACG)(2), d(CGATCG)(2) and d(CGCGCG)(2) shows a strong but reversible binding to G: *C* base pairs. *Bioorganic Medicinal Chemistry, 20*(24), 6979–6988.

McLeod, H. L., Cassidy, J., Powrie, R. H., Priest, D. G., Zorbas, M. A., Synold, T. W., et al., (2000). Pharmacokinetic and pharmacodynamic evaluation of the glycinamide ribonucleotide formyltransferase inhibitor AG2034. *Clinical Cancer Research, Clinical Trials, 6*(7), 2677–2684.

Momparler, R. L., Karon, M., Siegel, S. E., & Avila, F., (1976). Effect of adriamycin on DNA, RNA, and protein synthesis in cell-free systems and intact cells. *Cancer Research, 36*(8), 2891–2895.

Newlands, E. S., Stevens, M. F., Wedge, S. R., Wheelhouse, R. T., & Brock, C., (1997). Temozolomide: a review of its discovery, chemical properties, pre-clinical development, and clinical trials. *Cancer Treatment Reviews, 23*(1), 35–61.

Nitiss, J. I., (2009). Targeting DNA topoisomerase II in cancer chemotherapy. *Nature Reviews, Cancer 9*(5), 338–350.

Offermanns, S., & Rosenthal, W., (2008). *Antineoplastic Agents–In Encyclopedia of Molecular Pharmacology* (2nd edn., Vol. 1). Springer.

Oncea, I., & Duley, J., (2008). In: Brunton, L. L., Lazo, J. S., & Parker, K., (eds.), *Pharmacogenetics of Thiopurines* (11th edn.). Goodman & Gilman's the pharmacological basis of therapeutics. McGraw-Hill's Access, Chapter 38.

Pang, B., Qiao, X., Janssen, L., Velds, A., Groothuis, T., Kerkhoven, R., et al., (2013). Drug-induced histone eviction from open chromatin contributes to the chemotherapeutic effects of doxorubicin. *Nature Communications, 4,* 1908.

Peters, G. J., Van der Wilt, C. L., Van Moorsel, C. J., Kroep, J. R., Bergman, A. M., & Ackland, S. P., (2000). Basis for effective combination cancer chemotherapy with antimetabolites. *Pharmacological Therapy, 87*(2/3), 227–253.

Pommier, Y., (2006). Topoisomerase I inhibitors: Camptothecins and beyond. *Nature Reviews Cancer, 6*(10), 789–802.

Pratt, W. B., Ruddon, R. W., Ensminger, W. D., & Maybaum, J., (1994). *The Anticancer Drugs* (2nd edn., p. 141). Oxford University Press Publishers, (Chapter "Covalent DNA-Binding Drugs").

Qian, L., Zheng, J., Wang, K., Tang, Y., Zhang, X., Zhang, H., Huang, F., Pei, Y., & Jiang, Y., (2013). Cationic core-shell nanoparticles with carmustine contained within O6-benzylguanine shell for glioma therapy. *Biomaterials, 34*(35), 8968–8978.

Rai, K. R., Peterson, B. L., Appelbaum, F. R., Kolitz, J., Elias, L., Shepherd, L., et al., (2000). Fludarabine compared with chlorambucil as primary therapy for chronic lymphocytic leukemia. *New England Journal of Medicine, 343,* 1750–1757.

Ravina, E., (2011). *The Evolution of Drug Discovery: From Traditional Medicines to Modern Drugs* (1st edn., p. 157). Weinheim: Wiley-VCH.

Sahasranaman, S., Howard, D., & Roy, S., (2008). Clinical pharmacology and pharmacogenetics of thiopurines. *European Journal of Clinical Pharmacology, 64*(8), 753–767.

Sauter, C., Lamanna, N., & Weiss, M. A., (2008). Pentostatin in chronic lymphocytic leukemia. *Expert Opinion on Drug Metabolism and Toxicology, 4*(9), 1217–1222.

Schellens, J. H. M., McLeod, H. L., & Newell, D. R., (2005). *Cancer Clinical Pharmacology* (p. 95). (Chapter "Alkylating Agents"). Oxford University Press.

Siddik, Z. H., (2005). *Mechanisms of Action of Cancer Chemotherapeutic Agents: DNA-Interactive Alkylating Agents and Antitumor Platinum-Based Drugs.* John Willey & Sons.

Skeel, R. T., (2003). *Handbook of Cancer Chemotherapy* (6th edn.). Lippincott Williams & Wilkins.

Slatter, M. A., Rao, K., Amrolia, P., Flood, T., Abinun, M., Hambleton, S., et al., (2011). Treosulfan-based conditioning regimens for hematopoietic stem cell transplantation in children with primary immunodeficiency: United Kingdom experience. *Blood, 117*(16), 4367–4375.

Sobell, H., (1985). Actinomycin and DNA transcription. *Proceedings of the National Academy of Sciences of the United States of America, 82*(16), 5328–5331.

Speck-Planch, A., Kleandrova, V. V., Luana, F., & Cordeiro, M. N. D. S., (2012). Rational drug design for anti-cancer chemotherapy: Multi-target QSAR models for the *in silico* discovery of anti-colorectal cancer agents. *Bioorganic & Medicinal Chemistry, 20*(15), 4848–4855.

Sudhakar, A., (2009). History of cancer, ancient, and modern treatment methods. *Journal of Cancer Science & Therapy, 1*(2), 1–4.

Tacar, O., Sriamornsak, P., & Dass, C. R., (2013). Doxorubicin: An update on anticancer molecular action, toxicity, and novel drug delivery systems. *The Journal of Pharmacy and Pharmacology, 65*(2), 157–170.

Tageja, N., & Nagi, J., (2010). Bendamustine: Something old, something new. *Cancer Chemotherapy and Pharmacology, 66*(3), 413–423.

Takimoto, C. H., (1996). New antifolates: Pharmacology and clinical applications. *Oncologist, 1*(1/2), 68–81.

Takimoto, C. H., & Calvo, E., (2008).–Principles of oncologic pharmacotherapy. In: Pazdur, R., Wagman, L. D., Camphausen, K. A., & Hoskins, W. J., (eds.), *Cancer Management: A Multidisciplinary Approach* (11ᵗʰ edn.).

Tang, M., Yu, X., Jiang, Y., Shi, Y., Liu, X., Li, W., & Cao, Y., (2014). Common methods used for the discovery of natural anticancer compounds, methods in pharmacology and toxicology. In: Ann, M. B., & Zigang, D., (eds.), *Cancer Prevention: Dietary Factors and Pharmacology* (pp. 33–52).

Timar, M., Gedrikh, I., Vrezhoiu, G., & Peushesku, E., (1969). Mechanism of action of 5-mercapto-6-azauracil. *Farmakologiia i Toksikologiia, 32*(5), 602–604.

Tomasz, M., (1995). Mitomycin C: Small, fast, and deadly (but very selective). *Chemistry and Biology, 2*(9), 575–579.

Trzaska, S., (2005). Cisplatin. *Chemical & Engineering News, 83*(25), 52.

Tyagi, A., Kapoor, P., Kumar, R., Chaudhary, K., Gautam, A., & Raghava, G. P., (2013). *In silico* models for designing and discovering novel anticancer peptides. *Scientific Reports, 18*, p. 2984.

Willmore, E., De Caux, S., Sunter, N. J., Tilby, M. J., Jackson, G. H., Austin, C. A., & Durkacz, B. W., (2004). A novel DNA-dependent protein kinase inhibitor, NU7026, potentiates the cytotoxicity of topoisomerase II poisons used in the treatment of leukemia. *Blood, 103*(12), 4659–4665.

Wright, J. C., Prigot, A., & Wright, B. P., (1951). An evaluation of folic acid antagonists in adults with neoplastic diseases. A study of 93 patients with incurable neoplasms. *Journal of National Medical Association, 43*, 211–240.

Zhu, Q., (2010). *QSAR for Anticancer Activity by Using Mathematical Descriptors*. University of Minnesota, MSc Thesis.

Zucchi, R., Yu, G., Ghelardoni, S., Ronca, F., & Ronca-Testoni, S., (2000). Effect of MEN 10755, a new disaccharide analog of doxorubicin, on sarcoplasmic reticulum Ca^{2+} handling and contractile function in rat heart. *British Journal of Pharmacology, 131*, 342–348.

CHAPTER 5

Cell Cycle

GRDISA MIRA[1] and ANA-MATEA MIKECIN[2]

[1]*Rudjer Boskovic Institute, 10000 Zagreb, Croatia, E-mail: grdisa@irb.hr*

[2]*Rudjer Boskovic Institute, Division of Molecular Medicine, Bijenicka 54, 10000 Zagreb, Croatia, E-mail: Ana-Matea.Mikecin@irb.hr*

5.1 DEFINITION

The cell cycle (CC) or cell-division cycle is a series of events that take place in a cell leading to its division and duplication (replication) what results with the production of two daughter cells. In prokaryotes, which lack a cell nucleus, the CC occurs via a process termed binary fission. In eukaryotic cells, which have a nucleus, the CC is divided into three parts: interphase, the mitotic (M) phase, and cytokinesis. During interphase, the cells growth, accumulate the nutrients necessary for mitosis, and prepare for DNA duplication as well as cell division. Within the mitotic phase, the cell splits itself into two assigned daughter cells. In the course of the final stage, cytokinesis, the new cell is completely divided. To provide the correct division of the cell, there are control mechanisms known as CC checkpoints. After cell division, each of the daughter cells begins the interphase of a new cycle. Although the various stages of interphase are usually not morphologically discernible, each phase of the CC has a defined set of specialized biochemical processes that prepare the cell for initiation of cell division. Cell division is a vital process by which skin, hair, blood cells and certain internal organs are renewed. With that process also fertilized egg develops into mature organisms.

5.2 HISTORICAL ORIGIN(S)

In early 1900s scientists have observed that the cells divide, but they have not understood the mechanisms of it (Nurse, Masui and Hartwell, 1998). In the last 40 years, scientists have explained many events in the molecular mechanism of CC control. These explanations have had a significant impact on the understanding of various human disorders, such as cancer and other diseases. However, many scientists have supposed that the understanding of the basics events of the CC, such as DNA replication and chromosome segregation couldn't be possible before the understanding of these regulations (Garber, 2008).

Nobel Prize in physiology and medicine in 2001 was given to three scientists for these discoveries of the regulation of CC. Each scientist used a unique organism for the investigation of CC regulation. Leland Hartwell and Paul Nurse conducted their studies in yeast, which share the advantage of being easily genetically manipulated. Also, yeast completes their life cycle quickly, making it much easier to look for mutants that failed to regulate their CCs correctly. Using their different mutants, Hartwell and Nurse discovered the proteins required for CC progression. On the other hand, Tim Hunt used the biochemistry for discovering the key proteins important for the progression of the CC. Hunt and his collaborators discovered the protein cyclin, working with sea urchin embryos. Production and degradation of cyclin regulate CC progression. His work and the work of other scientists (Yoshio Masui, Marc Kirschner, and John Gerhart) on frog eggs have indicated that a clock mechanism existing in the cytoplasm could control CC progression. These different studies were completed when Lee Hartwell suggested thinking about the CC progression in terms of 'checkpoints.' The fundamental idea of the checkpoint model was that a control system in the cell would trigger an alarm that would delay the cycle's progression until the previous event was completed. To test the checkpoint model Lee Hartwell and Ted Weinert had conducted critical experiments, using yeast mutants which died when their DNA was damaged. It was unusual that these mutants have not been arrested in the CC after their DNA was damaged. That demonstrated that these cells had not started the alarm to indicate the needing of repair of their DNA like a normal cell would. They anyway divided into two daughter cells with damaged DNA, which could not survive. This experiment resulted in the revelation of important proteins that acted as a "checkpoint" or "stop button" to make sure that all DNA is repaired before the cell division. It was shown that the same proteins were important also in humans.

5.3   NANO-SCIENTIFIC DEVELOPMENT(S)

The CC is controlled by numerous mechanisms that ensure accurate cell division. These mechanisms include the regulation of cyclin-dependent kinases (CDK) by cyclins, CDK inhibitors (CKI), phosphorylating events, as well as the activation of checkpoints after DNA damage.

Cell division comprises of two successive processes characterized by DNA replication and segregation of replicated chromosomes into two separate cells. It was divided into two stages: interphase, which includes G1, S, and G2 phases and it is the interlude between two *M* phases. The second stage is mitosis (M), the process of nuclear division, which includes prophase, metaphase, anaphase, and telophase. (Norbury and Nurse, 1992). During G1 phase the cells grow in size and prepare for DNA synthesis. Replication of DNA occurs in a specific part of the interphase called *S* phase. During the G2 phase, the cell prepares for mitosis (Figure 5.1). Before commitment to DNA replication, the cells in G1 can enter a resting state called G0. The majority of the non-growing, non-proliferating cells in the human body are in G0 state.

The transition from one CC phase to another occurs in an orderly fashion and is regulated by different cellular proteins. Key regulatory proteins are the CDK, a family of serine/threonine protein kinases that are activated at specific points of the CC (Morgan, 1995; Pines, 1995). CDK protein levels constantly rest during the CC, in contrast to the cyclins, their activating proteins. Cyclin protein levels rise and fall during the CC, and in this way, they periodically activate CDK (Evans et al., 1983; Pines, 1991). When CDK is activated, its target proteins become phosphorylated and physiologically relevant for CC progression (Buchkovich et al., 1989; Brehm et al., 1998; Montagnoli et al., 1999).

CDK activity can be suppressed by CC inhibitory proteins, called CKI. They regulate CDK activity by binding to CDK alone or to the CDK-cyclin complex (Sherr and Roberts, 1995; Carnero and Hannon, 1998; El Deiry et al., 1993; Reynisdottir et al., 1995).

The restriction point (R) is defined as a point of no return in G1, following which the cell is committed to entering the CC. To ensure the accurate sequence of events in the CC, the additional controls or checkpoints exist in the CC (Hartwell and Weinert, 1989).

DNA damage checkpoints are positioned before the cell enters *S* phase (G1-S checkpoint) or after DNA replication (G2-M checkpoint) and there appear to be DNA damage checkpoints during *S* and *M* phases.

5.4 NANO-CHEMICAL APPLICATION(S)

Uptake of nanoparticles by cells is influenced by the phases of CC. It was shown that cells in different phases of the CC internalize nanoparticles at similar rates (Kim et al., 2012). After 24 *h* the concentration of nanoparticles in the cells according to the different phases could be arranged: G2/M > *S* > G0/G1. After internalization by the cells, the nanoparticles are not exported from the cells. They are split between daughter cells when the parent cell divides. The future studies on the uptake of nanoparticle should consider the CC because the dose of internalized nanoparticles varies through the CC.

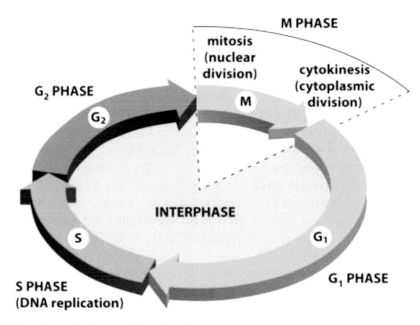

FIGURE 5.1   The phases of the cell cycle.

5.5   MULTI-/TRANS-DISCIPLINARY CONNECTION(S)

Discovering of the proteins which regulate the progression of CC could help in developing specific drugs for better treating of various diseases (cancer, immunological disorders, etc.).

5.6   OPEN ISSUES

Nanoparticles are considered as a primary vehicle for targeted therapies. They can pass the biological barriers and by energy-dependent pathways enter and distribute within the cells. The properties of nanoparticles such as size and surface can influence on their internalization by the cells.

KEYWORDS

- **cell cycle**
- **checkpoints**
- **cyclin-dependent kinase**
- **cyclin-dependent kinases inhibitors**
- **cyclins**

REFERENCES AND FURTHER READING

Buchkovich, K., Duffy, L. A., & Harlow, E., (1989). The retinoblastoma protein is phosphorylated during specific phases of the cell cycle. *Cell, 58*, 1097–1105.

Carnero, A., & Hannon, G. J., (1998). The INK4 family of CDK inhibitors. *Current Topics in Microbiology and Immunology, 227*, 43–55.

El Deiry, W. S., Tokino, T., Velculescu, V. E., Levy, D. B., Parsons, R., Trent, J. M., et al., (1993). WAF1, a potential mediator of p53 tumor suppression. *Cell, 75*, 817–823.

Evans, T., Rosenthal, E. T., Youngblom, J., Distel, D., & Hunt, T., (1983). Cyclin: A protein specified by maternal mRNA in sea urchin eggs that is destroyed at each cleavage division. *Cell, 33*, 389–396.

Hartwell, L. H., & Weinert, T. A., (1989). Checkpoints: Controls that ensure the order of cell cycle events. *Science, 246*, 629–634.

Kim, J. A., Aberg, C., Salvati, A., & Dawson, K. A., (2012). Role of cell cycle on the cellular uptake and dilution of nanoparticles in a cell population. *Nature Nanotechnology, 7*, 62–68.

Montagnoli, A., Fiore, F., Eytan, E., Carrano, A. C., Draetta, G. F., Hershko, A., & Pagano, M., (1999). Ubiquitination of p27 is regulated by CDK-dependent phosphorylation and trimeric complex formation. *Genes Development, 13*, 1181–1189.

Morgan, D. O., (1995). Principles of CDK regulation. *Nature, 374*, 131–134.

Norbury, C., & Nurse, P., (1992). Animal cell cycles and their control. *Annual Review of Biochemistry, 61*, 441–470.

Pines, J., (1991). Cyclins: Wheels within wheels. *Cell Growth Differentiation, 2*, 305–310.

Pines, J., (1995). Cyclins and cyclin-dependent kinases: Theme and variations. *Adv. Cancer Res., 66*, 181–212.

Reynisdottir, I., Polyak, K., & Iavarone, M. J., (1995). Kip/cip and INK4 CDK inhibitors co-operate to induce cell cycle arrest in response to TGF-b. *Genes Dev., 9*, 1831–1845.

Sherr, C. J., (1994). G1 phase progression: Cycling on cue. *Cell, 79*, 551–555.

CHAPTER 6

Classification Models Using Decision Tree, Random Forest, and Moving Average Analysis

ROHIT DUTT,[1] HARISH DUREJA,[2] and A. K. MADAN[3]

[1]*School of Medical and Allied Sciences, GD Goenka University, Gurugram–122103, India*

[2]*Department of Pharmaceutical Sciences, M.D. University, Rohtak–124001, India*

[3]*Faculty of Pharmaceutical Sciences, Pt. B.D. Sharma University of Health Sciences, Rohtak–124001, India, Tel.: +91-98963 46211, E-mail: madan_ak@yahoo.com*

ABSTRACT

The use of machine learning techniques in *in-silico* assessment of biological activity has gained paramount importance in the modern era of the drug discovery process. A diverse range of statistical techniques is available in the literature for performing both regression and classification analysis. Depending on the methodology used, either quantitative (regression) or classification prediction is feasible for even vast and diverse chemical data sets. Though correlation models far outnumber classification models in the area of computer-aided drug designing but the significance of classification models for the development of potential therapeutic agents can't be underestimated. With the aid of classification techniques, the chemical space can be partitioned into regions of higher and lower "fitness," i.e., each chemical molecule gets attributed to a predefined class. This enables the identification of a "fitness landscape" having the potential for obtaining desired outcomes. The basic methodology and applications of three classification techniques, i.e., decision tree (DT), random forest (RF), and moving average analysis (MAA) have been briefly reviewed in this work.

6.1   INTRODUCTION

Due to recent advances in chemical, biological, and combinatorial technologies, it is now possible to rapidly synthesize or generate a large number of chemical compounds within no time. Nowadays, it is in routine practice to predict the common properties like lipophilicity, binding affinity, etc., through swift biological tests with novel molecular agents. Accordingly, the scientific world is confronted with vast amounts of physicochemical and biological data (Schneider, 2000). Use of statistical as well as computational techniques is now typically used as a tool to schedule experimental framework at each phase of the drug discovery and development process (Clark, 2006). The later a chemical agent fails in the discovery phase, the more exorbitant it has been, hence envisaging this drug attrition in initial phases is particularly desirable (Cross and Cruciani, 2010). Machine learning depicts a popular family of algorithmic methods based on conventional statistical theories and able to handle ever-increasing chemical database. The application of statistical methods to experimental chemical datasets is an essential part of the (quantitative) structure-activity relationship ((Q)SAR) (Michielan and Moro, 2010). (Q)SAR models can be principally split into two categories: regression/correlation and classification. A substantial amount of statistical/machine learning techniques are reported in the literature for carrying out regression/correlation as well as classification. Some of these are suited for only classification modeling, whereas others can perform both classifications as well as regression (Dutt and Madan, 2012).

The initial QSAR technique, developed by Hansch and Fujita (1964), assumed that there exists a relationship between the molecular properties and its chemical structure, and it was possible to establish simple mathematical equations for the depiction of the desired property shared by a set of compounds which had a definite chemical diversity but some extent of structural similarity (Yang and Huang, 2006). Literature reveals a vast amount of information that overviews the statistical method for establishing correlation among molecular properties of available chemical structural datasets. However, among quantitative, qualitative, and categorical activities, only classification algorithms that build a predictive model by splitting a dataset into predefined classes can be used. Such classification outcome of physicochemical and biological activity defines the diversity among the compounds as well as with respect to their variables (Varma et al., 2015).

In the past decade, several nonlinear modeling methods have emerged as robust alternative tools to the conventional linear statistical methods (Durant et al., 2002) and these are increasingly being used to model a wide range of physicochemical/biological/toxicological properties. Originally developed in the domain distant from chemical or pharmaceutical sciences, the techniques like decision tree (DT) or classification trees (Breiman et al., 1984), ensemble methods (Svetnik et al., 2003; Breiman, 2001) and moving average analysis (MAA) (Gupta et al., 2001, 2003; Dureja et al., 2008) are found effective for data mining large databases as well as to solve both regression and classification problems (Fox and Kriegl, 2006). Every technique has its own advantages and limitations. The selection of appropriate machine learning method is chiefly relying upon the quantity of the physicochemical data and the type of the activity being measured (numerical vs. categorical data).

In this review, a brief summary and applications of three such machine learning techniques, i.e., DT, random forest (RF), and MAA in modeling diverse biological properties have been presented.

6.2 METHODOLOGICAL OVERVIEW

(Q)SAR models which classify the available data into two or more predefined classes (i.e., biologically active or inactive, soluble or insoluble, bio-available or bio-unavailable, mutagenic or non-mutagenic, permeable or impermeable, toxic or non-toxic) are known as classification models. Classification modeling is applicable for both *predictive* (allocating a predefined class) and *descriptive* (to discriminate between the entity of divergent class) purpose. A classification technique (or classifier) is a methodical approach to build a classification model from the chemical dataset. Each classifier employs a suitable learning algorithm to build a model that optimally describes the relationship between the chemical compounds and activity/property of the input data set. The outcomes of the classification can be organized into the so-called confusion or contingency matrix, where the rows depict the reference or reported classes (experimental data), and the columns depict the predicted class assigned during classification (Das and Sengur, 2010). The main diagonal cell represents the cases where the reference class concurs with the predicted class, i.e., the number of compounds classified correctly in each class, whereas the non-diagonal cells illustrate misclassifications or compounds not predicted correctly (Massarelli et al., 2011).

6.2.1 *DECISION OR CLASSIFICATION TREE*

Decision or classification tree is a well-known machine learning statistical algorithm for data mining and pattern recognition that yield classification models by clustering data according to similar attributes or characteristics. Breiman et al., (1984) had a pioneering impact both in bringing the classification scheme to the notice of statisticians and in tendering new algorithms for building classification trees. A wide variety of composite problems can sometimes be easily addressed by recursively splitting them down into compact subsets that are easily resolved. The ability to translate the classification tree into a set of predictive rules in Boolean logic make DT a distinct method in comparison to other statistical techniques (Almuallim and Dietterich, 1994). Moreover, it efficiently governs the high dimensional data, ignores extraneous descriptors, models nonlinear problems well, manages multi-target activities, and is easily acquiescent to model prediction (Breiman, 2001). DT uses recursive segregation to appraise every attribute of the input data and grade them as per their ability to split the remaining data, thereby creating a tree-like framework (Li et al., 2006). A branch of the DT corresponds to a group of classes, and a leaf represents a specific class (see Figure 6.1) (Guha, 2005).

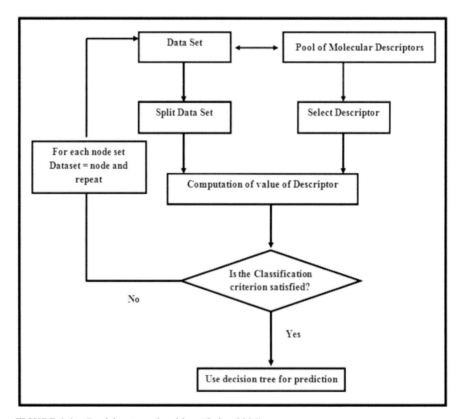

FIGURE 6.1   Decision tree algorithm (Guha, 2005).

An example of data classification into biologically active and inactive by a conventional DT is illustrated in Figure 6.2 (Dutt and Madan, 2012a).

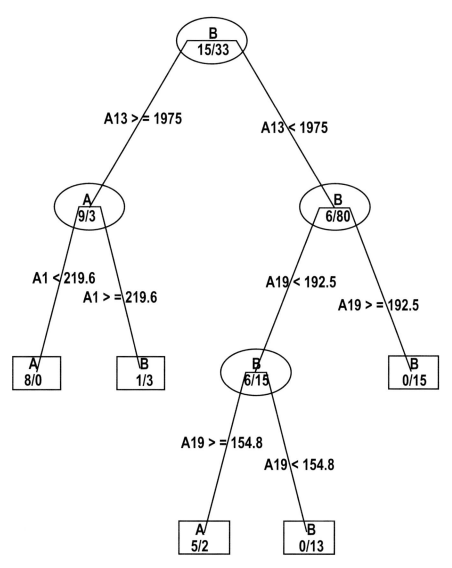

FIGURE 6.2 The decision tree for distinguishing active (A) from inactive (B) (Reproduced with permission from Dutt and Madan, 2012a).

Every predefined class (active and inactive) occupies their specific area in the chemical space (Figure 6.2). Such a graphical presentation can be very useful for a drug designer in assessing the selected molecular descriptor and complementing other feature extraction and classification techniques reported in the literature.

6.2.2   RANDOM FOREST (RF)

Taking the tree paradigm a step further, Breiman (2001) initiated the concept of RFs. This method is based on an ensemble of a large number of classification trees called "forests." This grouping continues until each tree grows on the value of provided random vector, singly instituted and with the similar allocation for the entire range of trees in a forest (Svetnik et al., 2003; Meng et al., 2009). A RF is built by a bootstrap specimen of the input data and random feature extraction in tree induction (Figure 6.3) (Guha, 2005). Therefore, all classification trees in a forest are divergent. The predictive outcome is assessed by a majority vote (or average) of the independent trees (Polishchuk et al., 2009). The graphical representation of RF is similar to DT. The classification outcome of RF is generally shown in terms of confusion matrix as exemplified in Table 6.1 for prediction of cannabinoid receptor-1 (CB-1) antagonistic activity (Dutt and Madan, 2010).

6.2.3   MOVING AVERAGE ANALYSIS (MAA)

MAA model is principally a classification model based upon the maximization of moving average of the input data. For the assortment and analysis of predefined ranges specific attributes, unique activity ranges were located from frequency distribution of biological response level and subsequently assigning the active range by evaluating the outcome by maximization of moving average with regard to active compounds ($<35\%$ = inactive, 35–65% = transitional and $>65\%$ = active) (Gupta et al., 2001, 2003; Dureja et al., 2008). MAA-based classification models are exclusive and vary widely from traditional regression models. Both systems of modeling have their specific advantages and limitations. However, MAA-based modeling system has a clear cut advantage of identification of narrow but vital biologically active range(s), which may be erroneously skipped during routine regression analysis (Dureja and Madan, 2006). An illustrative example of various classification ranges of MAA-based models in the form of a histogram is shown in Figure 6.4 (Dutt and Madan, 2012).

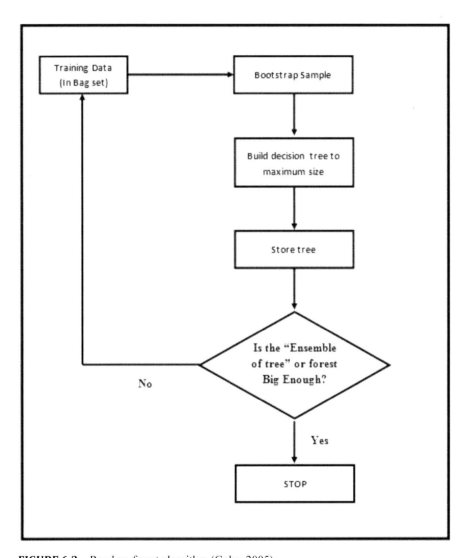

FIGURE 6.3   Random forest algorithm (Guha, 2005).

Based on various ranges of MAA-based model, it is feasible to compile sets of molecules that fall into the "active ranges/drug" area of the chemical space. *Active ranges of the MAA-based models can be easily exploited*

TABLE 6.1 Confusion Matrix for Anti-Proliferative Activity and Recognition Rate of Model Based on Random Forest (Reproduced with Permission from (Dutt and Madan, 2012)

Description	Ranges	Number of compounds predicted		Sensitivity (%)	Specificity (%)	OOB error (%)	Overall accuracy of prediction (%)	MCC
		Active	Inactive					
Random forest	Active	15	05	75	76	24	76	0.50
	Inactive	07	22					

for providing lead structures either through virtual screening of the ever-increasing chemical database or reverse engineering.

The application of DT, RF, and MAA in predicting diverse biological activities is many folds and resulted in a considerable number of predictive models in the last decade. An overview of the applications of these techniques reported in literature for diverse range of therapeutic molecules/activities is summarized in Table 6.2.

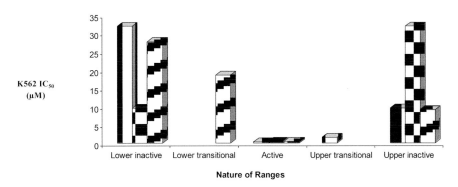

FIGURE 6.4   Average K562 IC_{50} (μM) values of correctly predicted analogs in various ranges of MAA-based model (Reproduced with permission from Dutt and Madan, 2012).

6.3   CONCLUSION AND FUTURE PERSPECTIVE

During the past few decades, various machine learning techniques or classification algorithms have emerged as an effective *in silico* alternative to cope up with the non-linear properties of chemical compounds. Since the precise experimental values are not needed in the classification techniques, therefore, these techniques will naturally be much more suitable to study multiple mechanisms and structural classes. Increasing applications of DT, RF, and MAA has encouraged researchers to utilize these techniques for instant virtual screening. The classification models described here can be used as filters in early stages of a drug discovery project. Simultaneous use of diverse classification techniques/models for discrimination of biologically active or inactive compounds prior to their synthesis will naturally accelerate the drug discovery processes in a rapid and cost-effective manner.

TABLE 6.2   Classification Models for Prediction of Biological Activities

Technique	Chemical class/Source	Number of compounds	Biological activity	Overall accuracy of prediction (%)	References
MAA	5- vinyl pyrimidines nucleoside analogs	100	Antiviral activity	83	Mendiratta and Madan, 1994
	Hydantoin/thiohydantoin analogs	82	Anticonvulsant activity	85	Dang and Madan, 1994
	Pyrazole carboxylic acid hydrazide analogs	76	Anti-inflammatory activity	75–89	Goel and Madan, 1995
	Diverse set	83	Antitumor activity	91	Goel and Madan, 1996
	N-aryl isoxazole carboxamide/benzamide analogs	69	Antiepileptic activity	91	Goel and Madan, 1997
	Piperidinyl methyl and methylene methyl ester analogs	94	Analgesic activity	79–86	Sharma et al., 1997
	4-substituted-2-guanidino thiazoles	128	Gastric H+,K+-ATPase enzyme inhibitory activity	82	Gupta et al., 1999
	N-benzyl imidazoles	81	Angiotensin II receptor antagonistic activity	80	Gupta et al., 2000
	Sulfamoyl benzoic acid derivatives	68	Diuretic activity	82–90	Sardana and Madan, 2001
	HEPT derivatives	107	Anti-HIV activity	> 90	Gupta et al., 2001
	Dihydroseselin analogs	48	Anti-HIV activity	93–99	Gupta et al., 2000a
	5-vinyl pyrimidines nucleosides	104	Anti-HSV-1 activity	74–83	Gupta et al., 2001a
	Pyrazole carboxylic acid hydrazide analogs	76	Anti-inflammatory activity	84–90	Gupta et al., 2002
	Quinolone analogs	52	Anti-mycobacterial activity	81–85	Sardana and Madan, 2002
	TIBO derivatives	121	Anti-HIV activity	84–86	Sardana and Madan, 2002a
	Substituted benzamides/benzylamines	41	Anticonvulsant activity	88–97	Sardana and Madan, 2002b
	Benzimidazoles	81	Antihypertensive activity	82–87	Sardana and Madan, 2003
	Substituted (aminosulfonyl) ureas	41	ACAT inhibitory activity	83–91	Lather and Madan, 2003
	Substituted carboxamides	73	Dopamine D3 receptor binding affinity	82–86	Lather and Madan, 2004

TABLE 6.2 (Continued)

Tech-nique	Chemical class/Source	Number of com-pounds	Biological activity	Overall accuracy of prediction (%)	References
	N-arylanthranilic acids	112	Anti-inflammatory activity	83–90	Bajaj et al., 2004
	Phenethylthiazolethiourea (PETT) compounds	62	Anti-HIV activity	90	Bajaj et al., 2004a
	Substituted-pyridinium-azole compounds	42	Carbonic anhydrase activators	91–100	Bajaj et al., 2004b
	4-amino[1,2,4]triazolo[4,3-a]quinoxalines	138	Adenosine receptors binding activities	80–90	Lather and Madan, 2004a
	Arylpiperazines	30	Adrenoceptor antagonistic activity	90–93	Kumar and Madan, 2004
	2, 3-diaryl–1, 3-thiazolidine-4-one	31	Anti-HIV activity	81–90	Kumar et al., 2004
	Aminothiazoles	54	CDK–2 inhibitory activity	80–86	Kumar and Madan, 2004a
	N-arylanthranilic acids	112	Anti-inflammatory activity	88–92	Bajaj et al., 2005
	4-Benzyl pyridinone derivatives	32	Anti-HIV activity	87–100	Bajaj et al., 2005a
	HEPT analogs	33	Anti-HIV activity	88	Bajaj et al., 2005b
	Acylthiocarbamates	61	Anti-HIV activity	95–98	Bajaj et al., 2005c
	Indole–2-ones	67	CDK2 inhibitory activity	88–90	Dureja and Madan, 2005
	3-Aminopyrazoles	42	Antitumor activity	86–89	Bajaj et al., 2005d
	N-phenylphenylglycine analogs	39	CRF receptor antagonizing activity	82	Bajaj et al., 2005e
	Aloisines	51	CDK–1 inhibitory activity	82–84	Lather and Madan, 2005
	Tetrahydropyrimidine–2-one analogs	80	HIV-protease inhibitory activity	86–88	Lather and Madan, 2005a
MAA	Dihydro (alkylthio) (naphthylmethyl) oxopyrimidines	67	Anti-HIV activity	86–89	Lather and Madan, 2005b
	Pyrrolopyrimidine	82	MRP–1 inhibitory activity	88	Lather and Madan, 2005c
	N-[4-(4-Arylpiperazin–1-yl)butyl]aryl carboxamides	37	Dopamine D3 receptor binding affinity	89–92	Lather and Madan, 2005d

TABLE 6.2 (Continued)

Technique	Chemical class/Source	Number of compounds	Biological activity	Overall accuracy of prediction (%)	References
	Mercaptoacyl dipeptides	39	Dopamine D4 receptor binding affinity	78–81	
			NEP inhibitory activities	82–89	Lather and Madan, 2006
			ACE inhibitory activities	77–90	
	Substituted thiadiazolidinones	28	Glycogen synthase kinase–3b inhibitory activity	83–87	Kumar and Madan, 2005
	1,2-Bis(sulfonyl)–1-methylhydrazines	61	Anti-neoplastic activity	86–88	Bajaj et al., 2006
	1-alkoxy–5-alkyl–6-(arylthio)uracils	36	Anti-HIV-1 activity	81–94	Bajaj and Madan, 2007
	Arylindoles	31	h5-HT$_{2A}$ receptor antagonistic activity	81–84	Dureja and Madan, 2006
	2-pyridinone derivatives	72	Anti-HIV activity	81–89	Bajaj et al., 2006b
	Propan–2-ones	44	Cytosolic phospholipase A$_2$ inhibitory activity	84–88	Kumar and Madan, 2006
	Hydantoins and related non-hydantoins	50	Voltage-gated sodium channel binding activity	>99	Gupta and Madan, 2007
	Thiophenes	59	Agonist allosteric enhancer activity	84–91	Kumar and Madan, 2007
	Flavonoids	30	Telomerase inhibitory activity	80–83	Dureja and Madan, 2007
	Indole–1-ones	41	Poly(ADP-Ribose) polymerase inhibitory activity	83–85	Dureja and Madan, 2007a
	4-oxopyrimido [4, 5-b] quinolines	49	Anti-allergic activity	86–88	Kumar and Madan, 2007a
	Sulfonamides	34	Carbonic anhydrase (CA) inhibitory activities	85–91	Kumar and Madan, 2007b
	6-arylbenzonitriles	81	Anti-HIV-1 activity	81	Dureja et al., 2008

TABLE 6.2 *(Continued)*

Technique	Chemical class/Source	Number of compounds	Biological activity	Overall accuracy of prediction (%)	References
	α,γ–Diketo acids	30	Hepatitis *C* virus inhibitory activity	90–93	Bajaj et al., 2008a
	N-(phenylsulfonyl)benzamides	62	Anti-tumor activity	80–84	Dureja and Madan, 2008
	3-ethyl–1*H*-indoles	26	5-HT$_6$ binding affinity	81–84	Dureja and Madan, 2009
	Dimethylaminopyridin–2-ones	103	Anti-HIV activity	81–85	Dureja and Madan, 2009a
	Dihydro-alkoxy-benzyl-oxopyrimidine	53	Anti-HIV activity	83–96	Dutt et al., 2009
	Biarylpyrazolyl oxadiazoles	76	Cannabinoid–1 receptor antagonistic activity	95–96	Dutt and Madan, 2010
	2,6-Disubstituted pyrazines	56	BRAF inhibitory activity	91–93	Dutt et al., 2010
	5'-*O*-[(*N*- Acyl)sulfamoyl]adenosines	31	Antitubercular activity	90–91	Goyal et al., 2010
	2-Chloro-N^6-substituted-4'-thioadenosine–5'-uronamides	42	A$_3$ adenosine receptor binding affinity	81–97	Goyal et al., 2010a
	Analogs of pyridoimidazolones	43	^{V600E}BRAF inhibitory activity	≥90	Dutt and Madan, 2010a
	3-aminoindazole analogs	42	Receptor tyrosine kinase inhibitory activities	88	Gupta et al., 2011
	Bisphosphonates	65	Anti-tumor activity	87–89	Goyal et al., 2011
	2-aryl benzimidazoles	47	Checkpoint kinase inhibitory activity	90–95	Gupta et al., 2012
	Carbonitriles (check for dual activity)	44	Anti-inflammatory activity	87–93	Goyal et al., 2012
	N-Benzoylated phenoxazines and phenothiazines	53	Anti-proliferative activity	86–90	Dutt and Madan, 2012
	Substituted pyrazines	48	H-CRF–1 receptor binding affinity	≤85	Dutt and Madan, 2012a
	Substituted quinolone carboxylic acids	50	HIV integrase inhibitory activity	96–98	Gupta and Madan, 2012

TABLE 6.2 *(Continued)*

Tech-nique	Chemical class/Source	Number of com-pounds	Biological activity	Overall accuracy of prediction (%)	References
	Aza and diazabiphenyl analogs of PA–824	121	Antitubercular activity	81–99	Marwaha and Madan, 2012
MAA	Isatins	65	hCE1 and hiCE inhibitory activities	81–94	Dutt et al., 2012
	4- substituted piperidin–4-ol derivatives	56	Melanocortin–4 receptor agonist activity	78–98	Gupta and Madan, 2013
	Substituted 4-aminomethylene isoquinoline–1, 3-diones	52	CDK4 inhibitory activity	90–96	Gupta and Madan, 2013a
	Indanylacetic acid derivatives	73	PPARα agonistic activity PPARγ agonistic activity PPARδ agonistic activity PPAR pan (α+γ+δ) agonistic activity		Dutt and Madan, 2013
	Sulfonylurea derivatives	49	H₃ receptor antagonist activity	88–95	Khatri and Madan, 2014
	3-(*1H*-imidazol–1-yl) propyl thiourea derivatives	45	human glutaminyl cyclase (hQC) inhibitory activity	86–95	Gupta et al., 2014
	6-(phenylthio)thymines	107	Anti- HIV activity	81–89	Gupta et al., 2003
	Substituted aminohydantoin analogs	42	β-secretase (BACE1) inhibitory activity	>92	Gupta and Madan, 2014
	Indomethacin analogs	38	Cycloxygenase–2 (COX–2) inhibitory activity	81–97	Singh and Madan, 2014
	N,N-disubstituted indol–3-yl-glyoxylamides	62	TPSO receptor binding affinity	77	Dutt and Madan, 2014
	4-morpholinopyrrolopyrimidines	39	mTOR and PI3Kα and PC3 cells inhibitory activities	87–99	Gupta and Madan, 2014a

TABLE 6.2 *(Continued)*

Tech-nique	Chemical class/Source	Number of com-pounds	Biological activity	Overall accuracy of prediction (%)	References
	Aza and diazabiphenyl analogs of PA–824	121	Antitubercular activity	79–90	Marwaha and Madan, 2014
	Analogs of cysmethynil	46	Icmt inhibitory activity	87–91	Marwaha and Madan, 2014a
	1-aryl–3-(1-acyl)piperidin–4-yl) urea analogs	52	Human sEH inhibitory activity	90–93	Mahajan et al., 2014
	Thiourea derivatives	121	Anti-hepatitis *C* virus activity	89–98	Khatri et al., 2015
	Benzyl phenyl ether derivatives	62	Anti-parasitic activities	88–95	Khatri et al., 2015a
	Purine nucleoside analogs	36	Anti-HIV activity	>91	Khatri et al., 2015b
DT	2,4,6-trisubstituted 1,3,5-triazines	58	Cannabinoid receptor–2 (CB–2) agonistic activity	93	El-Atta et al., 2014
	4-aminomethylene isoquinoline–1,3-(2H,4H)-dione	52	CDK4 inhibitory activity	96 (T) and 69 (CV)	Gupta and Madan, 2013a
	Analogs of pyridoimidazolones	43	V600EBRAF inhibitory activity and melanoma cells growth inhibitory activities	98 (T) and 79 (CV)	Dutt and Madan, 2010
	Substituted pyrazines	48	H-CRF–1 receptor binding affinity	92 (T) and 71 (CV)	Dutt and Madan, 2012a
	2,6-Disubstituted pyrazines	56	BRAF inhibitory activity	96 (T) and 79 (CV)	Dutt et al., 2010
	N-Benzoylated phenoxazines and phenothiazines	53	Anti-proliferative activity	96 (T) and 73 (CV)	Dutt and Madan, 2012
	N,N-disubstituted indol–3-yl-glyoxylamides	62	TPSO receptor binding affinity	94 (T) and 61 (CV)	Dutt and Madan, 2014
	2-aryl benzimidazoles	47	Checkpoint kinase inhibitory activity	96 (T) and 77 (CV)	Gupta et al., 2012

TABLE 6.2   *(Continued)*

Technique	Chemical class/Source	Number of compounds	Biological activity	Overall accuracy of prediction (%)	References
	Quinolone carboxylic acid	50	HIV integrase inhibitory activity	98 (T) and 96 (CV)	Gupta and Madan, 2012
	4-morpholinopyrrolopyrimidines	39	mTOR and PI3Kα cells inhibitory activities	98 (T) and 90 (CV)	Gupta and Madan, 2014b
	3-aminoindazole analogs	42	Receptor tyrosine kinase inhibitory (KDR) activity	88 (T)	Gupta et al., 2011
	Substituted aminohydantoin analogs	42	β-secretase (BACE1) inhibitory activity	98 (T) 81 (CV)	Gupta and Madan, 2014
	4- substituted piperidin–4-ol derivatives	56	Melanocortin–4 receptor agonist activity	98 (T) 96 (CV)	Gupta and Madan, 2013
	3-(1H-imidazol–1-yl) propyl thiourea derivatives	45	Human glutaminyl cyclase (hQC) inhibitory activity	96 (T) 84 (CV)	Gupta et al., 2014
DT	Dihydro-alkoxy-benzyl-oxopyrimidine	53	Anti-HIV-1 activity	98 (T) 77 (CV)	Dutt et al., 2009
	Thiourea derivatives	121	Anti-hepatitis C virus activity	>99 (T) 87 (CV)	Khatri et al., 2015
	Purine nucleoside analogs	36	Anti-HIV activity	>99 (T) 89 (CV)	Khatri et al., 2015b
	Aza and diazabiphenyl analogs of PA–824	121	Antitubercular activity	98 and 83 (T) 72 and 75 (CV)	Marwaha and Madan, 2014
	Analogs of cysmethynil	46	Icmt inhibitory activity	96 (T) 80 (CV)	Marwaha and Madan, 2014a
	Indomethacin analogs	38	COX–2 inhibitory activity	94 (T) 86 (CV)	Singh and Madan, 2014

TABLE 6.2 *(Continued)*

Tech-nique	Chemical class/Source	Number of com-pounds	Biological activity	Overall accuracy of prediction (%)	References
	Diverse range	261	P-gp (P-glycoprotein) substrate specificity	79 (T) 69 (CV)	Joung et al., 2012
	Collected from literatures	163	P-glycoprotein substrates and non-substrates	88 (T) 88 (CV)	Li et al., 2007
	Collected from literatures	377	Aurora-A kinase inhibitory activity	84–87	Morshed et al., 2011
	Collected from 105 literatures	1273	P-gp inhibitors and non-inhibitors	84 and 67	Chen et al., 2011
	9-fluorenyl methoxy carbonyls (peptides)	60	Antimicrobial activity	70	Lira et al., 2013
	Hydantoin derivatives	287	Anticonvulsant activity	75	Sutherland and Weaver, 2003
	1-aryl-3-(1-acylpiperidin-4-yl) urea analogs	52	Human sEH inhibitory activity	96 (T) 79 (CV)	Mahajan et al., 2014
RF	2,4,6-trisubstituted 1,3,5-triazines	58	Cannabinoid receptor-2 (CB-2) agonistic activity	100	El-Atta et al., 2014
	4-aminomethylene isoquinoline-1,3-(2H,4H)-dione	52	CDK4 inhibitory activity	92	Gupta and Madan, 2013a
	N-Benzoylated phenoxazines and phenothiazines	53	Anti-proliferative activity	76	Dutt and Madan, 2012
	N,N-disubstituted indol-3-yl-glyoxylamides	62	TPSO receptor binding affinity	77	Dutt and Madan, 2014
	2-aryl benzimidazoles	47	Checkpoint kinase inhibitory activity	83	Gupta et al., 2012
	Substituted aminohydantoin analogs	42	β-secretase (BACE1) inhibitory activity	93	Gupta and Madan, 2014
	4- substituted piperidin-4-ol derivatives	56	Melanocortin-4 receptor agonist activity	96	Gupta and Madan, 2013
	3-(1H-imidazol-1-yl) propyl thiourea derivatives	45	Human glutaminyl cyclase (hQC) inhibitory activity	93	Gupta et al., 2014

TABLE 6.2 (Continued)

Technique	Chemical class/Source	Number of compounds	Biological activity	Overall accuracy of prediction (%)	References
	Dihydro-alkoxy-benzyl-oxopyrimidine	53	Anti-HIV-1 activity	85	Dutt et al., 2009
	4-morpholinopyrrolopyrimidines	39	mTOR and PI3Kα cells inhibitory activities	> 92	Gupta and Madan, 2014b
	Purine nucleoside analogs	36	Anti-HIV activity	83	Khatri et al., 2015b
	Thiourea derivatives	121	Anti-hepatitis C virus activity	84	Khatri et al., 2015
	Aza and diazabiphenyl analogs of PA–824	121	Antitubercular activity	>80	Marwaha and Madan, 2014
	Analogs of cysmethynil	46	Icmt inhibitory activity	83	Marwaha and Madan, 2014a
	Indomethacin analogs	38	COX–2 inhibitory activity		Singh and Madan, 2014
	Diverse scaffolds	~ 3500	EGFR inhibitors and non-inhibitors	84	Singh et al., 2015
	Diverse scaffolds	91	MMP–3 inhibitors and non-inhibitors	> 99	Li et al., 2015
	Diverse scaffolds	120	MMP–9 inhibitors and non-inhibitors	> 99	Li et al., 2015
	1-aryl–3-(1-acylpiperidin–4-yl) urea analogs	52	Human sEH inhibitory activity	89	Mahajan et al., 2014

KEYWORDS

- **carbonic anhydrase**
- **cyclooxygenase-2**
- **decision tree**
- **moving average analysis**
- **phenylethylthiazolethiourea**
- **random forest**

REFERENCES

Almuallim, H., & Dietterich, T. G., (1994). Learning Boolean concepts in the presence of many irrelevant features. *Artificial Intell.*, *69*, 279–306.

Bajaj, S., & Madan, A. K., (2007). Topochemical models for anti-HIV activity of 1-alkoxy-5-alkyl-6-(arylthio) uracils. *Chem. Papers, 61*, 127–132.

Bajaj, S., Sambi, S. S., & Madan, A. K., (2004a). Topological models for prediction of anti-inflammatory activity of n-arylanthranilic acids. *Bioorg. Med. Chem.*, *12*, 3695–3701.

Bajaj, S., Sambi, S. S., & Madan, A. K., (2004b). Prediction of anti-HIV activity of phenyl-ethylthiazolethiourea (PETT) analogs using wiener's topochemical index. *J. Mol. Struct., 684*, 197–203.

Bajaj, S., Sambi, S. S., & Madan, A. K., (2004c). Prediction of carbonic anhydrase activation of tri/tetra substituted-pyridinium-azole compounds-computational approach using novel topochemical descriptor. *QSAR Comb. Sci., 23*, 506–514.

Bajaj, S., Sambi, S. S., & Madan, A. K., (2005a). Prediction of anti-inflammatory activity of N-arylanthranilic acids: Computational approach using Zagreb topochemical indices. *Croat. Chem. Acta*, *78*, 165–174.

Bajaj, S., Sambi, S. S., & Madan, A. K., (2005b). Topochemical models for prediction of anti-HIV activity of 4-benzyl pyridinone derivatives. *Drug Develop. Indust. Pharm.*, *31*, 1041–1051.

Bajaj, S., Sambi, S. S., & Madan, A. K., (2005c). Topochemical models for prediction of anti-HIV activity of hept analogs. *Bioorg. Med. Chem. Lett.*, *15*, 467–469.

Bajaj, S., Sambi, S. S., & Madan, A. K., (2005d). Topological models for prediction of anti-HIV activity of acylthiocarbamates. *Bioorg. Med. Chem., 13*, 3263–3268.

Bajaj, S., Sambi, S. S., & Madan, A. K., (2005e). Topochemical approach of predicting anti-tumor activity of 3-aminopyrazoles. *Chem. Pharm. Bull.*, *53*, 611–615.

Bajaj, S., Sambi, S. S., & Madan, A. K., (2005f). Topochemical model for prediction of corticotropin-releasing factor antagonizing activity of n-phenylphenylglycines. *Acta Chim. Slov., 52*, 292–296.

Bajaj, S., Sambi, S. S., & Madan, A. K., (2006a). Models for prediction of anti-neoplastic activity of 1,2-bis(sulfonyl)–1-methylhydrazines: computational approach using Wiener's indices. *Commun. Math. Comput. Chem.*, *56*, 193–204.

Bajaj, S., Sambi, S. S., & Madan, A. K., (2006b). Model for prediction of anti-HIV activity of 2-pyridinone derivatives using novel topological descriptor. *QSAR Comb. Sci.*, *25*, 813–823.

Bajaj, S., Sambi, S. S., & Madan, A. K., (2008). Topochemical models for prediction of activity of α,γ–Diketo acids as inhibitors of hepatitis *C* virus NS5b RNA-dependent RNA polymerase. *Pharm. Chem. J.*, *40*, 650–654.

Breiman, L., (2001). Random forests. *Mach. Learn.*, *45*, 5–32.

Breiman, L., Friedman, J. H., Olshen, R. A., & Stone, C. J., (1984). *Classification and Regression Trees* (Vol.1, pp. 18–55). Boca Raton, Florida: Chapman & Hall/CRC.

Chen, L., Li, Y., Zhao, Q., Peng, H., & Hou, T., (2011). ADME evaluation in drug discovery. 10. predictions of p-glycoprotein inhibitors using recursive partitioning and naive Bayesian classification techniques. *Mol. Pharm.*, *8*, 889–900.

Clark, D. E., (2006). What has computer-aided molecular design ever done for drug discovery? *Expert Opin. Drug Discov.*, *1*, 103–110.

Cross, S., & Cruciani, G., (2010). Molecular fields in drug discovery: Getting old or reaching maturity? *Drug Discov. Today*, *15*, 23–32.

Dang, P., & Madan, A. K., (1994). Structure-activity study of anticonvulsant (thio) hydantoins using molecular connectivity indices. *J. Chem. Inf. Comput. Sci.*, *34*, 1162–1166.

Das, R., & Sengur, A., (2010). Evaluation of ensemble methods for diagnosing of valvular heart disease. *Expert Syst. Appl.*, *37*, 5110–5115.

Durant, J. L., Leland, B. A., Henry, D. R., & Nourse, J. G., (2002). Reoptimization of MDL keys for use in drug discovery. *J. Chem. Info. Comput. Sci.*, *42*, 1273–1280.

Dureja, H., & Madan, A. K., (2005). Topochemical models for prediction of cyclin-dependent kinase 2 inhibitory activity of indole–2–ones. *J. Mol. Model.*, *11*, 525–531.

Dureja, H., & Madan, A. K., (2006). Models for the prediction of h5-HT$_{2A}$ receptor antagonistic activity of arylindoles: Computational approach using topochemical descriptors. *J. Mol. Graph. Mod.*, *25*, 373–379.

Dureja, H., & Madan, A. K., (2006). Prediction of 852 h5-HT2A receptor antagonistic activity of arylindoles: Computational approach using topochemical descriptors. *J. Mol. Graph. Model.*, *25*, 373–379.

Dureja, H., & Madan, A. K., (2007a). Topochemical models for prediction of telomerase inhibitory activity of flavonoids. *Chem. Biol. Drug Des.*, *70*, 47–52.

Dureja, H., & Madan, A. K., (2007b). Topochemical models for the prediction of poly (ADP-ribose) polymerase inhibitory activity of indole–1–ones. *Med. Chem. Res.*, *16*, 15–27.

Dureja, H., & Madan, A. K., (2008a). Prediction of antitumor activity of N-(phenylsulfonyl) benzamides: Computational approach using topochemical descriptors. *Leonardo J. Sci.*, *7*, 214–231.

Dureja, H., & Madan, A. K., (2009a). Topochemical model for the prediction of 5-HT$_6$ binding affinity of 3-ethyl–1H-indoles. *Leonardo J. Sci.*, *8*, 112–123.

Dureja, H., & Madan, A. K., (2009b). Predicting anti-HIV activity of dimethylamino pyridine–2-ones: Computational approach using topological descriptors. *Chem. Biol. Drug Des.*, *73*, 258–270.

Dureja, H., Gupta, S., & Madan, A. K., (2008a). Predicting anti-HIV-1 activity of 6-arylbenzonitriles: Computational approach using super-augmented eccentric connectivity topochemical indices. *J. Mol. Graph. Model.*, *26*, 1020–1029.

Dureja, H., Gupta, S., & Madan, A. K., (2008b). Topological models for prediction of pharmacokinetic parameters of cephalosporins using random forest, decision tree and moving average analysis. *Sci. Pharm.*, *76*, 377–394.

Dutt, R., & Madan, A. K., (2010a). Models for the prediction of cannabinoid–1 receptor antagonistic activity of biarylpyrazolyl oxadiazoles. *In Silico Biol.*, *10*, 247–263.

Dutt, R., & Madan, A. K., (2010b). Models for prediction of V^{600E} BRAF and melanoma cells growth inhibitory activities of pyridoimidazolones. *Arch. Pharm.*, *343*, 664–679.

Dutt, R., & Madan, A. K., (2012a). Classification models for anticancer activity. *Curr. Top. Med. Chem.*, *12*, 2705–2726.

Dutt, R., & Madan, A. K., (2012b). Predicting biological activity: Computational approach using novel distance-based molecular descriptors. *Comput. Biol. Med.*, *42*, 1026–1041.

Dutt, R., & Madan, A. K., (2013). Models for prediction of PPARs agonistic activity of indanylacetic acid derivatives. *Med. Chem. Res.*, *22*, 3218–3228.

Dutt, R., & Madan, A. K., (2014). Models for prediction of TPSO receptor binding affinity of N,N-disubstituted indol–3yl-glyoxylamides. *J. Eng. Sci. Manage. Educ.*, *7*(III), 166–177.

Dutt, R., Dureja, H., & Madan, A. K., (2009). Models for prediction of anti HIV-1 activity of 5-alkyl–2-alkylamino–6-(2,6-difluorophenylalkyl)–3,4-dihydropyrimidin–4(3*H*)-ones. *J. Comput. Method Sci. Eng.*, *9*, 95–114.

Dutt, R., Dureja, H., & Madan, A. K., (2010). Decision tree and moving average analysis based models for prediction of BRAF inhibitory activity of 2,6-disubstituted pyrazines. *Int. J. Chem. Model.*, *3*, 73–97.

Dutt, R., Singh, M., & Madan, A. K., (2012). Improved superaugmented eccentric connectivity indices for QSAR/QSPR Part II: Application in development of models for prediction of hiCE and hCE1 inhibitory activities of isatins. *Med. Chem. Res.*, *21*, 1226–1236.

El-Atta, A. H. A., Moussa, M. I., & Hassanien, A. E., (2014). Predicting biological activity of 2,4,6-trisubstituted 1,3,5-triazines using random forest. In: Kömer, P., Abraham, A., & Snášel, V., (eds.), *Proceedings of the Fifth International Conference on Innovations in Bio-Inspired Computing and Applications-IBICA 2014* (Vol. 303, pp. 101–110). Springer International Publishing: Switzerland.

Fox, T., & Kriegl, J. M., (2006). Machine learning techniques for *in silico* modeling of drug metabolism. *Curr. Top. Med. Chem.*, *6*, 1579–1591.

Goel, A., & Madan, A. K., (1995). Structure-activity study on anti-inflammatory pyrazole carboxylic acid hydrazide analogs using molecular connectivity indices. *J. Chem. Inf. Comput. Sci.*, *35*, 510–514.

Goel, A., & Madan, A. K., (1996). Structure-activity study on antitumor/carcinogenic agents using valence molecular connectivity index. *Pharm. Pharmacol. Lett.*, *6*, 70–72.

Goel, A., & Madan, A. K., (1997). Structure-activity study on antiepileptic *N*-aryl isoxazole carboxamide/ N-isoxazolylbenzamide analogs using topological indices. *Struct. Chem.*, *8*, 155–159.

Goyal, R. K., Dureja, H., Singh, G., & Madan, A. K., (2010a). Models for antitubercular activity of 5'-O-[(N- Acyl)sulfamoyl]adenosines. *Sci. Pharm.*, *78*, 791–820.

Goyal, R. K., Dureja, H., Singh, G., & Madan, A. K., (2010b). Topological models for the prediction of A3 adenosine receptor binding affinity of 2-chloro-N–6-substituted-4´-thioadenosine–5´-uronamides. *Int. J. Chem. Model.*, *2*, 397–416.

Goyal, R. K., Dureja, H., Singh, G., & Madan, A. K., (2012). Models for anti-inflammatory activity of 8-substituted-4-anilino-6-aminoquinoline–3-carbonitriles. *Med. Chem. Res.*, *27*, 1044–1055.

Goyal, R. K., Singh, G., & Madan, A. K., (2011). Models for anti-tumor activity of bisphosphonates using refined topochemical descriptors. *Naturwissenschaften*, *98*(10), 871–887.

Guha, R., (2005). Methods to improve the reliability, validity, and interpretability of QSAR models, pp. 42–43, PhD Thesis, Department of Chemistry: The Pennsylvania State University, The United States.

Gupta, M., & Madan, A. K., (2007). Topochemical models for the prediction of voltage-gated sodium channel binding activity of hydantoins and related non-hydantoins. *J. Mol. Model.*, *13*, 137–145.

Gupta, M., & Madan, A. K., (2012). Diverse models for the prediction of HIV integrase inhibitory activity of substituted quinoline carboxylic acids. *Arch. Pharm.*, *345*, 989–1000.

Gupta, M., & Madan, A. K., (2013a). Models for the prediction of melanocortin–4 receptor agonist activity of 4- substituted piperidin-4-ol. *Int. J. Comput. Biol. Drug Des.*, *6*, 294–317.

Gupta, M., & Madan, A. K., (2013b). Models for the prediction of CDK4 inhibitory activity of substituted 4-aminomethylene isoquinoline–1, 3-diones. *J. Chem. Sci.*, *125*, 483–493.

Gupta, M., & Madan, A. K., (2014a). Detour cum distance matrix based topological indices for QSAR/QSPR Part-II: Application in development of models for the prediction of BACE 1 inhibitory activity of substituted aminohydantoins. *Lett. Drug Des. Discov.*, *11*, 864–876.

Gupta, M., & Madan, A. K., (2014b). Diverse models for the prediction of dual mTOR and PI3Kα inhibitory activities of substituted 4-morpholinopyrrolopyrimidines. *Lett. Drug Des. Discov.*, *11*, 454–473.

Gupta, M., Dureja, H., & Madan, A. K., (2011). Models for the prediction of receptor tyrosine kinase inhibitory activity of substituted 3-aminoindazole analogs. *Sci. Pharm.*, *79*, 239–257.

Gupta, M., Gupta, S., Dureja, H., & Madan, A. K., (2012). Superaugmented eccentric distance sum connectivity indices: Novel highly discriminating topological descriptors for QSAR/QSPR. *Chem. Biol. Drug Des.*, *79*, 38–52.

Gupta, M., Jangra, H., Bharatam, P. V., & Madan, A. K., (2014). Relative eccentric distance sum/product indices for QSAR/QSPR: Development, evaluation, and application. *ACS Comb. Sci.*, *16*, 101–112.

Gupta, S., Singh, M., & Madan, A. K., (1999). Superpendentic index: a novel topological descriptor for predicting biological activity. *J. Chem. Inf. Comput. Sci.*, *39*, 272–277.

Gupta, S., Singh, M., & Madan, A. K., (2000a). Connective eccentricity index: a novel topological descriptor for predicting biological activity. *J. Mol. Graphics Model.*, *18*, 18–25.

Gupta, S., Singh, M., & Madan, A. K., (2000b). Eccentric distance sum: a novel graph invariant for predicting biological and physical properties. *J. Math. Anal. Appl.*, *275*, 386–401.

Gupta, S., Singh, M., & Madan, A. K., (2001a). Predicting anti-HIV activity: Computational approach using novel topological descriptor. *J. Comput. Aided Mol. Des.*, *15*, 671–678.

Gupta, S., Singh, M., & Madan, A. K., (2001b). Application of graph theory–relationship of molecular connectivity index and atomic molecular connectivity index with anti-HSV activity. *J. Mol. Struct.*, *571*, 147–152.

Gupta, S., Singh, M., & Madan, A. K., (2002). Application of graph theory–relationship of eccentric connectivity index and wiener's index with anti-inflammatory activity. *J. Math. Anal. Appl.*, *266*, 259–268.

Gupta, S., Singh, M., & Madan, A. K., (2003). Novel topochemical descriptors for predicting anti-HIV activity. *Indian J. Chem.*, *42A*, 1414–1425.

Hansch, C., & Fujita, T., (1964). p-σ-π Analysis. A method for the correlation of biological activity and chemical structure. *J. Am. Chem. Soc.*, *86*, 1616–1629.

Joung, J. Y., Kim, H., Kim, H. M., Ahn, S. K., Nam, K. Y., & No, K. T., (2012). Prediction models of p-glycoprotein substrates using simple 2d and 3d descriptors by a recursive partitioning approach. *Bull. Korean Chem. Soc.*, *33*, 1123–1127.

Khatri, N., & Madan, A. K., (2014). Models for H3 receptor antagonist activity of sulfonylurea derivatives. *J. Mol. Graph. Model.*, *48*, 87–95.

Khatri, N., Dutt, R., & Madan, A. K., (2015). Role of moving average analysis for development of multi-target (Q) SAR models. *Mini Rev. Med. Chem.*, *15*, 659–674.

Khatri, N., Lather, V., & Madan, A. K., (2015a). Diverse classification models for anti-hepatitis *C* virus activity of thiourea derivatives. *Chemometr. Intell. Lab.*, *140*, 13–21.

Khatri, N., Lather, V., & Madan, A. K., (2015b). Diverse models for anti-HIV activity of purine nucleoside analogs. *Chem. Cent. J.*, *9*, 29.

Kumar, V., & Madan, A. K., (2004a). Topological model for the prediction of alpha–1 adrenoceptor antagonistic activity of arylpiperazines. *J. Theor. Comput. Chem.*, *3*, 245–255.

Kumar, V., & Madan, A. K., (2004b). Topological models for the prediction of cyclin-dependant kinase 2 inhibitory activity of aminothiazoles. *Comm. Math. Comput. Chem.*, *(MATCH)*, *51*, 59–78.

Kumar, V., & Madan, A. K., (2005). Application of graph theory: Prediction of glycogen synthase kinase–3 inhibitory activity of thiadiazolidinones as potential drugs for the treatment of Alzheimer's disease. *Eur. J. Pharm. Sci.*, *24*, 213–218.

Kumar, V., & Madan, A. K., (2006). Application of graph theory: Prediction of cytosolic phospholipase A2 inhibitory activity of propan–2-ones. *J. Math. Chem.*, *39*, 511–521.

Kumar, V., & Madan, A. K., (2007a). Prediction of agonist allosteric enhancer activity of thiophenes at human A1 adenosine receptors using topological indices. *Pharm. Chem. J.*, *41*, 140–145.

Kumar, V., & Madan, A. K., (2007b). Predicting anti-allergic activity of 4-oxopyrimido [4, 5-b] quinolines: Computational approach using topochemical indices. *Med. Chem. Res.*, *16*, 88–99.

Kumar, V., & Madan, A. K., (2007c). Topological models for the prediction of carbonic anhydrase inhibitory activity of sulfonamides. *J. Math. Chem.*, *42*, 925–940.

Kumar, V., Sardana, S., & Madan, A. K., (2004). Predicting anti-HIV activity of 2, 3-diaryl–1, 3-thiazolidin-4-ones: Computational approach using reformed eccentric connectivity index. *J. Mol. Mod.*, *10*, 399–407.

Lather, V., & Madan, A. K., (2003). Predicting acyl-coenzyme A: Cholesterol O-acyltransferase inhibitory activity: Computational approach using topological descriptors. *Drug Des. Discov.*, *18*, 117–122.

Lather, V., & Madan, A. K., (2004a). Predicting dopamine d_3 receptor binding affinity of n-(ω-(4-(2-methoxyphenyl)piperazin–1-yl)alkyl) carboxamides: Computational approach using topological descriptors. *Comm. Math. Comput. Chem.*, *(MATCH)*, *52*, 65–89.

Lather, V., & Madan, A. K., (2004b). Models for the prediction of adenosine receptors binding activity of 4-amino[1,2,4]triazolo[4,3-a]quinoxalines. *J. Mol. Struct.*, *678*, 1–9.

Lather, V., & Madan, A. K., (2005a). Application of graph theory: Topological models for the prediction of CDK–1 inhibitory activity of aloisines. *Croat. Chem. Acta*, *78*, 55–61.

Lather, V., & Madan, A. K., (2005b). Topological models for the prediction of HIV-protease inhibitory activity of tetrahydropyrimidin–2-ones. *J. Mol. Graph. Model.*, *23*, 339–345.

Lather, V., & Madan, A. K., (2005c). Topological models for the prediction of anti-HIV activity of dihydro (alkylthio) (naphthylmethyl) oxopyrimidine. *Bioorg. Med. Chem.*, *13*, 1599–1604.

Lather, V., & Madan, A. K., (2005d). Topological models for the prediction of MRP1 inhibitory activity of pyrrolopyrimidines and templates derived from pyrrolopyrimidine. *Bioorg. Med. Chem.*, *15*, 4967–4972.

Lather, V., & Madan, A. K., (2005e). Predicting dopamine receptors binding affinity of N-[4-(4-arylpiperazin–1-yl)butyl]aryl carboxamides: Computational approach using topological descriptors. *Curr. Drug Discov. Technol., 2*, 115–121.

Lather, V., & Madan, A. K., (2006). Topological models for the prediction of neutral endopeptidase and angiotensin-converting enzyme inhibitory activity of mercaptoacyldipeptides. *J. Theor. Comput. Chem., 5*, 565–577.

Li, B. K., He, B., Tiana, Z. Y., Chen, Y. Z., & Xue, Y., (2015). Modeling, predicting, and virtual screening of selective inhibitors of MMP–3 and MMP–9 over MMP–1 using random forest classification. *Chemometr. Intell. Lab. Sys., 147*, 30–40.

Li, H., Ung, C. Y., Yap, C. W., Xue, Y., Li, Z. R., & Chen, Y. Z., (2006). Prediction of estrogen receptor agonists and characterization of associated molecular descriptors by statistical learning methods. *J. Mol. Graph. Model., 25*, 313–323.

Li, W. X., Li, L., Eksterowicz, J., Ling, X. B., & Cardozo, M., (2007). Significance analysis and multiple pharmacophore models for differentiating P-glycoprotein substrates. *J. Chem. Inf. Model., 47*, 2429–2438.

Lira, F., Perez, P. S., Baranauskas, J. A., & Nozawa, S. R., (2013). Prediction of antimicrobial activity of synthetic peptides by a decision tree model. *Appl. Environ. Microbiol., 79*, 3156–3159.

Mahajan, N., Lather, V., Sambi, S. S. A. K., & Madan, A. K., (2014). Models for human soluble epoxy hydrolase inhibitory activity of 1-aryl-3-(1-acylpiperidin-4-yl) urea analogs. *Int. J. Chem. Model., 6*, 41–63.

Marwaha, R. K., & Madan, A. K., (2012). Fourth generation detour matrix based topological descriptors for QSAR/QSPR Part-II: Application in development of models for prediction of biological activity. *Int. J. Comput. Biol. Drug Des., 5*, 335–360.

Marwaha, R. K., & Madan, A. K., (2014a). Fourth generation detour matrix-based topological descriptors for QSAR/QSPR Part-II: Application in development of models for prediction of biological activity. *Int. J. Comput. Biol. Drug Des., 7*, 1–30.

Marwaha, R. K., & Madan, A. K., (2014b). Path eccentricity based highly discriminating molecular descriptors for QSAR/QSPR Part-II. Application in development of models for prediction of a biological activity. *Int. J. Chem. Model., 6*, 557–579.

Massarelli, I., Imbriani, M., James, T. L., Mundula, T., & Bianucci, A. M., (2011). Development of classification model batteries for predicting inhibition of tubulin polymerization by small molecules. *Chemomet. Intell. Lab., 107*, 206–214.

Mendiratta, S., & Madan, A. K., (1994). Structure-activity study on antiviral 5-vinyl nucleoside analogs using wiener's topological index. *J. Chem. Inf. Comput. Sci., 34*, 867–871.

Meng, Y. A., Yu, Y., Cupples, L. A., Farrer, L. A., & Lunetta, K. L., (2009). Performance of random forest when SNPs are in linkage disequilibrium. *BMC Bioinformatics, 10*, 78.

Michielan, L., & Moro, S., (2010). Pharmaceutical perspectives of nonlinear QSAR strategies. *J. Chem. Inf. Model., 50*, 961–978.

Morshed, M. N., Cho, Y. S., Seo, S. H., Han, K. C., Yang, E. G., & Pae, A. N., (2011). Computational approach to the identification of novel Aurora-A inhibitors. *Bioorg. Med. Chem., 19*, 907–916.

Polishchuk, P. G., Muratov, E. N., Artemenko, A. G., Kolumbin, O. G., Muratov, N. N., & Kuzmin, V. E., (2009). Application of random forest approach to QSAR prediction of aquatic toxicity. *J. Chem. Inf. Model., 49*, 2481–2488.

Sardana, S., & Madan, A. K., (2001). Application of graph theory–relationship of molecular connectivity index, wiener's index and eccentric connectivity index with diuretic activity. *J. Math. Comp. Chem., 43*, 85–98.

Sardana, S., & Madan, A. K., (2002a). Application of graph theory: Relationship of antimycobacterial activity of quonolone derivatives with eccentric connectivity index and zagreb group parameter. *J. Math. Comput. Chem.*, *45*, 35–53.

Sardana, S., & Madan, A. K., (2002b). Predicting anti-HIV activity of TIBO derivatives: Computational approach using novel topological descriptor. *J. Mol. Model.*, *8*, 258–265.

Sardana, S., & Madan, A. K., (2002c). Predicting anticonvulsant activity of benzamides/benzylamines: Computational approach using topological descriptors. *J. Comput. Aided Mol. Des.*, *16*, 545–550.

Sardana, S., & Madan, A. K., (2003). Topological models for prediction of antihypertensive activity of substituted benzylimidazoles. *J. Mol. Struct.*, *638*, 41–49.

Schneider, G., (2000). Neural networks are useful tools for drug design. *Neural Netw., 13*, 15–16.

Sharma, V., Goswamy, R., & Madan, A. K., (1997). Eccentric connectivity index: a highly discriminating topological descriptor for structure-property and structure-activity studies. *J. Chem. Inf. Comput. Sci.*, *37*, 273–282.

Singh, H., Singh, S., Singla, D., Agarwal, S. M., & Raghava, G. P. S., (2015). QSAR-based model for discriminating EGFR inhibitors and non-inhibitors using Random forest. *Biol. Direct, 10*, 10.

Singh, M., & Madan, A. K., (2014). Detour matrix based adjacent path eccentric distance sum indices for (Q)SAR/QSPR Part II: Application in development of models for COX–2 inhibitory activity of indomethacin derivatives. *Int. J. Comput. Biol. Drug Des.*, *7*, 319–340.

Sutherland, J. J., & Weaver, D. F., (2003). Development of quantitative structure-activity relationships and classification models for anticonvulsant activity of hydantoin analogs. *J. Chem. Inf. Comput. Sci.*, *43*, 1028–1036.

Svetnik, V., Liaw, A., Tong, C., Culberson, J. C., Sheridan, R. P., & Feuston, B. P., (2003). Random forest: A classification and regression tool for compound classification and QSAR modeling. *J. Chem. Inf. Comput. Sci., 43*, 1947–1958.

Varma, S., Fisher, L., Lyons, T., & Chen, D., (2015). *Creating a Smart Virtual Screening Protocol, Part II: Recursive Partitioning for Sequential Screening.*

Yang, G. F., & Huang, X., (2006). Development of quantitative structure-activity relationships and its application in rational drug design. *Curr. Pharm. Des., 12*, 4601–4611.

CHAPTER 7

Colorants, Pigments, and Dyes

SERHAT VARIŞ and LEMI TÜRKER

Department of Chemistry, Middle East Technical University, 06800, Ankara, Turkey

7.1 DEFINITION

Colorant is the term that color scientists use for the entire range of coloring materials including pigments and dyes (Gottsegen, 1993). So, colorants are classified as either pigments or dyes depending on the medium they are involved. Pigments are insoluble materials (in water, oils, and resins) having fine sized particles. They are used in coatings for one or more of five reasons: to provide color, to hide substrates, to modify the performance properties of films, and/or to reduce cost. Pigments and dyes are often derived from the same basic building blocks. Pigments are practically insoluble in the medium in which they are incorporated; however, dyes dissolve in a proper solvent during application, losing their crystal or particulate structure in the process (see Figure 7.1). The difference between pigments and dyes is therefore due to physical characteristics rather than main chemical composition, such as phthalocyanine dyes and pigments. Pigments seem the colors they have because they selectively reflect and absorb certain wavelengths of visible light. White light is an approximately equal mixture of the whole spectrum of visible light with a wavelength in a range from about 375 or 400 nm to about 760 or 780 nm. When this light comes on a pigment, some parts of the spectrum are absorbed by the chemical bonds of conjugated systems and other components of the pigment. Some other wavelengths or parts of the spectrum are reflected or scattered. The new reflected light spectrum creates the appearance of a color. For example, a blue pigment absorbs red and green light in the daylight, but reflects blue, creating the color blue (Griffiths, 1976; Fabian and Hartmann, 1980).

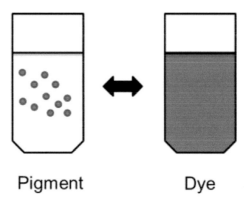

FIGURE 7.1 The visual description of pigments and dyes.

Pigments are divided into four broad classes: white, color, inert, and functional pigments. White (hiding) pigments are the ones that contribute light-scattering properties to coatings. They act by scattering all wavelengths of light, owing to their relatively high refractive index, so they are perceived as white by the human eye. They are also named as hiding pigments because the scattering of light reduces the probability that light will penetrate through a pigmented film to the substrate. A paint film of sufficient thickness and concentration of light-scattering pigment is truly opaque, hiding the substrate. The whiteness and opacity contributed by this class of pigments make them among the most extensively used pigments for coatings. The most extensively used white pigment is the rutile crystal form of titanium dioxide (TiO_2). Rutile has the highest index of refraction (2.76) of any material that can be manufactured in pigment form at a reasonable cost, making it the most efficient white pigment available. Another crystal form of TiO_2, *anatase*, is sometimes used in coatings, but its lower index of refraction (2.55) makes it a less optically efficient pigment. Furthermore, surface-treated TiO_2 in its rutile form yields coatings that are more durable to exterior exposure than are equivalent *anatase* pigments. TiO_2 pigments are used in very high volume worldwide, especially in the so-called trade sales market, which includes retail, architectural, and contractor markets. In these applications, light, pastel, and white coatings predominate, thus the demand for TiO_2. Other white pigments are zinc oxide (ZnO), zinc sulfide (ZnS), and lithopone, a mixture of barium sulfate ($BaSO_4$) and ZnS. The earliest commercial white pigment was "white lead," basic lead carbonate ($Pb(OH)_2 \cdot PbCO_3$), which was widely used until about 1925–30. Since the compound is water soluble and toxic, its use in coatings has been restricted since the 1960s.

Its commercial use actually stopped much earlier; because of its low index of refraction (1.94), the white lead had been replaced by titanium dioxide, which is more than eight times as efficient in hiding power. Nevertheless, the presence of old, peeling paint containing white lead pigment continues to be a health hazard in older buildings that are poorly maintained.

Color pigments act by absorbing certain wavelengths of visible light and transmitting or scattering the other wavelengths. There are a large number of color pigments, both organic and inorganic, that allow paint users to create films of almost all the colors in the visible spectrum. Some commonly used color pigments are copper phthalocyanine-based greens and blues, quinacridone red, iron oxide red, iron oxide yellow, diarylide yellow, and perinone orange.

Inert pigment is a paint additive that does not change the shade or hue, but extends or otherwise imparts a special working quality to the paint. Inert pigments absorb little light and have refractive indexes close enough to those of binders that they give little light scattering when used as pigments. Inert fillers and extenders are synonymous terms used for inert pigments. They are quite cheap and reduce the cost of coating. The main function of inert pigment is to occupy volume in a film. The other functions are to adjust the rheological properties of fluid coatings and mechanical properties of films. The chalk (calcium carbonate, $CaCO_3$), magnesium silicate, gypsum, barium white, alumina white are best known inert pigments.

Functional pigments are the ones that are used to modify the application characteristics, appearance or film properties of coatings. Corrosion inhibiting pigments take place in this class. Complex zinc chromate pigments, strontium chromate, barium phosphosilicates, barium borosilicates, red lead and zinc phosphates are used in primers to inhibit corrosion of steel by passivation of anodic area. In contrast to other pigments, they must be somewhat soluble in water to function.

As for dyes, they are soluble in the medium they involved during application, losing their crystal or particulate structure in the process contrary to pigments. The difference between pigments and dyes is, therefore, due to physical characteristics. The dyes used to fall into many categories: basic dyes, fat-soluble dyes, metal complex dyes, infrared absorbing dyes, etc. (Gordon and Gregory, 1983; Matsuoka, 1990).

Basic dyes are cationic dyes which are highly soluble in polar solvents such as alcohols, glycols, and water. They are used by the printing ink industry with lacking agents such as tannic acid to produce clean, bright shades. However, their poor light-fastness limits their usefulness.

Fat-soluble dyes include nonionic, metal-free azo and anthraquinone dyes, which are highly soluble in less polar solvents, such as aromatic and aliphatic hydrocarbons. Azo dyes are mainly used for aliphatic/aromatic solvent based wood stains and in the coloration of styrene polymers. Anthraquinone dyes are used much more widely in the coloration of plastics and fibers on account of their much wider range of resistance properties.

Metal complex dyes are mainly anionic chromium and cobalt complexes of azo dyes. The cation is either a sodium ion or a substituted ammonium ion. Substituted soluble phthalocyanines also fall into this category. These dyes are normally soluble in alcohols, glycol ethers, ketones, and esters.

The molecular orbital essentials of color chemistry, chromophores, auxochromes, etc., have been discussed (Dewar and Dougherty, 1975). The differences between photochemistry and color chemistry have been emphasized theoretically (Turro, 1991).

7.2 HISTORICAL ORIGIN(S)

Textile dyeing dates back to the Neolithic period. Scarce dyestuffs that produced brilliant and enduring colors such as the natural invertebrate dyes, Tyrian purple, and Crimson Kermes were highly prized luxury items in the ancient and medieval world. Plant-based dyes such as woad, indigo, saffron, and madder were raised commercially and were important trade goods in the economies of Asia and Europe. Across Asia and Africa, patterned fabrics were produced using resist dyeing techniques to control the absorption of color in piece-dyed cloth. Dyes from the New World such as cochineal and logwood were brought to Europe by the Spanish treasure fleets, and the dyestuffs of Europe were carried by colonists to America. Dyed flax fibers have been found in the Republic of Georgia in a prehistoric cave dated to 36,000 BP. Archaeological evidence shows that, particularly in India and Phoenicia, dying has been widely carried out for over 5,000 years. The dyes were obtained from animal, vegetable or mineral origin, with none to very little processing. By far the greatest source of dyes has been from the plant kingdom, notably roots, berries, bark, leaves, and wood, but only a few have ever been used on a commercial scale. The discovery of man-made synthetic dyes late in the 19th century ended the large-scale market for natural dyes (Booth et al., 2000).

The earliest known pigments were natural minerals. Natural iron oxides give a range of colors and are found in many Paleolithic and Neolithic cave paintings. Two examples include Red Ochre, anhydrous Fe_2O_3, and the hydrated Yellow Ochre ($Fe_2O_3.H_2O$) Charcoal, or carbon black, has also been used as a black pigment since prehistoric times (Garfield, 2000). Archaeologists have discovered that pigments and paint grinding equipment believed to be between 350,000 and 400,000 years old have been reported in a cave at Twin Rivers, near Lusaka, Zambia. Biological pigment, e.g., Tyrian Purple, is made from the mucus of Murex snail. Production of Tyrian purple for use as a fabric dye began as early as 1200 BC by the Phoenicians, and was continued by the Greeks and Romans until 1453 CE, with the fall of Constantinople (Kassinger, 2003). Since the pigment was quite expensive, it became associated with power and wealth. Two of the first synthetic pigments were white lead (basic lead carbonate ($Pb(OH)_2$. $PbCO_3$ and blue frit (Egyptian Blue). White lead is made by combining lead with vinegar (acetic acid, CH_3COOH) in the presence of CO_2. Some others are Cobalt Blue (1802), Synthetic Ultramarine Blue (1828), Viridian Green (1838), Cadmium Yellow (1846), and Zinc White (1751) (Gottsegen, 1993). Blue frit is calcium copper silicate and was made from glass colored with a copper ore, such as malachite. These pigments were used as early as the second millennium BCE. The Industrial and Scientific Revolutions brought a huge expansion in the range of synthetic pigments, pigments that are manufactured or refined from naturally occurring materials, available both for manufacturing and artistic expression. Prussian blue was the first modern synthetic pigment, discovered by accident in 1704. By the early 19th century, synthetic, and metallic blue pigments had been added to the range of blues, including French ultramarine, a synthetic form of lapis lazuli, and the various forms of Cobalt and Cerulean Blue. In the early 20th century, organic chemistry added Phthalo Blue (copper phthalocyanine), a synthetic, organometallic pigment with overwhelming tinting power.

Discoveries in color science produced new industries and caused changes in fashion and taste. The discovery in 1856 of Mauveine, the first aniline dye, was a forerunner for the development of hundreds of synthetic dyes and pigments like azo and diazo compounds which are the source of a wide spectrum of colors. Mauveine was discovered by an 18-year-old chemist named William Henry Perkin, who went on to exploit his discovery in the industry and become wealthy. His success attracted a generation of followers, as young scientists went into organic chemistry to pursue riches. Within a few years, chemists had synthesized a substitute for Madderin the production of Alizarin Crimson. By the closing decades of the 19th century, textiles, paints, and other commodities in colors such as red, crimson, blue, and purple had become affordable (Garfield, 2000).

Development of chemical pigments and dyes helped bring new industrial prosperity to Germany and other countries in northern Europe, but it brought dissolution and decline elsewhere. In Spain's former New World Empire, the production of cochineal colors employed thousands of low-paid workers. The Spanish monopoly on cochineal production had been worth a fortune until the early 19th century, when the Mexican War of Independence and other market changes disrupted production. Organic chemistry delivered the final blow for the cochineal color industry. When chemists created inexpensive substitutes for carmine, an industry and a way of life went into steep decline (Greenfield, 2006). Today, the synthetic organics make up the largest share of colorants used in industry.

7.3   NANO-SCIENTIFIC DEVELOPMENT(S)

A pigment should not react chemically or physically with other paints or supplementary materials to which it is exposed. These materials include the binder, varnishes, thinner or solvents, glaze or painting mediums, grounds, and other pigments. Pigments should not alter in hue (color), chroma (the relative intensity of color) or value (the relative lightness or darkness of the color) when exposed to normal conditions of light over a long period of time. Also, a pigment should not react to changes in normal atmospheric conditions. It should form a good film with the binder which binds the pigment particles on the surface (Gottsegen, 1993).

Nano-scientific developments in the area of colorants occur mainly in pigment industry. Nano-pigments are organic or inorganic substances; insoluble, chemically, and physically inert into the substrate or binders, with a particle size less than 100 nm (Cain and Morrell et al., 2001). Particle size affects color strength, transparency or opacity, exterior durability, solvent resistance and other properties of pigments. Nano-sized pigments have quite high surface area, which assures higher surface coverage, and high number of reflectance points, hence improved scattering. Compared to the effects of conventional pigments, the use of nano-pigments, especially nano-fillers, improves resistance to scratch, mar, abrasion, heat, radiation, and swelling resistance, decrease water permeability and increase hardness, weatherability, modulus, and strain to failure while maintaining toughness (Fasaki et al., 2012). Although nano-pigment term is new, nano-pigments have been used in coating for many years. For instance, high strength channel blacks have particle diameters in the range 5 to 15 nm. Some clay can also be considered as nano-pigments, as they can be separated into sheets on the order of 1 nm thick (Ahmad et al., 2010). Mica-based pigments (particle size about 20 nm) with pearlescent effect are used in cosmetics, automobile coatings, plastics, etc. In cosmetic applications, both doped TiO_2 and ZnO are being developed for use in sunscreens, in order to avoid the skin damage by sunlight radiation. Novel nano-sized reflecting powders, providing a broad spectrum protection against UV radiation, are more acceptable from the cosmetic viewpoint because they are flesh-toned and turn invisible when applied. Jet-color is dedicated in producing nano-dispersed pigment for use in digital inkjet inks, paints, and coatings. Jet colors nano-dispersed pigments have ultrafine particle size and wide color gamut. With nano-dispersed pigments, users can easily make high-quality inkjet inks, paints, and coatings.

Another application is nano-pigment screen, i.e., a new style phosphor screen applied to cathodic ray tube TVs, exploiting nano-pigment to improve contrast, color gamut and body color without additional process or cost. Using smaller pigment particles in the liquid crystal display (LCD) technology can improve not only the stability of pigment in the dispersion medium, but also color strength, contrast, and transmittance. Also traditional inorganic pigments, including titanium dioxide, zinc oxide, silicon oxide, and magnesium oxide, are being available in a nano-size range for use in rubber and plastics, e.g., polyethylene (PE), poly(ethylene-vinyl acetate) (PEVA), and poly(vinyl chloride) (PVC) (Cavalcante and Dondi et al., 2009).

7.4 NANO-CHEMICAL APPLICATION(S)

Current nano-chemical applications of pigments of variant areas in the literature are summarized below.

Iron oxide (Fe_2O_3) red is technologically an important pigment and has superior character in non-toxicity, chemical stability, and durability and low costs. It is widely applied as pigments in the building industry, inorganic dyes, ceramics, and adsorbents in the paper industry, lacquers or plastics. Natural iron oxides are derived from hematite, which is a red iron oxide mineral; limonites, which vary from yellow to brown, such as ochers, siennas, and umbers; and magnetite, which is black iron oxide. Synthetic iron oxide pigments are produced from basic chemicals. The three major methods for the manufacture of synthetic iron oxides are thermal decomposition of iron salts or iron compounds, precipitation of iron salts usually accompanied by oxidation, and reduction of organic compounds by iron. Lately, the synthesis of nano-iron red oxide pigment by cyanided tailings via ammonia process with urea has been published. The particle size of iron oxide crystal prepared on different temperature and pH conditions showed different color shades (Dengxin et al., 2008).

Metal nanoparticle pigments are employed to give yellow to red coloration on glass and transparent glazes. Color also depends on particle size, particles inter-distance, alternating dielectric, and metal particles. Color is developed through the mechanism of surface plasmon resonance (SPR), which is due to charge-density oscillations confined to metal nanoparticles. The excitation of surface plasmons by an electric field at an incident wavelength, where resonance occurs, results in strong light scattering, appearance of intense SPR absorption bands, and enhancement of the local electromagnetic field. The frequency (i.e., absorption maxima and therefore color) and intensity of SPR bands are characteristic of each metal and highly sensitive to particle size and shape. Ceramic nano-pigments have been recently developed for ink-jet decoration of ceramic tiles using quadrichromic technology (cyan, magenta, yellow, and black colors). The coloring mechanisms and performance of $CoAl_2O_4$, Au, (Ti, Cr, Sb)O_2, and $CoFe_2O_4$ nano-pigments were investigated (Cavalcante et al., 2009).

Recently, a simple microwave-assisted route for producing gold, silver, and Au-Ag structures in the form of stable nano-suspensions as ceramic colorant has been published. Control of particle size and colloidal stability was pursued through accurate reaction optimization combined with microwave heating. The synthesis of bimetallic Au-Ag nano-pigments provided a rather wide range of particle size and optical properties. It is a flexible route which allows to change particles size easily and composition only by tuning the process parameters. Nano-pigments synthesized in this way are suitable for application as ceramic inks for digital decoration. A set of red colors with different shades can be obtained changing the percent composition of silver and the particle size Ag-Au structure (Blosi et al., 2012).

Inorganic pigments synthesized in nano-size are an integral part of many decorative and protective coatings and used for the coloration of ceramic materials including glazes, clay bodies, and porcelain enamels. Nano-sized pigments have a considerable market potential due to their high surface area, which assures higher surface coverage, and high number of reflectance points, hence improved scattering. Cobalt aluminate ($CoAl_2O_4$) has received significant attention as a coloring agent in glaze and bulk tile compositions due to its superior properties, such as high refractive index, chemical, and thermal stability, and color stability. Recently, ceramic blue nano-pigments of cobalt aluminate have been synthesized using an ultrasonic-assisted hydrothermal method. Nano-sized cobalt aluminate is expected to have a spinel-type structure, with Co^{2+} situated at the tetrahedral site. The transition for Co^{2+} ions in a tetrahedral site shows that the two first spin-allowed bands fall in the infrared region (ca. 1400 and 1600 nm), and only the third one is present in the visible region, usually as a triple band around 540 nm (green region), 590 nm (yellow-orange region), and 640 nm (red region), which induces blue color. The UV–visible spectra indicate that the absorption behavior might be affected by the crystallite and the particle size (Kim and Cho et al., 2012). Nano-sized cobalt aluminate has also been obtained by hydrothermal method and been employed in color TV tubes as contrast-enhancing luminescent (Chen et al., 2002). In another study, different types of nano-sized cobalt aluminate pigments were prepared by co-precipitation process. The cetyltrimethylammonium bromide (CTAB) and polyvinyl-pyrrolidone (PVP) as double capping agents have been employed in the co-precipitation process. The particle size (about 50 nm) and its distribution of the particles have been controlled by varying the amount of CTAB and PVP. Homogeneous and stable cyan ink, without any visible particles, have been obtained with dispersion of the produced nano-$CoAl_2O_4$ particles by an itaconic acid-co-acrylic acid dispersant. Nano-sized cobalt aluminate is employed in direct ceramic ink-jet printing (DCIJP) applications (Peymanniaa et al., 2014). Another ceramic ink-jet blue ceramic nano-pigment (24–35 nm), $Co_xMg_{1-x}Al_2O_4$, has recently been produced as a solid solution from two mixed phases $CoAl_2O_4$ and $MgAl_2O_4$ spinels and 3-methyl-pyrazole-5-one (3MP5O) followed by calcination. The study shows that the variation of calcination temperatures from 500 to 1200°C affects the color and particle size (Ahmed et al., 2008).

Magnetic nano-powder of spinel ferrite (MFe_2O_4; where M may be Mg, Fe, Mn, Co, Ni, Zn, Cu, and Cd) is a technologically important material which was first synthesized 50 years ago. Special physicochemical properties of these materials result different application such as radio frequency coils, transformer, magnetocaloric refrigeration, contrast enhancement in magnetic resonance imaging (MRI), magnetically guided drug delivery, ferrofluid, microwave absorbing material, sensor, as radar absorbing materials (RAM) in military to increase the invisibility to radar and heat-resistant pigments. The high electrical resistivity and excellent magnetic properties make this ferrite a good candidate as a core material for power transformers in telecommunication and electronic applications in megahertz frequency regions. Magnetic inorganic pigments have been used in high technology applications because of some of their unique properties. Formerly, nickel zinc ferrites have been prepared by conventional methods, which have disadvantages such as high period heating which may result in Zn volatilization and change of the stoichiometry. In recent years, researchers are concerned with the synthesis of spine based brown pigments, which has been widely studied during last decades to improve color hue. Nano-crystalline magnetic pigment has been synthesized via polymeric precursor method and followed by calcination. The average crystallite size using Scherrer's equation confirms crystal growth with enhancement of calcination temperature. Increase of calcination temperature causes more saturation of magnetization. Colorimetric results indicate that the color of the nano-sized pigment changes from reddish-brown to dark-brown depending on the calcination temperature (Moeen et al., 2010).

The near infra-red (NIR) reflecting inorganic pigments absorbs in the visible region and reflect the NIR portion of incident radiation. They have been widely employed in the defense, construction, plastic, and ink industries. NIR reflective pigments coated on the external walls and roofs of a building have been reported to reduce energy consumption by making the interior of the building cooler. There is a strong incentive to develop novel colored, NIR reflecting inorganic pigments that are less hazardous to health and environment. Recently, NIR reflective nano-crystalline Y_2BaCuO_5 green pigment (50 nm) with impressive NIR reflectance (61% at 1100 nm) has been obtained using the nano-emulsion method at relatively lower temperatures as compared to micron-sized pigment. The coloring performance of the designed pigment has been evaluated by applying on to roofing materials like concrete cement blocks and plastic materials. Notably, the developed green nano-pigment sample exhibits NIR solar reflectance, when coated on cement concrete block, thus having a great potential in applications such as cool materials used for buildings with energy saving performance. The current green nano-inorganic pigment also possesses high chemical and thermal stability. The potential utility of the synthesized nano-pigment formulations has also been evaluated by coating on polymer materials like PVC with the goal to develop polymer NIR coatings for use in automobile and electrical cable industries (Jose et al., 2014). Similarly, another nano-sized orange, yellow pigment, Cr, and Sb co-doped TiO_2, has been synthesized and its NIR-reflectance has been determined as % 52. Also, it has good cooling properties (Zou et al., 2014). Besides, a series of near-infrared reflective nano-pigments based on $Co_{(1-x)}Zn_xCr_{(2-y)}Al_yO_4$ ($x = 0–1$ and $y = 0–2$) have been synthesized via Pechini-type sol-gel method. The synthesized pigments have been reported to be new candidates for use as cool pigments owing to their good NIR reflectance (%50) (Hedayati et al., 2015). In addition, the synthesis of nickel titanate ($NiTiO_3$) yellow nano-pigment has been reported. $NiTiO_3$ nanoparticles (30–60 nm) were prepared by tetra-n-butyl titanate and nickel(II) acetate in polyacrylic acid and followed by pyrolyzing the polymer precursor at a moderate temperature. It has been reported to serve as an excellent cool pigment for building roofs and facades due to its 62% NIR reflectance (Wang et al., 2013).

Inorganic nano-pigments are extensively used in a variety of applications such as paints, plastics, coatings, glasses, and especially anticorrosion application. The majority of inorganic pigments used in the anticorrosion application are derived from toxic metals such as cadmium, lead, chromium or cobalt. Recently, the anticorrosive cerium zinc molybdate nano-pigment (about 27 nm) has been synthesized by the reaction of zinc oxide, cerium nitrate, sodium molybdate and nitric acid using ultrasound method. This synthesis enables to produce anticorrosion pigment without toxic metals (Patel et al., 2013).

Polymers are generally known to be good insulating materials due to their stable physical and chemical properties. Both mechanical and electrical properties, however, can be further improved or modified with the addition of inorganic filler type nano-pigments as demonstrated by the increase in the mechanical strength of the composite and changes in the electrical properties. Recent studies showed that nano-sized CoO/MgO/kaolin mixed pigments improves the mechanical, rheological, and dielectric properties of styrene-butadiene rubber composites (Ahmed et al., 2011).

7.5 MULTI-/TRANS-DISCIPLINARY CONNECTION(S)

Nano-pigments are widely employed in various disciplines. They are used in cosmetic applications due to absorption of UV characteristic of some nano-pigments. Another application is nano-pigment screen employed in cathodic ray tube TVs. Ceramic nano-pigments have been used for ink-jet decoration of ceramic tiles using quadrichromic technology. The NIR reflecting inorganic nano-pigments are widely employed in the defense, construction, plastic, and ink industries. They are also employed in NIR coatings for use in automobile and electrical cable industries. Inorganic nano-pigments are extensively used in indoor and outdoor paints, plastics, coatings, glasses, and especially anticorrosion application. Metal nano-pigments represent an efficient way to give yellow to red coloration on glass and transparent glazes. They are also employed in different application such as radio frequency coils, transformer, magnetocaloric refrigeration, contrast enhancement in MRI, magnetically guided drug delivery, ferrofluid, microwave absorbing material, sensor, and RAM in military area.

7.6   OPEN ISSUES

As the particle diameters of pigments are reduced, especially below 10 nm, the physical properties of the materials change, causing various complications (Wicks et al., 2007). No matter what kind of methods is used to fabricate the nanoparticles, the nanoparticles will aggregate again due to effects of Coulomb electrostatic forces and van der Waals forces. Many mechanical processing methods have been successfully developed for dispersing agglomerated particles in liquids, including agitator discs, colloid mills, high-pressure homogenizers, triple roller mills, ball mills, sand mills, and beads mills. In particular, bead mills are used in industrial processing for grinding and dispersing agglomerated particles with primary particles in the submicrometer size range as milling systems have used to prepare photoresist pigment of color filter in LCDs industry. It is necessary to form a stable colloidal suspension for the nanoparticle applications (Chiu et al., 2010).

KEYWORDS

- **coating**
- **color**
- **dispersion**
- **dyes**
- **pigments**

REFERENCES AND FURTHER READING

Ahmad, W. Y. W., Ahmad, M. R., Hamid, H. A., Kadir, M. I. A., Ruznan, W. S., & Yusoh, M. K. M., (2010). Nano-sized natural colorants from rocks and soils. International conference on advancement of materials and nanotechnology: (ICAMN—2007), Langkawi, Kedah (Malaysia). *AIP Conf. Proc., 1217*, 515.

Ahmed, I. S., Ali, A. A., & Dessouki, H. A., (2008). Synthesis and characterization of new nanoparticles as blue ceramic pigment. *Spectrochimica Acta Part-A, 71*, 616–620.

Ahmed, N. M., El-Messieh, S. L. A., & El-Nashar, D. E., (2011). Utilization of new micronized and nano-CoO-MgO/kaolin mixed pigments in improving the properties of styrene-butadiene rubber composites. *Materials and Design, 32*, 170–182.

Blosi, M., Albonetti, S., Baldi, G., Dondi, M., & Gatti, F., (2012). Au-Ag nanoparticles as red pigment in ceramic inks for digital decoration. *Dyes and Pigments, 94*, 355–362.

Booth, G., McLaren, K., Sharples, W. G., Weswell, A., & Zollinger, H., (2000). Dyes, general survey. In: *Ullmann's Encyclopedia of Industrial Chemistry*, Wiley-VCH.

Cain, M., & Morrell, R., (2001). Nanostructured ceramics: A review of their potential. *Applied Organometallic Chemistry*, *15*, 321–330.

Cavalcante, P. M. T., Baldi, G., Dondi, M., Guarini, G., & Raimondo, M., (2009). Color performance of ceramic nano-pigments. *Dyes and Pigments*, *80*, 226–232.

Chen, Z., Li, W., Shi, E., Zheng, Y., & Zhong, W., (2002). Hydrothermal synthesis and optical property of nano-sized $CoAl_2O4$ pigment. *Materials Letters, 55*, 281–284.

Chiu, H., Chang, C., Chiang, T., Huang, Y., & Kuo, M., (2010). Preparation, particle characterizations and application of nano-pigment suspension. *Polymer-Plastics Technology and Engineering*, *49*(15), 1552–1562.

Dengxin, L., Chong, J., Fanling, M., & Guolong, G., (2008). Preparation of nano-iron oxide red pigment powders by use of cyanided tailings. *Journal of Hazardous Materials, 155*, 369–377.

Dewar, M. J. S., & Dougherty, R. C., (1975). *The PMO Theory of Organic Chemistry*. Plenum-Rosetta Edition, NY.

Eastaugh, N., Chaplin, T., Siddall, R., & Walsh, V., (2004). *The Pigment Compendium A Dictionary of Historical Pigments.* Elsevier Butterworth-Heinemen.

Fabian, J., & Hartmann, H., (1980). *Light Absorption of Organic Colorants*. Springer-Verlag, Berlin.

Fasaki, I., Arabazis, I., Arin, M., Glowacki, B. A., Hopkins, S. C., Lommens, P., Siamos, K., & Van Driessche, I., (2012). Ultrasound-assisted preparation of stable water-based nanocrystalline TiO_2 suspensions for photocatalytic applications of inkjet-printed films. *Applied Catalysis A–General*, *411/412*, 60–69.

Garfield, S., (2000). *Mauve: How One Man Invented a Color That Changed the World*. Faber and Faber. ISBN 0–393–02005–3.

Gordon, P. F., & Gregory, P., (1983). *Organic Chemistry in Color*. Springer-Verlag, Heidelberg.

Gottsegen, M. D., (1993). *The Painter's Handbook.* Watson-Guptill Pub. NY.

Greenfield, A. B., (2006). *A Perfect Red: Empire, Espionage, and the Quest for the Color of Desire.* HarperCollins, NY, USA.

Griffiths, J., (1976). *Color and Constitution of Organic Molecules*. Academic Press, London.

Hedayati, H. R., Alvani, A. A. S., Moosakhani, S., Salimi, R., Sameie, H., Tabatabaee, F., & Zarandi, A. A., (2015). Synthesis and characterization of Co(1-x)ZnxCr(2-y)AlyO4 as a near-infrared reflective color tunable nano-pigment. *Dyes and Pigments, 113*, 588–595.

Jose, S., Laha, S., Natarajan, S., Prakash, A., & Reddy, M. L., (2014). Green colored nano-pigments derived from Y2BaCuO5: NIR reflective coatings. *Dyes and Pigments, 107*, 118–126.

Kassinger, R. G., (2003). *Dyes: From Sea Snails to Synthetics*, 21st Century Books. USA, ISBN 0–7613–2112–8.

Kim, J. H., Cho, W., Hwang, K., Kim, U., Noh, H., Son, B., & Yoon, D., (2012). Characterization of blue CoAl2O4 nano-pigment synthesized by ultrasonic hydrothermal method. *Ceramics International*, *38*, 5707–5712.

Matsuoka, M., (1990). *Infrared Absorbing Dyes*. Plenum Press, NY.

Moeen, S. J., Vaezi, M. R., & Yousefi, A. A., (2010). Chemical synthesis of nano-crystalline nickel-zinc ferrite as a magnetic pigment. *Prog. Color Colorants Coat*, *3*, 9–17.

Patel, M. A., Bhanvase, B. A., & Sonawane, S. H., (2013). Production of cerium zinc molybdate nano pigment by innovative ultrasound assisted approach. *Ultrasonics Sonochemistry, 20*, 906–913.

Peymanniaa, M., Ghaharib, M., Najafi, F., & Soleimani-Gorgania, A., (2014). Production of a stable and homogeneous colloid dispersion of nano CoAl2O4 pigment for ceramic ink-jet ink. *Journal of the European Ceramic Society, 34*, 3119–3126.

Rossotti, H., (1983). *Color: Why the World Isn't Grey*. Princeton, NJ: Princeton University Press. ISBN 0–691–02386–7.

Turro, N. J., (1991). *Modern Molecular Photochemistry*. University Science Books, Sausalito, California.

Wang, J., Byon, Y., Li, Y., Mei, S., & Zhang, G., (2013). Synthesis and characterization of NiTiO3 yellow nano pigment with high solar radiation reflection efficiency. *Powder Technology, 235*, 303–306.

Wicks, Z. W., Jones, J. F. N., Pappas, S. P., & Wicks, D. A., (2007). *Chapter 20: Pigments in Organic Coatings: Science and Technology* (pp. 418–434). Wiley-VCH, ISBN: 9780471698067.

Zou, J., Liu, C., Peng, Y., & Zhang, P., (2014). Highly dispersed (Cr, Sb)-co-doped rutile pigments of cool color with high near-infrared reflectance. *Dyes and Pigments, 109*, 113–119.

CHAPTER 8

Cuboctahedron Model of Atomic Mass Based on a Dual Tetrahedral Coordinate System

JAMES GARNETT SAWYER II[1] and MARLA JEANNE WAGNER[2]

[1]241 Lexington Avenue, Buffalo, NY 14222, USA

[2]407 Norwood Avenue, Buffalo, NY 14222, USA

8.1 DEFINITION

The periodic of the elements (PTE) includes a fundamental periodic order formula constant related to wavelength, speed of light, and frequency for each of the 92 elements. In challenging the Heisenberg Uncertainty Principle (Heisenberg, 1927), we concur with Einstein that there exists a uniform electromagnetic and quantum gravity (QG) field (Einstein, 1927). The QG field was not found at the Planck scale. We contend that the QG field exists within a cuboctahedron model as a dual coordinate system of electron orbital, and quantum loop in three and six dimensions as dual tetrahedral coordinate system (TDCS), between spinning electrons radiating from the nucleus on four and six axes of symmetry, and returning to the nucleus on three axes of symmetry (x, y, z) as QG. We contend that polyhedral order can be proven to describe the geometry of the nucleus, based upon the integration of Hawking's twelve spinning quarks (Hawking, 2001) with the cuboctahedron's six axes of symmetry model (r, s, t, u, v, w). Six dimension theory integrates charge with spinning quarks (instead of neutrons and protons) to present a new vision for quantum space. We contend that there exists a common six-dimensional (6D) polyhedral geometry related to the atomic mass of quantum quarks for individual atoms. The Planck spinning sphere theory of Michael Sarnowski connects cuboctahedron packing of quarks with Buckminster Fuller's vector equilibrium model (Sarnowski, 2015). The electrons are

spinning as gyroscopes in one direction; however, two opposing electrons on an axis of symmetry are spinning in opposite directions relative to the central nucleus. The electromagnetic forces of the spinning electrons are radiating from the nucleus in four, eight, and twelve directions on four and six axes of symmetry and returning to the nucleus in six directions on three axes of symmetry, in a continuous unified loop. This TDCS geometry is revealed in the self-assembly of Clathrate polyhedrons.

8.2   HISTORICAL ORIGINS

The geometric polyhedrons of nanocrystals exhibit a fundamental order of atomic force fields based on three, four, and six axes of symmetry of polyhedrons such as octahedron, cuboctahedron, and truncated octahedron. The early history of chemistry was founded on principles of geometric properties of liters as cubic decimeters, counting atoms by Amedeo Avogadro (1776–1856) as a mole, Mendeleev PTE, and Robert Mosley's deterministic x-ray frequency ratio for the 92 atomic elements. Linus Pauling demonstrated the four axes of symmetry tetrahedral bonding of the carbon atom. The counting and weighting of elements revealed a geometric shell layer pattern concerning the positions of the atomic elements in space. Humphry Davy first discovered electricity as electrons that bind atomic matter, which is now presented as electronegativity by Linus Pauling. Albert Einstein in general relativity (GR) applied a dual coordinate system to define the geodesic fields of astronomical gravity and positional Galilean geometry in space. The application of the nanocrystal polyhedral geometry, spinning electron field shielding effect, and DCCMAM to the nobel gases (NG) predicts the location in space (noted by gas constant) of spinning electrons within the three-dimensional (3D) fields of QG in a state of equilibrium with the stronger bonding 6D fields of energy called electronegativity.

8.3   NANO-SCIENTIFIC DEVELOPMENTS

Henry Moseley used the cathode ray tube to establish that the periodic of 92 elements was classified as atomic number z and based on a frequency constant of the square (Moseley, 1913). The periodic order must be derived from the spinning frequency constant of an inner electron and nucleon found in all atomic elements. We contend that the resultant frequency and number

of electrons per shell present a geometric pattern for the positions of all atomic elements in three and six dimensions. Three new atomic elements were predicted by Moseley and were later discovered as Tc 43, Pm 61, Re 75 (Moseley, 1913). This set of new elements is a column in the Quark 6D of Atomic Mass (8.1).

The inverse meter electron volt is in a constant ratio with the atomic wavelength of the specific element. All of the elements are related to Planck's constant in the following formula: electron volts multiplied by wavelength λ equals 1.2398 constant (MIT, 2012). Atomic z as a unit of frequency energy is proportional to the quantum number of spinning electrons for a specific atomic mass. There are exceptions for atomic mass as weight (Cobalt 27 has a greater mass then Nickel 28), which present evidence for the existence of a dual coordinate system. This spinning electron energy constant must originate from a specific atomic nuclear isotope. The frequency of spinning electron rest energy has a nucleon with an equivalent spinning frequency. The sum of the energy for a given atomic element (z) is the sum of spinning energy for all subatomic particles. The 92 atomic elements (z) are theoretical atomic motors which spin at different frequencies based on the following formula (which is related to λ, the speed of light, and frequency for each atomic element) (MIT, 2012):

$$\lambda \ (nm) = c/v = hc/E = 1.238/E \ (keV)$$

The nucleon-electron force must equal the theoretical photon force (c) that positions an electron on an axis of symmetry. The nucleon force liberating the spinning electron releases at the speed of light and thus becomes a photon. For example, the s inner electron spinning at a specific frequency of revolution is held by a force equivalent to an electron traveling at the speed of light. When an electron is liberated by another electron, it travels at the constant speed of light (c) and has a frequency based on the revolutions of the spinning electron. The photo polarization of light presents light in the rotation as either a right or left the circular rotation.

We contend that the order of atomic wavelength, a specific number of units per sell (Magic Numbers), and Mosley's proportion for atomic number reveal a progressive polyhedral order for electron shell layers related to an increase in frequency. We contend that the frequency waves are, in actuality, 360° rotations of spinning electrons on axes of symmetry. The speed of light (c) is equivalent to the energy needed by the nucleus to release an electron from its rest mass.

TABLE 8.1 Mosely's X-Ray Frequencies with Bohr Model for Atomic Elements (Van Grieken)

Z		Kev	λ	Z		Kev	λ	Z		Kev	λ
H	1	0	Infinity	Ge	32	9.80631	0.12643	Eu	63	39.22523	0.031608
He	2	0.0102	121.502	As	33	10.44918	0.11865	Gd	64	40.50077	0.030613
Li	3	0.04082	30.3755	Se	34	11.11246	0.11157	Tb	65	41.79671	0.029664
Be	4	0.09184	13.5022	Br	35	11.79614	0.10511	Dy	66	43.11306	0.028758
B	5	0.16327	7.59388	Kr	36	?	?	Ho	67	44.44982	0.027893
C	6	0.25511	4.86008	Rb	37	13.22474	0.09375	Er	68	45.80699	0.027066
N	7	0.36735	3.37506	Sr	38	13.9697	0.08875	Tm	69	47.18457	0.026276
O	8	0.50001	2.47963	Y	39	14.7349	0.08414	Yb	70	48.58255	0.02552
F	9	0.65307	1.89847	Zr	40	15.5207	0.07988	Lu	71	50.00095	0.024796
Ne	10	?	?	Nb	41	16.32684	0.07594	Hf	72	51.43975	0.024103
Na	11	1.02043	1.21502	Mo	42	17.15338	0.07228	Ta	73	52.898961	0.023438
Mg	12	1.23478	1.00415	Tc	43	18.00034	0.06888	W	74	54.378581	0.0228
Al	13	1.46442	0.84376	Ru	44	18.8677	0.06571	Re	75	55.878609	0.022188
Si	14	1.72452	0.71895	Rh	45	19.75547	0.06276	Os	76	57.399046	0.0216
P	15	2.00004	0.61991	Pd	46	20.66365	0.06	Ir	77	58.939892	0.021035
S	16	2.29596	0.540009	Ag	47	21.59224	0.05742	Pt	78	60.501146	0.020492
Cl	17	2.61229	0.474617	Cd	48	22.54124	0.055	Au	79	62.082809	0.0199707
Ar	18	?	?	In	49	23.51065	0.05273	Hg	80	?	?
K	19	3.30619	0.375006	Sn	50	24.50046	0.0506	Ti	81	65.30736	0.018985
Ca	20	3.68374	0.336571	Sb	51	25.51068	0.0486	Pb	82	66.95025	0.0185188
Sc	21	4.08171	0.303755	Te	52	26.54132	0.04671	Bi	83	68.61354	0.0180699
Ti	22	4.50008	0.275514	I	53	27.59235	0.04493	Po	84	?	?
V	23	4.93887	0.251037	Xe	54	?	?	At	85	?	?
Cr	24	5.39806	0.22968	Cs	55	29.75566	0.041667	Rn	86	?	?
Mn	25	5.87766	0.21094	Ba	56	30.86743	0.040165	Fr	87	?	?
Fe	26	6.37767	0.1944	La	57	32.0006	0.038744	Ro	88	?	?
Co	27	6.89809	0.17974	Ce	58	33.15369	0.037396	Ac	89	?	?
Ni	28	7.43892	0.16667	Pr	59	34.32718	0.036118	Th	90	80.828062	0.0153392
Cu	29	8.00015	0.15498	Nd	60	35.52108	0.034904	Po	91	?	?
Zn	30	8.58179	0.14447	Pm	61	36.73539	0.03375	U	92	84.5016	0.0146723
Ga	31	9.18385	0.135	Sm	62	37.97012	0.032653				

8.4 NANO-CHEMICAL APPLICATIONS

The 6D periodic polyhedral order found in the PTE is also documented in the geometry of the truncated octahedron as Brillouin Zone and Fermi surface of atomic copper. The present Fermi surface is reoriented according to the science of crystallography, whereby Px, Py, Pz are in alignment with (100) square planes as three axes of symmetry (x, y, z) and the (111) triangular-hexagonal planes as six axes of symmetry (r, s, t, u, v, w). The 6D cuboctahedron model with three and six axes of symmetry is the standard to unify the periodic order between the alignments of the spinning electrons with the spinning quarks in the nucleus. Octahedral ligand bonding functions internally on three axes of symmetry. The cuboctahedron model justifies the dual-polyhedral bonding found in the 12 transition elements $_{19}^{39}$K through $_{30}^{65}$Zn (Figures 8.1–8.3).

FIGURE 8.1   Six octahedral ligand bond locations found in the Fermi model of copper.

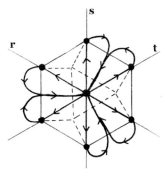

FIGURE 8.2   Cross-section through cuboctahedron of electromagnetic forces on (r, s, t) plane.

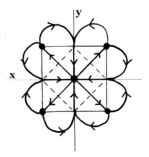

FIGURE 8.3   Cross-section through cuboctahedron of electromagnetic forces on (x, y) plane.

The Fermi model of copper is composed of an internal octahedron with six vertices and an external cuboctahedron (Choy, 2000), verifying a connection between ligand octahedral geometry in six locations and 6D positions of electrons in twelve locations in space on six axes of symmetry. The transition elements that form quasicrystals, K-Zn, include the *D* shell layer and ligand octahedral bonding, which we contend is the internal QG force location.

Figures 8.2–8.4 illustrate the location of quantum loop gravity theory by Lee Smolin (2001).

The *F* orbitals include hexagonal orbital patterns which map triangular and hexagonal planes (Atomic Orbitals, 2014), and have octahedral symmetry. Linus Pauling applied a theory of spherical packing to the PTE and discovered a *s* polyhedral order (Pauling, 1965). The stability of atomic nucleons (protons or neutrons) is directly related to the number of atomic nucleon particles in spherical polyhedral shell layers.

These complete shell layers of the atomic nucleus are packing in tetrahedral, octahedral, and cuboctahedral order. The fundamental polyhedral packing is called Magic Numbers and include 2, 8, 20, 28, 82, and 126 for nucleons. The successive layers include 12, 24, and 32 are related to the magic numbers of spinning electrons on the axis of symmetry. The magic numbers for NG (Neon, Argon, Krypton, Xenon, and Radon) with filled electron shells are presented as 2, 2 + 8, 18, 36, 54, and 86 (Teo, 1985).

We present polyhedral order related to nucleons as cuboctahedral, tetrahedral, and truncated octahedral symmetry. The model for Carbon 12 is a cuboctahedron (Pauling, 1968). The model for Neon 20 is a tetrahedron. Metal clusters such as Au and Pt form cuboctahedral spherical packing polyhedral structures in shell layers (Wilcoxon, 2006). Metals such as Co, Fe, and Ni exhibit properties of shell layer formations in truncated octahedral symmetry (Besley, 1995). The *F* orbitals include hexagonal orbital patterns which map triangular and hexagonal planes and atomic orbitals and have octahedral symmetry (Atomic Orbitals, 2014).

The interatomic forces at work in frustrated geometry (FG) and GR are directly related to triangular lattice planes with electromagnetic spins (Kogerler, 2009). Octahedral lattice systems with (111) triangular planes (Lapa, 2012) and cuboctahedral models are presented as the Calogero model (Hakobyan, 2009).

The small-network geometrically frustrated systems of Bilin Zhuang and Courtney Lammert connect the geometry of the tetrahedron, octahedron, and icosahedron to the dimensions of atomic space (Zhuang). The primary models presented are the cuboctahedron and the icosidodecahedron (the icosidodecahedron is an exception to most FG compounds).

The icosidodecahedron is the foundation of the geodesic all-space coordination system of R. Buckminster Fuller21. The model of Mo72Fe30 is a geodesic polyhedron based on the icosahedron (Schnack, 2013), (Fu). The gravitational wave experiment GR by Krishna Venkateswara updated Truncated Icosahedral Gravitational Antenna originally presented by Johnson and Merkowitz (Venkateswara, 2007). Venkateswara presented the Buckyball of R. Buckminster Fuller to model transducer positions on the TIGA geodesic sphere. The placement of optical transducers is in alignment with the geodesic pentagons of TIGA (Venkateswara, 2007). Piezoelectric compounds of FG and optical mechanisms are used as detectors of the theoretical gravity waves of GR. Also, the testing is conducted with a dual detector system which validates the need for a 6DDC system. GR meets FG in this icosahedral geodesic geometry by utilizing the piezoelectric compounds used in the transducers.

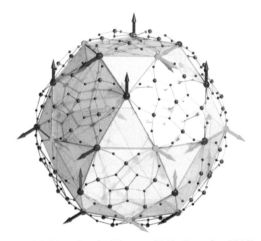

FIGURE 8.4 One Mo72Fe30 molecule (Zhuang, 2012; Kogerler, 2010).

The earth itself as related to the philosophy of R. Buckminster Fuller is a geodesic dome. The earth is technically almost a perfect geodesic sphere in a state of hydrostatic equilibrium (Figure 8.5).

FIGURE 8.5 Icosahedral model of the Earth.

The Euler formula $V + F = E + 2$ by Jeffery Ventrella digitally produces a spherical earth by utilizing the spherical cellular automata formula (Ventrella, 2009); the spherical surface of the earth was approximated into a 17 x 106 frequency geodesic dome. Geodesic frequency is the number of times an icosahedron edge is divided, which forms a triangular geodesic lattice system (1.5' x 1.5' x 1.5') triangle (Sawyer, 2014).

Michael Wysession, Professor of Earth and Planetary Science, mapped a gravitational anomaly called the Midcontinental Rift (Wysession, 2014). The ancient tectonic plate collision between the East and West coast of North America created Molton high-density copper and iron metal deposits. This hybrid rift alters the $F = G\,(M_1 M_2/r^2)$ of Einstein concerning the gravity field of the earth. Seismologists using an isomer for detecting sound waves through the earth discovered this gravitational rift. The isometer detected a greater gravity field under the laved based copper and iron atomic elements in the Midcontinental Rift. A lesser gravity rift is found in less dense sedimentary rocks. Earthquakes are detected as sound waves passing through the atomic mass of earth. Therefore, there is a direct connection to quantum atomic gravity in the geometry of the high-density packing of copper and iron metals. We see the twelve-fold bonding symmetry of the Fermi copper cuboctahedron, including three axes of internal symmetry, as a model connected to QG.

The NG includes a unique volume increase for a molar liter of atomic mass. A theoretical 3D cubic decimeter liter Avogadro mole of a gas at Standard Temperature and Pressure (STP) increases to 22.4 liters. The increase in the volume of a mole of any atomic mass of gas at STP is always a 22.4 liter constant. We contend this unique property of atomic gas is a geometric property of a dual coordinate system of forces between electronegativity and QG. The outer spinning electron shells are balanced in a quantum polyhedral geometry between four and six axes of symmetry and three axes of symmetry returning to the nucleus as QG. Henry Moseley noticed an increasing frequency per atomic shell layer with an increasing shielding effect, theoretically caused by electrons (Mosley, 1913).

The discovery of NG resulted in a new column in the PTE. The 2,240% increase in the volume of the NG must be a direct result of a dual coordinate system where the internal atomic bonding properties of atoms as TDCS is equalized by a spinning electron field shielding effect, which is maximized with full electron shells based on 6D polyhedrons. The full outer electron shells spinning electron field shielding effect of NG are predicted as spinning electrons in polyhedral geometry such as the octahedron, cuboctahedron, truncated octahedron, and truncated rhombic dodecahedron. The NG all resist the natural forces of GR gravity field of the earth by levitation. A Dual Coordinate system between the atomic bonding TDCS and QG must be the mechanism where atomic mass transforms into an atomic gas (Figures 8.6 and 8.7).

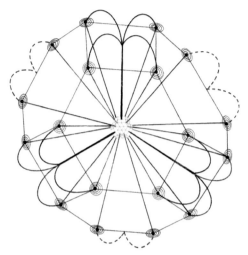

FIGURE 8.6 Continuous force field of a truncated octahedron and central Xenon with 116 nuclear mass units (18 of 24 locations of force fields radiating from nucleus are shown, and three of six locations of quantum loop force fields returning to the nucleus are shown).

FIGURE 8.7 Truncated octahedron with central cuboctahedron. (see 8.2 for refined nucleus of Xe.).

Figures 8.6 and 8.7 present a uniform, continuous field between the nucleus and outer electrons for Xenon gas. The force fields are radiating from the nucleus to 24 locations at the vertices of the truncated octahedron; there are four unique sets of six axes of symmetry for the force fields of Xenon gas. The force fields are returning to the nucleus on three axes of symmetry through the centroids of six square faces of the truncated octahedron (Figure 8.8 and 8.9).

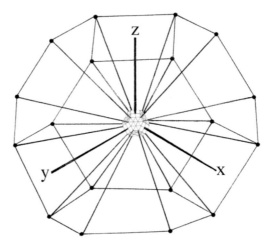

FIGURE 8.8   Nucleus of Xenon gas as a truncated octahedron with 24 spinning electrons in the outer shell (18 of 24 electrons are shown) and (x, y, z) axes of symmetry.

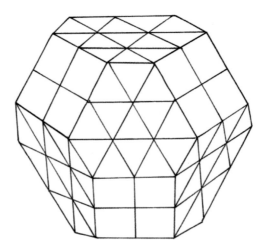

FIGURE 8.9   Nucleus of Xenon gas with 116 atomic mass spherical units at each intersection of a tetrahedron-octahedron truss, modeled in six-dimensional geometry.

The unification of nucleons and electrons produces a theoretical model of xenon gas with an exterior shell of 24 spinning electrons in the geometry of a truncated octahedron, which is in perfect alignment with the 116 nuclear mass model on twelve axes of symmetry.

The following model of radon as a truncated rhombic dodecahedron is based on an outer shell of 32 spinning electrons. The nucleus is based on 116 nucleons for the atomic mass of xenon. It is assumed that the nucleus of xenon is inside the full nucleus of radon with a larger atomic mass of 222 nucleons (Figures 8.10a and 8.10b).

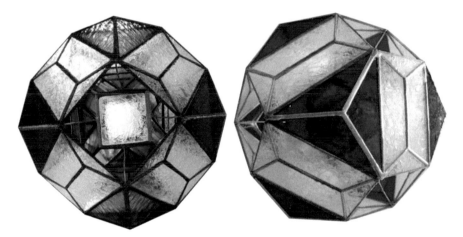

FIGURES 8.10a,b Radon gas model as truncated rhombic dodecahedron with Xenon nucleus.

The 6D geometry of a truncated rhombic dodecahedron is out of alignment with the parallel edges of the tetrahedron, octahedron, cuboctahedron, and truncated octahedron. The 32 vertices of the truncated rhombic dodecahedron model for Radon 86 are revealed in Figures 8.10a and 10.b. The six squares on three axes of symmetry (x, y, z) are 45° out of phase with the six squares formed by the 24 vertices of the truncated octahedron model for Xenon 54 (Radon 86–Xenon 54 = 32 vertices). The age of the radioactive elements 87–92 began with transmutation of the elements by Marie Curie and Lise Meitner. We predict that the electrons of elements Francium 87 through Uranium 92 (beyond the electrons of Radon 86 as a truncated rhombic dodecahedron) are on the same (x, y, z) axes of symmetry as the QG locations on (x, y, z), resulting a destabilization of this group of six natural elements.

This series includes elements with a high-frequency blue light on the electromagnetic spectrum (note the fact that the element Francium 87 liberates heat naturally). The geometric evidence predicts that elements Francium–Uranium are uns polyhedrons utilizing the TDCS. The age of the atom bomb with radioactive elements or man-made elements include: Neptunium 93 – Copernicum 112 and atomic numbers 113–118. The element Copernicum has a shelf life of about 29 seconds; therefore, it is extremely uns and radioactive (we only represent the natural 92 elements in our PTE). We contend that there is enough potential energy in water molecules for generating electrical energy by utilizing TDCS and DCCMAM applied to the PTE. We do not need to use a potential runaway $E = mc^2$ reaction on Uranium atoms to boil water for electrical energy.

The electrons for xenon rearrange into a truncated octahedron. We contend that the instability of uranium is a direct result of the six spinning electrons on the same axes as the six opposing quantum loop gravity locations in space. The man-made elements 93–110 must contribute an increased geometric destabilization for elements beyond Uranium by producing an environment with a reduced QG effect on (x, y, z) axes of symmetry. Therefore, quantum loop gravity is extremely important for the stability of the 92 natural elements.

The interatomic distances and electron shell structures follow the geometries of tetrahedron, octahedron, cuboctahedron, truncated octahedron, icosahedron, and rhombic dodecahedron. Fritsche and Benfield developed a coordination quantum of atoms which could be applied to quantum atomic electrons in the PTE (Fritsche, 1993) (Figure 8.11).

FIGURE 8.11   Octahedral salt formed in zero gravity (NASA, 2014).

An experiment conducted by NASA formed octahedral salt crystals in zero gravity (NASA, 2014). In outer space, the atomic gravity forces on three axes of symmetry are weaker than the electronegativity forces on six axes of symmetry of sodium chloride (NaCl) crystals. On earth, atomic gravity forces are stronger on three axes of symmetry, forming cubic salt crystals. This supports the existence of TDCS between GR and QG.

The addition of a triangular coordinate system to the present square coordinate system will advance the study of atomic space for the scientific community. The resolution of the concept of atomic mass, spinning electrons and photons, and the quantum quark nucleus can be determined by understanding the full dimensions of Planck space, recognized as 6D String theory. We present the cuboctahedron polyhedron as a geometric model for defining atomic space based on the integration of 6D triangular space with 3D square space as a new form of mathematics. This dual coordinate system is referenced in the writings of Albert Einstein in general relativity as a balance between curved geodesic space and 3D square space (Einstein, 1927). Triangular space is a more efficient system to describe multidimensional polyhedrons as geodesic spheres.

Quarks are an elementary particle and constituent of matter including protons and neutrons. The integration of 6D String Theory, quantum loop gravity, and spinning quarks with the geometry of the cuboctahedron determines the position and forces of atomic particles such as electrons, nucleons, and gravitons. The electrons are spinning on two, four, and six axes of symmetry where atomic forces are radiating from the nucleus, balanced with forces returning to the nucleus on three axes of symmetry, called quantum loop gravity, ligand bonding in octahedral symmetry or QG.

Stephen Hawking defines protons and neutrons as triple combinations of up and down quarks (Hawking, 2001). We contend that at each of twelve cuboctahedron vertices are twelve spinning quarks: six pairs of two quarks spinning on six axes of symmetry through opposite vertices. The current scientific process of observing quarks in groups of three causes the twelve quarks to be represented as protons and neutrons. 6D theory presents up quarks as spinning clockwise and down quarks as spinning counterclockwise (Figures 8.12–8.17).

Linus Pauling constructed a Boron molecule as four icosahedrons in tetrahedral symmetry (Pauling, 1964). Six carbon has two AMO units and four sets of three spinning quarks with cuboctahedral symmetry, based on the twelve vertices of a cuboctahedron. There must be a synergetic effect between the AMO units (as triple quark polyhedrons) and actual variations in atomic mass numbers, based on TDCS and QG.

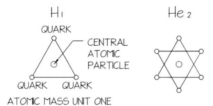

FIGURE 8.12 Hydrogen atom contains one set of spinning quarks with a central atomic particle.

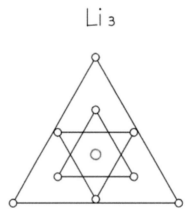

FIGURE 8.13 Helium contains two atomic mass one (AMO) units.

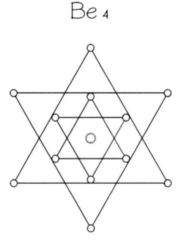

FIGURE 8.14 Lithium nucleus contains two internal AMO units and a second shell with a set of three spinning quarks.

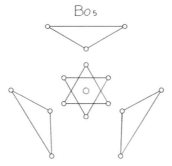

FIGURE 8.15   Beryllium contains two AMO units and two sets of three spinning quarks.

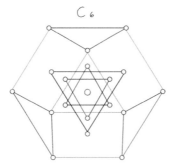

FIGURE 8.16   Boron contains two AMO units and three sets of three spinning quarks with icosahedral symmetry.

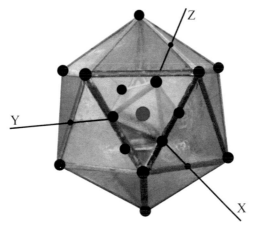

FIGURE 8.17   Nuclear model of Boron as an icosahedron with central octahedron, connected by quantum gravity on (x, y, z), with two sets of triple quarks at the vertices of the octahedron, and three sets of triple quarks at nine vertices of the icosahedron (central atomic particle in red).

When the quarks are in groups of three, there is one triangle of three clockwise quarks, one triangle of three counterclockwise quarks, three triangles of two clockwise quarks and one counterclockwise quark (protons), and three triangles of two counterclockwise quarks and one clockwise quark (neutrons). When the quarks are in groups of four (Aaig, 2014), there are three squares of three counterclockwise quarks and one clockwise quark, and three squares of three clockwise quarks and one counterclockwise quark (Figures 8.18 and 8.19).

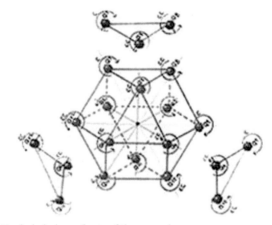

FIGURE 8.18   Exploded view of sets of three quarks.

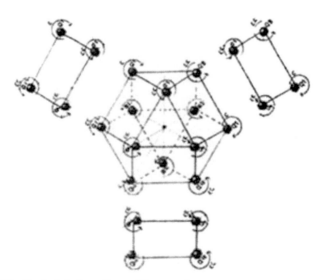

FIGURE 8.19   Exploded view of sets of four quarks.

The QG force functions between the three axes of symmetry at the centroids of six square vortexes of the cuboctahedron model (Feynman). The energy returning to the nucleus is directed through the centroids of groups of four quarks on (x, y, z) axes of symmetry as quantum loop gravity. Larger groups of quarks can be visualized in the geometry of our quark 6D of atomic mass. Unifying cuboctahedral geometry of three and four quarks for atomic mass units produces a tetrahedral-octahedral building system with three and six axes of symmetry. Multiple cuboctahedron models produce the generic integration of protons and neutrons as nucleons, representing individual nuclear mass and isotope units for the periodic (Figure 8.20).

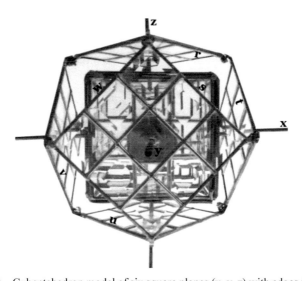

FIGURE 8.20 Cuboctahedron model of six square planes (x, y, z) with edges (r, s, t, u, v, w).

Many possible combinations of quarks are presented as triangular planes in a tetrahedral-octahedral matrix pattern. A theoretical nuclear mass PTE is developed to exhibit the geometry of the nuclear particles such as generic quarks. Unifying the concept of spinning quarks in triple patterns produces protons and neutrons which results in the normalization of the nucleus for atomic mass units. Carbon with atomic mass of twelve is represented as a cuboctahedron (Feynman) (8.2).

We present a mechanism theory for Lise Meitner's experiment on fission. The fission of Uranium 92 produces Barium 56 and Krypton 36 with a mass of approximately 92 units; Uranium has 92 electrons. We theorize that the slow spinning alpha-type atomic nucleon particle destabilizes the high-frequency Uranium nucleus. The outer 92 AMO units and 92 electrons separate from the Uranium nuclear core of 146 nucleons. The 92 quark units re-combine into a Krypton 36 polyhedral nuclear mass units (see #92 in Box 1). Barium's atomic mass of approximately 146 atomic mass units accepts the 56 electrons. The nuclear AMO must be equivalent to 92 separate Hydrogen 1 atoms. This fission theory recognizes the destabilized 56 uranium electrons as they return to a remaining 146 nucleus as a transmutation of Barium 56. We contend that a fundamental geometric mechanism produced barium and krypton in Lise Meitner's atom-splitting experiment (Sime, 1998). The destabilization mechanism involved in the fission of uranium is related to individual Q units of mass:

$$^{238}\text{Ur 92 electrons fission} = {}^{92}\text{Kr 36 electrons} + {}^{146}\text{Ba 56 electrons}$$
$$^{235}\text{Ur 92 electrons fission} = {}^{90}\text{Cs 37 electrons} + {}^{143}\text{Rb 55 electrons}$$

Box 1   Quark Six-Dimensional for Atomic Mass

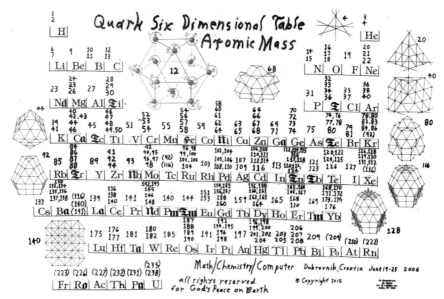

For both reactions, the number of electrons is extremely close to the atomic mass number for both ^{90}Kr and ^{90}Cs. The destabilization of ^{235}Ur with 92 electrons produces 92 AMO and one nucleus composed of 146 nucleons of Barium 56. The uranium nucleus with 92 spinning triple sets of quarks and 92 complementary electrons are all simultaneously released as 92 AMO, which rearrange as a nucleus with an atomic mass of approximately 90 (^{90}Kr or ^{90}Cs with a reconfigured 36 or 37 electrons per shell layer). The remaining electrons and uranium nucleons form ^{146}Ba or ^{143}Rb, respectively. We contend the slower spinning helium nucleus destabilizes the uranium nucleus by reducing the velocity of spin, thereby releasing all 92 AMO for Uranium. The remaining 146 nucleons become a new slower spinning nucleus of ^{146}Ba with 56 electrons or ^{143}Rb with 55 electrons and energy.

The cuboctahedron model for quarks is most useful to describe triple sets of quarks which compact into a uniform tetrahedron-octahedron geometry. This formula may vary in number of neutral quarks related to the geometry of nuclear packing of isotopes.

The theoretical mechanism for the nuclear fusion of a supernova by a nucleosynthesis produces Nobel gas $_{54}$^{116}Xe + meteorites + energy. The atomic structure of meteorites reveals the process for a supernova event. The achondrite family of elements has an interatomic relationship in the early stages of a supernova event. The Willamette meteorite in the Hayden Planetarium has a nickel-iron composition with large holes where nuclear fusion of Fe 26 + Ni 28 could have yielded Xe 54 in a supernova star.

The nuclear fusion of four hydrogen atoms into one helium molecule is the fuel for young stars. The chemical elements aluminum, silicon, oxygen, and manganese are part of star formation that potentially leads to a super-nova event. We predict that during a supernova event, Xe 54 is produced at the center from the fusion of Fe 26 and Ni 28. Gas formation results in the 22.4 volume increase which explains the huge force behind a supernova event. The cuboctahedron model could complete the explanation of nucleo-synthesis, the strong force between protons and neutrons or quarks. The nuclear fusion sequence utilizes the alpha process for a supernova event and evolves from the formation of a star core primarily based on silicon and sulfur. The $^{4}_{2}$He fusion continues to fuse in the following sequence (Wikipedia, 2015):

$$^{28}_{14}\text{Si} + {}^{4}_{2}\text{He} \rightarrow {}^{32}_{16}\text{S}$$

$$^{32}_{16}\text{S} + {}^{4}_{2}\text{He} \rightarrow {}^{36}_{18}\text{Ar}$$

$$^{36}_{18}\text{Ar} + {}^{4}_{2}\text{He} \rightarrow {}^{40}_{20}\text{Ca}$$

$$^{40}_{20}\text{Ca} + {}^{4}_{2}\text{He} \rightarrow {}^{44}_{22}\text{Ti}$$

$$^{44}_{22}\text{Ti} + {}^{4}_{2}\text{He} \rightarrow {}^{48}_{24}\text{Cr}$$

$$^{48}_{24}\text{Cr} + {}^{4}_{2}\text{He} \rightarrow {}^{52}_{26}\text{Fe}$$

$$^{52}_{26}\text{Fe} + {}^{4}_{2}\text{He} \rightarrow {}^{56}_{28}\text{Ni}$$

The final two stages of the helium fusion produces $^{52}_{26}\text{Fe}$ and $^{56}_{28}\text{Ni}$. Nucleosynthesis involves a massive collapse of $^{52}_{26}\text{Fe}$ and $^{56}_{28}\text{Ni}$., yielding $^{116}_{54}\text{Xe}$ + meteorites:

$$^{52}_{26}\text{Fe} + {}^{56}_{28}\text{Ni} + 2\,^{4}_{2}\text{He} \rightarrow {}^{116}_{54}\text{Xe} + \text{meteorites} + \text{energy}$$

8.5 MULTI-/TRANS-DISCIPLINARY CONNECTIONS

Advancement to geometric understanding of atomic structure in six dimensions provides the opportunity for development of triangular software pixels, highly efficient motors designed in 6D geometry, and for solving the geometric mechanism for cancer cells. The nanotechnology of potential energy and dynamic engineering are found in 6D geometry. We predict that the impact reaction in the LHC at CERN of Fe and Ni will produce xenon gas and a huge amount of energy.

The 6D triangular theory to model space is based on a system of four coordinated triangulated planes to describe the polyhedral geometry of atomic space. The internal structure of a cuboctahedron with six axes of symmetry accurately models the polyhedral geometry of FG compounds. GR utilizes a dual coordinate Cartesian-like system including a 3D Cartesian and a geodesic/icosahedral-type coordinated field equation system, including a time parameter to represent inertial mass.

A dual coordinate triangular and square system exists within the edges, vertices, and faces of a cuboctahedron to model atomic space by defining the dimensions of space with 3, 4 and 6 axes of symmetry. The 6D dual coordinate system includes six directional number lines as a paradigm shift to a 6D geometry, represented as (r, s, t, u, v, w) in triangular space.

The 6D dual coordinate system includes 6D polyhedral geometry theory and 3D Cartesian system which is based on the three axes of symmetry (x, y, z), intersecting between the centroids of square faces and nuclear center of cuboctahedron. The cuboctahedron model includes triangular and square lattice planes with three, four, and six axes of symmetry and one centroid as (0, 0, 0, 0, 0, 0).

We analyze the geometric properties of the standard models of particle physics within the polyhedral dimensions of atomic mass as space. FG and GR are thoroughly analyzed in relationship to triangular planes, polyhedrons, and including geodesic dome polyhedrons based on icosahedron.

GR was developed from special relativity and is based on the 3D Cartesian philosophy of space (Heisenberg, 1927). In order to geometrize gravity,

Einstein needed a dual coordinate system utilizing field equations in DC system of space. The dual Cartesian coordinate system of GR is composed of Koordinaten system K = Galilean based Cartesian-like (x, y, z) system, and Koordinaten system K' = Geodesic based Cartesian (x, y, z) system. The dual square Cartesian-like system presents the foundation for the curved geodesic space system found in GR (Einstein, 1927). In GR, gravity is a conceptual formatted as curved square lattice coordination system in time (Figure 8.21).

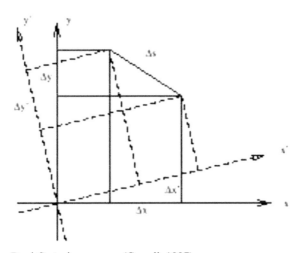

FIGURE 8.21   Dual Cartesian squares (Carroll, 1997).

The cuboctahedron as a model for a dual coordinate system is found in atomic models of FG compounds, Fermi 6D models of copper, Nanocrystals, and 6D PTE (Gielen, 2015).

FG meets GR philosophically between the definition of a triangular lattice plane and a square lattice plane within the polyhedral dimensions of atomic space. There are triangular lattice structures and icosahedral (geodesic classified) polyhedrons in FG, and there are Cartesian square lattice-based dual coordinate systems found in GR (Lehmkuhl, 2014; Einstein, 1914). There needs to be a coordinated system merge between the triangular lattice systems of FG and the dual Cartesian based Galilean field equations and geodesic coordinate systems in GR. The dual coordinate system we present represents all the laws of physics related to the triangular and square dimensional order of atomic elements in polyhedral space.

The group field theory of Steffen Gielen attempts to form a renormalized group based on GR with the Bose-Einstein condensates (as triangular lattice structures), and quantum loop gravity with the four triangular planes of the tetrahedron to represent the standard model of particle physics (Gielen, 2015). The standard model is a theory concerning the electromagnetic, weak, and strong nuclear interactions, of subatomic particles in atomic space. The group field theory attempts to solve Plato's problem of connecting the geometry of polyhedral space and mathematics by connecting the mathematical order of four axes of symmetry, triangular lattice of (Bose-Einstein condensates) and the tetrahedral geometry (Haggard, 2015). The (111) triangular lattice systems of FG is applied to GR. Our multi-dimensional philosophy of space needs to include the additional directions or dimensions of triangular/ hexagonal lattice systems to analyze the dynamic properties of the SM of atomic particles in space. The tetrahedron includes four axes of symmetry connecting vertices and centroids of opposite triangular planes through the center. FG includes the model cuboctahedron that in polyhedral geometry includes four triangular-hexagonal planes and six axes of symmetry.

We are now able to focus on the concept of atomic mass as related to triangular polyhedral geometry found in GR and FG. We are attempting "geometri-size" the dimensions of atomic mass as related to GR and atomic space into a form of platonic polyhedral geometry of three, four, five, and six axes of symmetry. The Platonic philosophy of space by Gödel and the poly-hedral geometry of matter must be found within the natural mathematical symmetry of space. The periodic order of the elements is applied to Gödel's philosophy and presents the important dimensional connection between polyhedral geometry of space and the natural properties of mathematical objects of mass in atomic space. The platonic solids (tetrahedron, octahe-dron, and icosahedron) are based on equilateral triangular planes, which are in need of number lines to form a 6D lattice coordinate system.

The use of the Cartesian coordinate system to analyze honeycomb, square, and triangular lattices6 needs to be upgraded to include 3D trian-gular plane and hexagonal lattice planes. The Bose-Einstein condensate experiment produced equilateral triangular arrays similar to vortex lattices in superconductors as 3D triangular planes (Conradson, 2013; Feder, 2001). This reveals the fundamental existence of triangular lattice structures in natures' atomic particles. Polyhedrons such as tetrahedron, octahedron, cuboctahedron, and icosahedron found in FG need 3D triangular planes with three directional number lines to understand the dynamic properties of atomic particles in space.

Supersymmetry is presented by Michael Kramer in snowflake crystals (Kramer, 2013). We see a mathematical understanding of 6D triangular symmetry within the six pinned connections seen in snow crystal image. See

Image of snow crystal. Note the six connections of the snow crystal growth are within the 120° triangular/hexagonal lattice structure. There is also a reversal of grain structure growth within the 6D triangular geometry.

Roderich Moessner presents crystal structure of spin ice based on pyrochlore lattice, consisting of corner-sharing tetrahedral (Moessner, 2005). The ising spins are forced to spin along local (111) axes for the four triangular/hexagonal lattices of the tetrahedron. This structure presents the need to describe the intersection of four hexagonal lattices found in the polyhedral geometry of the cuboctahedron.

Tsvi Piron uses the tetrahedron to describe the spherical collapse and emission of gravitational radiation wave packets in GR. A five tetrahedra set of simplex hedrons represents the 20 edges and 15 directions of a lattice system of icosahedron (Piron). (Simplex is a software term representing a tetrahedron.) The tetrahedral four axes of symmetry are defined by the (111) triangular planes in the field of crystallography. These (111) planes are parallel to the eight triangular faces of the octahedron and cuboctahedron. The Kagome lattice structures of FG are presented by Oleg Janson (et al.) compare the triangular lattice planes with the cuboctahedron. Their cuboctahedron model presents six axes of symmetry with twelve directions in space (Janson, 2012).

The triangular polyhedrons such as the tetrahedrons, octahedrons, and truncated octahedrons all have edges parallel to the six sets of four parallel edges of the cuboctahedron in the 6D dual coordinate system. Examples of these polyhedrons are presented in the nanoparticles of catalysis as tetrahedron, octahedron, cube, cuboctahedron, and including eight triangular hybrid planes in rotation forming icosahedron (Xiong, 2007).

FG and GR both exhibit properties of a dual-coordinate system based on kagome lattice, the cuboctahedron, icosidodecahedron, and icosahedron. The icosidodecahedron and icosahedron are the foundations of a triangular-based geodesic spherical space system (Zhuang, 2012).

The cuboctahedron as a vector equilibrium model with twelve flexible vertices transforms between icosahedron→cuboctahedron→icosahedron. Fuller presents many examples of geometry based on triangular lattice systems. The cuboctahedron has eight triangular planes which could be represented as the triangular lattice systems of FG (Fuller, 1979).

In *Structures of Nature is a Strategy for Design*, Peter Pierce notes a connection between the icosahedron and the truncated octahedron. Eight of the triangular faces of the icosahedron are parallel to the eight hexagonal faces of the truncated octahedron (Pierce). Twelve of the twenty-four vertices of the icosahedron are aligned in an icosahedral transformation based on triangular lattice systems, in rotation which illustrates the 6D dual coordinate system. This same polyhedral geometry is found in superconductive crystals, colossal magnetoresistance compounds (Choudhury, 2011), and the piezoelectric compounds (Figure 8.22).

FIGURE 8.22   Truncated octahedron (icosahedral triangles are in red).

The rare state of equilibrium balance occurs when forces out in 6D (r, s, t, u, v, w) equals forces in (x, y, z) within the spherical geodesic polyhedron with five-fold symmetry of icosahedron. The earth itself is in a state of hydrostatic equilibrium in GR.

8.6   OPEN ISSUES

Nature at an atomic Planck scale actually contains 92 separate coordinated systems based on PTE with three, four, and six primary axes of symmetry. The spinning of all atomic particles on these three, four, and six axes of symmetry produces the inertial force fields we refer to as atomic mass.

The spinning nucleons on four and six axes of symmetry as observed in 6D nuclear periodic represent force fields going out from the nuclear center hold the spinning electrons on four and six axes of symmetry for the appropriate periodic magic numbers of the PTE (see 6D electron). The three axes of symmetry ligand bonding is found primarily within the transition elements and related periodic columns which are synchronized with six squares or octahedron vertices of 6D dual coordinate cuboctahedron model.

The three axes of symmetry are the related to the atomic force fields returning to the nuclear center in a perfect QG loop. The atoms of earth are facing many different directions or spatial dimensions. GR actually averages the sum total average of all the directions of all the atoms on earth. The local forces of atomic bonding occur between specific atoms in synergetic order on four and six axes of symmetry based on specific magic numbered 6D polyhedrons and into the nucleus on three axes of symmetry located at six octahedral vertices locations in space. The atomic forces of quantum loop gravity are much greater than the cosmological measurement of gravity because the average force of gravity is much smaller proportionately cosmologically because atoms are directional with six dimensions out and three dimensions in. The *s* atoms on earth interact with each other to make *s* compounds, and molecules. Another characteristic property of the NG is the three axes of symmetry related to the six vertices of octahedron, or six square openings of appropriate 6D polyhedrons appear to geometrically block the internal ligand bonding of spinning electrons and thus violate the cosmological gravity constant of GR and levitate (see Box 1). The energy of the atomic mass of earth being in a state of hydrostatic equilibrium is much smaller proportionately at the cosmological GR scale called gravity vs. the Planck scale of atomic particles in FG called ligand and electron bonding.

The TDCS explains the self-assembly process found in bonding properties of nanoparticles of the atomic elements. Complicated polyhedral geometry has been found in the self-assembly of the clathrate family of crystals utilizing the TDCS. The bonding angles of these crystals are within 20 degrees of a tetrahedral angle. The building blocks of these clathrate crystals include truncated octahedrons, pentagonal dodecahedrons and various polyhedrons based on pentagons and hexagons. Various doping techniques produce similar polyhedral geometry, thus presenting a fundamental polyhedral order for the PTE based on TDCS.

KEYWORDS

- **frustrated geometry**
- **general relativity**
- **ligand quantum gravity**
- **self-assembly**
- **triangular planes**
- **triangular polyhedrons**

REFERENCES AND FURTHER READING

Aaij, R., (2014). Synopsis: Four-quark (FQ) state confirmed. *Rev. Letter, 112*, 222002.

Bentley, W. *Snowflake Photography Book.*

Besley, N. A., Johnston, R. L., Stace, A. J., & Uppenbrink, J., (1995). Theoretical study of the structures and stabilities of iron clusters. *Journal of Molecular Structure (Theochem.), 341,* 75–90. Elsevier.

Carroll, S. M., (1997). *Lecture Notes on General Relativity* (p. 5460). Enrico Fermi Institute. University of Chicago, S. Ellis Ave., Chicago, IL 60637. http://ned.ipac.caltech.edu/level5/March01/Carroll3/Carroll_contents.html [Accessed July 2, 2015].

Choudhury, N., Walizer, L., Lisenkov, S., & Bellaiche, L., (2011). Geometric frustration in compositionally modulated ferroelectrics. *Nature, 470,* 513–517. doi: 10.1038/nature09752.

Choy, T. S., Naset, J., Hershfield, S., & Stanton, C., (2000). *Periodic of the Fermi Surfaces of Elemental Solids.* Physics Department, University of Florida. http://www.phys.ufl.edu/fermisurface/periodic.pdf. [Accessed 20 February 2015].

Clark, D. L., Palmer, P. D., Tait, C., Drew, K. W., Condradson, S. D., & Donohoe, R. J., (2004). *Unusual Tetraoxo Coordination in Heptavalent Neptunium.* Actinide research quarterly, heavy element chemistry. Los Alamos National Laboratory, 1st Quarter. http://www.lanl.gov/orgs/nmt/nmtdo/AQarchive/04spring/index.shtml [Accessed 25 June 2015.]

Conradson, Steven, D., et al., (2013). Possible Bose-condensate behavior in quantum phase originating in a collective excitation in the chemically and optically doped Mott-Hubbard system UO2+x. *American Physical Society.* doi: 10.1103/PhysRevB.88.115135.

Einstein, A., (1914). *The Formal Foundation of the General Theory of Relativity.* Plenary Session of November 19. http://einsteinpapers.press.princeton.edu/vol6-trans/43.

Einstein, A., (1927). Formale beziehung des riemannschen krümmungstensors zu den feldgleichungen der gravitation formal relationship of the Riemannian curvature tensor to the field equations of gravity. "*The Foundation of the General Theory of Relativity.*"

Feder, D. L., & Clark, C. W., (2001). *Superfluid-to-Solid Crossover in a Rotating Bose-Einstein Condensate* (Vol. 87, Issue 19). Published November 01 2001. http://www.nist.gov/manuscript-publication-search.cfm?pub_id=840115. [Accessed 19 June 2015].

Feynman, R. *Richard Feynman's Story of Particle Physics.* Youtube.com/watch?v=kw6rR9h9vu8, time 12: 38/41: 30 [Accessed 10 January 2016].

Fritsche, H. G., & Benfield, R. E., (1993). Exact analytical formulae for mean coordination numbers in clusters. *Atoms, Molecules, and Clusters*, Supplement to *Z. Phys. D., 26,* 15–17. Springer-Verlag.

Fu, Z. Spin correlations and excitations in spin-frustrated molecular and molecule-based magnets. *Key Technologies* (Vol. 43). ISSN 1866–1807. ISBN 978–3–89336–797–9. http://publications.rwth-aachen.de/record/64085/files/4114.pdf [Accessed 25 June 2015].

Fuller, B., (1975). Synergetics. Exploration into the geometry of thinking. New York, London: MacMillan Publishers. *Synergetics, 1.*

Fuller, B., (1979). Synergetics. Exploration into the geometry of thinking. 2.5 octet truss, New York, London: MacMillan Publishers. *Synergetics, 2,* 124, 214.

Gielen, S., (2015). *Cosmology with Group Field Theory Condensates.* Imperial College London.

Haggard, H. M., Han, M., & Riello, A., (2015). *Encoding Curved Tetrahedra in Face Holonomies: A Phase of Shapes From Group-Valued Moment Maps.* Cornell University Library, Submitted 9 June 2015. http://arxiv.org/abs/1506.03053 [Accessed 7 July 7, 2015].

Hakobyan, T., (2015). *Cuboctahedric Higgs Oscillator from the Calogero Model.* Arxiv.org/pdf/0808.0430. Accessed June 21, 2015.

Hakobyan, T., Nersessian, A., & Yeghikyan, V., (2009). *Cuboctahedric Higgs Oscillator from the Rational Calogero Model*. arXiv: 0808.0430v3 [math-ph].

Hawking, S., & Mlodinow, L., (2010). The grand design. *Bantam Books Trade Paperbacks*. New York.

Hawking, S., (2001). *A Brief History of Time and The Universe in a Nutshell Illustrated Bantam Books* (p. 86, 89). New York City.

Heisenberg, W., (1927). "Über den anschulichen Inhalt der quantentheoretischen Kinematik und Mechanik." "*On the Perceptual Content of Quantum Theoretical Kinematics and Mechanics.*"

Hopkinson, J. M. Antiferromagnetic Ising Model on the sorrel net: A new frustrated corner-shared triangle lattice. *Department of Physics and Astronomy* (pp. R7A, 6A9). Brandon University, Brandon, Manitoba, Canada.

Hunpyo, L., (2009). *Hubbard Model with Geometrical Frustration* (p. 82). Rheinischen Friedrich-Wilhems-Universitat, Bonn.

Janson, O., Richter, J., Messio, L., Lhuillier, C., & Rosner, H., (2012). Phase diagram of the spin–1/2 Heisneberg J1—Jd model on the kagome lattice. *International Conference on Highly Frustrated Magnetism*. McMaster University, Hamilton, Ontario.

Kogerler, P., Tsukerblat, B., & Muller, A., (2009). Structure-related frustrated magnetism of nanosized polyoxometalates: Aesthetics and properties in harmony. *Electronic Supplementary Information for Dalton Transactions*. Copyright of the Royal Society of Chemistry.

Kogerler, P., Tsukerblat, B., & Muller, A., (2010). *Dalton Trans*, *39*, p. 21.

Kramer, M., (2013). *Searching for Supersymmetry: Some Frustration But no Despair*. The Guardian. Thursday. http://www.theguardian.com/science/life-and-physics/2013/mar/14/supersymmetry-spell. [Accessed June 19, 2015].

Lapa, M. F., & Henley, C. L., (2012). *Ground States of the Classical Antiferromagnet on the Pyrochlore Lattice*. arXiv: 1210.6810v1 [cond-mat.str-el].

Lehmkuhl, D., (2014). *Studies in the History and Philosophy of Modern Physics*, *Part B May, 46*, 316–326.

Maxwell, J. C., (1878). *A Treatise on Electricity and Magnetism*. Publication Oxford at the Claredon Press.

Meyer, P., (1999). *Lattice Geometries, Hermetic Systems*. http://www.hermetic.ch/compsci/lattgeom.htm. Written during 1999 CE, last revised 2000–01–17 CE. Published here 2001–02–17 CE (previously unpublished). Accessed June 19, 2015.

MIT OpenCourseWare. *Understanding X-Rays: The Electromagnetic Spectrum*. http://ocw.mit.edu/courses/earth-atmospheric-and-planetary-sciences/12–141-electron-microprobe-analysis-january-iap–2012/lecture-notes/MIT12_141IAP12_slides_day2.pdf. [Accessed 20 February 2015].

Moessner, R., (2005). *Geometry, Topology, and Frustration: The Physics of Spin Ice*. CNRS and PLT-ENS. Magdeburg.

Moseley, H., (1913). Philosophical magazine. "*The High-Frequency Spectra of the Elements.*"

NASA, About education. *Crystal Photo Gallery: Salt Crystals in Space*. http://chemistry.about.com/od/growingcrystals/ig/Crystal-Photo-Gallery/Salt-Crystals-in-Space.htm. Updated 28 August 2014. [Accessed 11 January 2010].

Pak, I., (2010). *Lectures on Discrete and Polyhedral Geometry, UCLA*, p. 153.

Pauling, L., & Hayward, R., (1964). *The Architecture of Molecules*. WH Freeman.

Pauling, L., (1965). The close-packed-spheron model of atomic nuclei and its relation to the shell model. *The National Academy of Sciences, 54*(4), 989–994.

Pauling, L., (1965). The close-packed-spheron theory and nuclear fission. *Science, 150,* 297–305.

Pauling, L., (1968). *Geometric Factors in Nuclear Structure* (pp. 83–88). Maria Sklodowska–Curie Centenary Lectures.

Pierce, P. *Structures in Nature is a Strategy for Design,* 177–178.

Piron, T. Numerical relativity from gravitational radiation. *Proceedings of 11th International Conference on General Relativity and Gravitation* (pp. 180–190). M.A.H. MacCallum.

Saghatelian, A., (2012). Constants of motion of the four-particle calogero model. ISSN 1063–7788, *Physics of Atomic Nuclei* (Vol. 75, No. 10, 1288–1293). Copyright Pleiados Publishing, Ltd.

Sarnowski, M. J. Vixra.org: Quantum gravity and string theory. Underlying cuboctahedron packing of Planck spinning spheres structure of the Hubble universe correlation with Higgs, W., Boson, Z. B. *Bottom Quark and Top Quark Masses a Structure Like Buckminster Fuller's Vector Equilibrium.* http://vixra.org/abs/1404.0035 [Accessed 8 April 8, 2015].

Sawyer, J., & Wagner, M., (2014). *De Revolution: Encyclopedia of Six Dimensions.* Self-published.

SCARC, (2012). *The Pauling Theory of Quasicrystals.* https://paulingblog.wordpress.com/2012/05/09/the-pauling-theory-of-quasicrystals/https://paulingblog.wordpress.com/2012/05/09/the-pauling-theory-of-quasicrystals/ Posted on May 9, 2012.

Schnack, J., (2013). *Exchange Randomness and Spin Dynamics in Frustrated Metagnetic Keplerate.* arXiv: 1304.2603v1[cond-mat.str-el] 9 April 2013.

Sime, R., (1998). Scientific American. *"Lise Meitner and the Discovery of Nuclear Fission."*

Smolin, L., (2001). *Three Roads to Quantum Gravity,* Basic Books, New York City.

Teo, Boon, K., & Sloane, N. J. A., (1985). "Magic numbers in polygonal and polyhedral clusters" (PDF). *Inorganic Chemistry, 24*(26), 4545–4558. doi: 10.1021/ic00220a025.

The Quantum Basis of Atomic Orbitals. http://chemistry.umeche.maine.edu/CHY251/Quantum.html. Updated 4 September 2014. [Accessed 20 February 2015].

Van Grieken, R. *Handbook of X-ray Spectography* (2nd edn., 101–118). (This Kev/wavelength has a +/- variation of 1% through interpolation based on many sources related to 1.238 ratio formula.)

Venkateswara, K., (2007). *Spherical Gravitational Wave Detector.*

Ventrella, J., (2009). *Spherical Cellular Automaton: Earth Day.* http://www.ventrella.com/earthday/earthday.html Accessed 25 June 2015.

Wikipedia. *Supernova Nucleosynthesis.* http://en.wikipedia.org/wiki/Supernova_nucleosynthesis. [Accessed 21 February 2015].

Wilcoxon, J. P., & Abrams, B. L., (2006). Synthesis, structure, and properties of metal clusters. *Chemical Society Reviews, 11,* 35, 1162–1194.

Wysession, M., (2014). *Gravity Map of the Midcontinental Rift.* Credit: Stein et al. Washington University, St. Louis MO.

Xiong, Y., Wiley, B. J., & Xia, Y., (2007). Nanocrystals with unconventional shapes: a class of promising catalysts. *Angew. Chem. Int. Ed., 46,* 7157–7159. Copyright Wiley-VCH Verlag GmbH & Co. KGaA, Weinheim.

Zhuang, B., & Lannert, C., (2012). Small-network approximations for geometrically frustrated ising systems. *Physical Review E., 85,* 031107. doi: 10.1103/PhysReg.85.031107.

CHAPTER 9

Diabetes Mellitus/Anti-DM Pharmacological Management

BOGDAN BUMBĂCILĂ,[1] CORINA DUDA-SEIMAN,[1]
DANIEL DUDA-SEIMAN,[2] and MIHAI V. PUTZ[1,3]

[1]*Laboratory of Computational and Structural Physical Chemistry for Nanosciences and QSAR, Biology-Chemistry Department, Faculty of Chemistry, Biology, Geography at West University of Timişoara, Pestalozzi Street No. 16, Timişoara, RO–300115, Romania*

[2]*Department of Medical Ambulatory, and Medical Emergencies, University of Medicine and Pharmacy "Victor Babes," Avenue C. D. Loga No. 49, RO–300020 Timisoara, Romania*

[3]*Laboratory of Renewable Energies-Photovoltaics, R&D National Institute for Electrochemistry and Condensed Matter, Dr. A. Paunescu Podeanu Str. No. 144, RO–300569 Timişoara, Romania, Tel.: +40-256-592-638, Fax: +40-256-592-620, E-mail: mv_putz@yahoo.com; mihai.putz@e-uvt.ro*

9.1 DEFINITION

Diabetes mellitus is a metabolic disorder characterized by the altered capacity of the body to process glucose intake correctly. Normally, sugars and carbohydrates are broken by the body to glucose, which is used by the cells for producing energy. A hormone, insulin, is the carrier of the glucose in the cell, where it will be burned to produce this energy. When the body does not secrete enough insulin, or it cannot properly use insulin for transporting glucose in the cells, high levels of glucose will be found in the circulatory system, the situation being diagnosed as diabetes. These increased levels of glucose will affect other biochemical processes leading to a wider metabolic disorder and in time, to life-threatening complications (WHO, 2013).

Three types of diabetes are today described: type I (insulin-dependent) – characterized by insufficient insulin production, type II (non-insulin-dependent) characterized by an inefficient insulin use by the body and gestational diabetes – having its onset during pregnancy and it is usually transitory.

The main goal of diabetes management is to rebalance the carbohydrate metabolism. In order to achieve that, changes in diet and the whole lifestyle should be adopted and of, course, proper pharmacotherapy.

9.2 HISTORICAL ORIGINS

The presence of sugar in the urine (glycosuria) was already associated with diabetes in the nineteenth century. Frederick Madison Allen (1879–1964) was a pioneer in the documentation of the disease. In 1913, he printed an almost 1200 pages book (*Studies Concerning Glycosuria and Diabetes*) including case studies and studies on sugar consumption. In 1921, he opened the first clinic for diabetes mellitus treatment in New Jersey, and he proposed a low-calorie diet to eliminate glycosuria. Till the discovery of insulin, it was the first attempt to manage the disease (Hockaday, 1981).

Insulin (named after the word "insula" in reference to the Langerhans islets of the pancreas producing this hormone) was already thought to be related to diabetes since 1889, when two physicians experimentally removed pancreases from dogs who immediately developed the signs of diabetes (Dallas, 2011).

The first pharmacological treatment of diabetes, with purified insulin extracted from bovine pancreases, happened in 1922. In 1923, Eli Lilly Pharmaceuticals Company started to produce insulin as a marketed drug.

The first clearly made and published the difference between type 1 diabetes and type 2 diabetes was made by Sir Harold Percival Himsworth (1936).

Trials to prepare orally administrated hypoglycemic agents ended successfully by first marketing of tolbutamide and carbutamide in 1955. (Ahmed, 2002) Experimental trials using sulfonylureas and biguanides were conducted locally in the worlds since 1920s, but they were not unanimously accepted. Some of the compounds proved to be toxic, and they were withdrawn from therapy. Afterward, as the research continued and the diabetes pathological mechanisms are more accurately understood, other oral antidiabetics were developed.

9.3 NANO-CHEMICAL IMPLICATIONS

Drugs used in diabetes target hyperglycemia. There are different pharmacological classes of antidiabetic drugs and their selection by the physician when prescribed depends on the type of diabetes, other chronic conditions of the patient, clinical parameters like weight, blood sugar, and hemoglobin A_{1C} values after the beginning of treatment (Mullan et al., 2009).

In type 1 diabetes, the pancreas of the patient produces little or no insulin, so a replacement therapy is needed, using insulin injections or an insulin pump. Insulin can be sometimes used in type 2 diabetes, too but in patients with great adherence to the treatment (self-monitoring blood sugar levels, self-administration of the parenteral treatment).

In type 2 diabetes, oral antidiabetic drugs are most commonly used.

9.3.1 ANTI DIABETIC DRUGS

9.3.1.1 INSULIN SENSITIZERS

Insulin sensitizers are treating insulin resistance, the main pathological mechanism of type 2 diabetes. Insulin resistance is a pathological condition when cells lose their normal response to the activity of insulin: the insulin receptors are either mutated and cannot recognize the hormone or their expression on the cell's surface is overwhelmed by the concentrations of the hormone, directly dependant by glucose concentration.

9.3.1.1.1 Biguanides

They reduce the amount of hepatic gluconeogenesis, and they increase the peripheral glucose uptake. The most commonly used drug is metformin, and its advantage is that it is a single drug that doesn't cause weight gain (Eurich et al., 2007).

It is possible that metformin's activity on the insulin's sensitivity is assigned to its effects on the insulin receptors' expression and the activity of the tyrosine kinase enzyme. All the present studies show that the metformin's primary function is to decrease hepatic glucose production, by inhibiting the gluconeogenesis process.

9.3.1.1.2 Thiazolidinediones

These molecules bind to receptor PPARγ, a regulatory protein that regulates the transcription of some genes involved in glucose and fat metabolisms. These PPARγ receptors act on some gene sequences named peroxisome

Main Pharma-cological Class	Pharmacological subclass	Agent	Mechanism of action	Other observations
Insulin sensitizers	Biguanides	Metformin	The direct mechanism is not fully understood. It is considered that increases insulin sensitivity in vivo, resulting in reduced plasma glucose concentrations, increased glucose uptake and decreased gluconeogenesis.	The drug is transported into hepatocytes mainly via OCT1 (Organic Cation Transporter 1, a protein), resulting in an inhibition of the mitochondrial respiratory chain (complex I) through a currently unknown mechanism(s). The resulting deficit in energy production is balanced by reducing the consumption of energy in the cell, particularly reduced gluconeogenesis in the liver. This is mediated in two main ways. First, a decrease in ATP and a concomitant increase in AMP concentration occur, which is thought to contribute to the inhibition of gluconeogenesis directly (because of the energy/ATP deficit). Second, increased AMP levels function as a key signaling mediator that has been proposed to (1) allosterically inhibit cAMP–PKA signaling through suppression of adenylate cyclase, (2) allosterically inhibit fructose–1,6-bisphosphatase, a key gluconeogenic enzyme, and (3) activates AMP-activated protein kinase. This leads to inhibition of gluconeogenesis (1 and 2) and lipid/cholesterol synthesis (3), which may contribute to the longer term metabolic and therapeutic responses to the drug. (Rena et al., 2013). Other analogs, like buformin and phenformin, were withdrawn from therapy due to their toxic effects, especially lactic acidosis. (Güthner et al., 2006)

(Continued)

Main Pharma- cological Class	Pharmacological subclass	Agent	Mechanism of action	Other observations
	Thiazolidinediones	Pioglitazone Rosiglitazone Lobeglitazone Troglitazone	They activate PPARG. When activated, the receptor binds to DNA in complex with another protein, the retinoid receptor *X* (RXR), regulating the transcriptions of some genes. The effect of this regulation is an increase in the storage of fatty acids in fatty cells, thereby decreasing the amount of fatty acids present in circulation. Thus cells become more dependent on the oxidation of carbohydrates for covering the necessary energy more specifically on glucose. (Plutzky, 2011)	They were first marketed as therapeutic agents in the late 1990's. They activate Peroxisome proliferator-activated receptors (PPAR) having specificity on PPAR gamma (PPARG), proteins which act as gene regulators. PPARG are regulating the expression of the genes implied in the lipid uptake and adipogenesis in fatty cells. They lower serum glucose without influencing the secretion of insulin in the pancreas. (Michalik et al., 2006).

(Continued)

Main Pharmacological Class	Pharmacological subclass	Agent	Mechanism of action	Other observations
				All these compounds have worldwide restrictions on their use. They proved to have serious side effects when administering them on the long term: increased risk of cancers, acute cardiovascular events, and hepatic toxicity. (Mannucci et al., 2008; Ferwana et al., 2013)
Insulin secretagogues	Sulfonylureas	First generation agents: Tolbutamide Acetohexamide Tolazamide Chlorpropamide Second generation agents: Glipizide Glibenclamide Glimepiride Gliclazide Gliquidone	They close the ATP-sensitive K-channels in the beta-cell plasma membrane, and so they initiate a chain of events which stimulates the insulin release (the closing of the channel depolarizes the beta-cells, opening the cells' calcium channels with a calcium influx that induces insulin secretion). The beta-cell ATP-sensitive K-channel is a complex of two proteins: a pore-forming subunit (Kir6.2) and a drug-binding subunit (SUR1). The SUR1 subunit is, in fact, the receptor for sulfonylureas. (Ashcroft, 1996)	They are used for type 2 diabetes mellitus, and they are ineffective for type 1 diabetes; (Greeley et al., 2010)
	Non-Sulfonylureas – Meglitinides	Repaglinide Nateglinide Mitiglinide	They close the ATP-sensitive K-channel present in the cell membrane of the beta-cells, like the sulfonylureas, but they	Their administration is co-dependent by the meals taken. If a meal is skipped, the meglinite dose will also be skipped because of the potentially induced hypoglycemia. (Rendell, 2004).

(Continued)

Main Pharma- cological Class	Pharmacological subclass	Agent	Mechanism of action	Other observations
			have a weaker binding affinity and a faster dissociation from the SUR1 binding site. (Blicklé, 2006)	
Alpha-Glucosi- dase Inhibitors		Miglitol Acarbose Voglibose	They are indirect hypoglycemic agents because they don't have a direct effect over the insulin secretion or sensitivity. They prevent the digestion of carbohydrates by inhibiting the maltase-glucoamylase enzyme present in the small intestine which catalyzes the transformation of starch and disaccharides to glucose. (Chiba, 1997)	The agents are modified saccharides which act as competitive inhibitors of the maltase-glucoamylase. (Bischoff, 1995)
Peptide Analogs	Injectable Incretin Mimetics	Glucagon-like Peptide 1 Gastric Inhibi- tory Peptide Injectable Glucagon-like peptide analogs and agonists: Exenatide Liraglutide Taspoglutide Lixisenatide	Glucagon-like peptide (GLP-1) analogs are a class of drugs used in the treatment of type 2 diabetes. They are administered by injection, and they regulate glucose levels by stimulating glucose-dependent insulin secretion and biosynthesis, suppressing glucagon secretion and delaying gastric emptying and promoting satiety. Shyangdan et al., 2010)	

(Continued)

Main Pharma-cological Class	Pharmacological subclass	Agent	Mechanism of action	Other observations
	Dipeptidyl Peptidase–4 Inhibitors: Vildagliptin Sitagliptin Saxagliptin Linagliptin Alogliptin Septagliptin Teneligliptin	By inhibiting the DPP4, they increase incretin levels thus inhibit glucagon release which in turn they increase insulin secretion, decrease gastric emptying and finally decrease blood glucose levels; (Dupre et al., 1995; Behme et al., 2003; McIntosh et al., 2005)	
	Injectable Amylin analogs	Pramlintide	An analog of amylin (peptide hormone, also released by beta-cells, like insulin, after a meal); (Jones, 2007).	By supplementing amylin, this peptide helps the cellular absorption of glucose and the regulation of the blood sugar by slowing the gastric emptying, promoting the satiety sensation via a hypothalamic pathway and by inhibiting the inappropriate secretion of glucagon; (Hollander et al., 2004)
Glycosurics	Gliflozins	Dapagliflozin Empagliflozin Canagliflozin Ipragliflozin Tofogliflozin Sergliflozin etabonate Remogliflozin etabonate Ertugliflozin	They are sodium/glucose cotransporter 2 (SGLT2) inhibitors	They block the re-uptake of glucose in the renal tubules, promoting loss of glucose in the urine; This causes both mild weight loss, and a mild reduction in blood sugar levels with little risk of hypoglycemia; (Dietrich et al., 2013)

proliferator hormone responsive elements (PPRE) which will influence further some insulin-sensitive genes. This process will enhance the production of mRNAs corresponding to insulin-dependent enzymes which translates to a better use of glucose by the cells (Michalik et al., 2006).

9.3.1.2 INSULIN SECRETAGOGUES

9.3.1.2.1 Sulfonylureas

Sulfonylureas are used for non-insulin dependent diabetes mellitus treatment. These drugs express their hypoglycaemic effects by stimulating insulin secretion from the pancreatic beta-cells (Proks et al., 2002).

The sulfonylureas can be classified as first and second generation agents. The second generation sulfonylurea hypoglycemics (glipizide, glyburide, and glimepiride) are the newer, "more potent" agents. The sulfonylureas produce their hypoglycemic actions through several mechanisms that can be broadly subclassified as pancreatic and extra-pancreatic. All sulfonylurea hypoglycaemics inhibit the efflux of $K+$ (from the $K+$ channel blockers) from pancreatic ß-cells via a sulfonylurea receptor which may be closely linked to an ATP-sensitive potassium-channel. The inhibition of efflux of $K+$ leads to depolarization of the ß-cell membrane and, as a consequence, voltage-dependent Ca^{2+}-channels on the ß-cell membrane then open to permit the influx of Ca^{2+}. The resultantly increased binding of Ca^{2+} to calmodulin results in activation of kinases associated with endocrine secretory granules thereby promoting the exocytosis of insulin granules. The sulfonylureas also reduce serum glucagon levels possibly contributing to its hypoglycemic effects (http://www.auburn.edu).

9.3.1.2.2 Non-Sulfonylureas

Meglitinides have similar mechanisms of action to the ones of sulfonylureas: they interact with binding sites on ATP-dependent potassium channels in the β-cell membrane. Still, these sites are distinct from those involved in sulfonylureas binding (http://www.auburn.edu).

9.3.1.3 ALPHA-GLUCOSIDASE INHIBITORS

One effective way of lowering postprandial blood glucose levels is represented by the delaying of glucose absorption. The idea was used to create competitive inhibitors of the intestinal alpha-glucosidases. By inhibiting these enzymes, the digestion of non-absorbable polysaccharides is prevented, and their split to absorbable monosaccharides (glucose, fructose) is inhibited (http://www.auburn.edu).

9.3.1.4 PEPTIDE ANALOGS

9.3.1.4.1 Incretin Analogs

This type of medication works by increasing the levels of hormones called 'incretins.' These hormones help the body produce more insulin only when needed and reduce the amount of glucose being produced by the liver when it's not needed. They reduce the rate at which the stomach digests food and empties, and can also reduce appetite.

Analogs of the incretins Glucagon-like peptide-1 (GLP-1), and Glucose-dependent insulinotropic peptide (GIP) have been developed to treat type 2 diabetes mellitus. They are protease resistant and have a longer biological half-life than the native peptides (Holscher, 2010).

1. Glucagon-Like Peptide Analogs

There are two naturally occurring incretin hormones that have a role in the maintenance of glycemic control: glucose-dependent insulinotropic polypeptide (GIP) and GLP-1, both of which have a short half-life because of their rapid inactivation by dipeptidyl peptidase-4 enzymes. In patients with type 2 diabetes, the incretin effect is reduced or, in some cases, absent. In particular, the insulinotropic action of GIP is lost in patients with type 2 diabetes. However, it has been shown that, after administration of pharmacological levels of GLP-1, the insulin secretory function can be restored in this population and thus GLP-1 has become an important target for research into new therapies for type 2 diabetes (Garber, 2011).

GLP-1 has multiple physiological effects that make it an attractive candidate for type 2 diabetes therapy. It increases insulin secretion while inhibiting glucagon release, but only when glucose levels are elevated thus offering the potential to lower plasma glucose while reducing the likelihood of hypoglycemia. Furthermore, gastric emptying is delayed, and food intake is decreased after GLP-1 administration (Garber, 2011).

Two classes of incretin-based therapy have been developed to overcome the clinical limitations of native GLP-1: GLP-1 receptor agonists (e.g., liraglutide, and exenatide), which exhibit increased resistance to DPP-4 degradation and thus provide pharmacological levels of GLP-1, and DPP-4 inhibitors (e.g., sitagliptin, vildagliptin, saxagliptin), which reduce endogenous GLP-1 degradation, thereby providing physiological levels of GLP-1 (Garber, 2011).

2. Dipeptidyl Peptidase–4 Inhibitors

Dipeptidyl-peptidase-4 (DPP-4) inhibitors inhibit the degradation of the incretins, GLP-1 and GIP. The first available DPP-4 inhibitors are sitagliptin and vildagliptin. These compounds are orally active and have been shown to be efficacious and well tolerated (Thornberry et al., 2009).

Dipeptidyl-peptidase IV is a complex enzyme that exists as a cell membrane peptidase and as a second smaller soluble form present in the circulation. Both GIP and GLP-1 are endogenous physiological substrates for DPP-4 and chemical inhibition of DPP-4 activity, or genetic inactivation of DPP-4 in rodents, results in increased levels of intact bioactive GIP and GLP-1. Sustained DPP-4 inhibition lowers blood glucose via stimulation of insulin and inhibition of glucagon secretion and is associated with preservation of β-cell mass in preclinical studies. Although DPP-4 cleaves dozens of regulatory peptides and chemokines in vitro, studies demonstrate that GIP and GLP-1 receptor-dependent pathway represent the dominant mechanisms transducing the glucoregulatory actions of DPP-4 inhibitors in vivo. The available preclinical data suggest that highly selective DPP-4 inhibition (because DPP-4 is a member of a complex gene family which encode other structural related enzymes) represents an effective and safe strategy for the therapy of type 2 diabetes (Drucker, 2007).

9.3.1.4.2 Amylin Analogs

Amylin, (islet amyloid polypeptide) or diabetes-associated peptide is co-secreted with insulin in the islet of Langerhans of diabetic patients in approximately 1:100, amylin-insulin ratio. Amylin is a small insoluble peptide with a molecular weight of only 3.9 kDa which has a relatively low tissue concentration. Amylin plays a role in glycemic regulation by slowing gastric emptying and promoting satiety. The actions of amylin analogs appear to be synergistic to that of insulin and can significantly reduce body weight, HbA1c values and even the dosage of insulin (Adeghate et al., 2011).

9.3.1.5 GLYCOSURIA

Glucosuria, the presence of glucose in the urine, has long been regarded as a consequence of uncontrolled diabetes. However, glucose excretion can be induced by blocking the activity of the renal sodium-glucose cotransporter 2 (SGLT-2). This mechanism corrects hyperglycemia independently of insulin, because SGLT2 is the major cotransporter involved in glucose reabsorption in the kidney. By lowering the renal threshold for glucose excretion, SGLT-2 inhibitors suppress renal glucose reabsorption (30–50% of the glucose filtered by the kidney) and thereby increase urinary glucose excretion. The advantages of this approach are reduced hyperglycemia without hypoglycemia, along with weight loss and blood pressure reduction (Chao, 2014).

9.4 TRANS-DISCIPLINARY CONNECTIONS

In 2014, a series of 2,4-thiazolidinedione derivatives were synthesized, and QSAR studied for their hypoglycemic activity. 3-(2,4-dimethoxyphenyl)–2-{4-[4-(2,4-dioxothiazolidin–5-yl-methyl)phenoxy]-phenyl}-acrylic acid methyl ester was found the most active compound in this series. Antihyperglycemic activity of these synthesized derivatives was described by five models in which the bi-parametric model-3 containing third-degree connectivity index (χ^3) and log P was found to be the best model (Chawla et al., 2014).

A study by Dixit et al. (2008) tried to assess the essential structural and physical-chemical characteristics for PPARγ agonistic activity of 23 compounds. It was found that the steric, electronic, and topological descriptors have the most important roles in determining the agonistic activity. The predicted activities using TATA-BioSuite software were in accordance with the observed activities.

In 2014, Balajee, R. and his colleagues performed a QSAR study on some compounds presenting a pharmacophore necessary for inhibiting DPP-4, using PHASE software for identifying the pharmacophore important for the inhibition and GLIDE software for molecular docking. After establishing the right pharmacophore and in silico attaching it to some compounds, the team evaluated the interaction between the optimized compounds and their specific receptor. The interactions were assessed, and a novel compound was found to be more effective in targeting DDP4 enzyme (Balajee et al., 2014).

9.5   OPEN ISSUES

Software-aided molecular design and QSAR studies have a great potential in designing molecules as therapeutic agents in diabetes mellitus. However, since type 2 diabetes mellitus is a complex disease that includes several biological targets, multi-target QSAR studies are recommended in the future to achieve efficient antidiabetic therapies.

KEYWORDS

- **alpha-glucosidase inhibitors**
- **glycourics**
- **insulin-secretagogue drugs**
- **insulin-sensitizing drugs**
- **peptide analogs**

REFERENCES AND FURTHER READING

Adeghate, E., & Kalasz, H., (2011). Amylin analogs in the treatment of diabetes mellitus: Medicinal chemistry and structural basis of its function. *The Open Medicinal Chemistry Journal, 5,* 78–81.

Ahmed, A. M., (2002). History of diabetes mellitus. *Saudi Medical Journal, 23*(4), 373–378.

Ashcroft, F. M., (1996). Mechanisms of the glycaemic effects of sulfonylureas. *Hormone and Metabolic Research, 28*(9), 456–463.

Balajee, E., & Dhanarajan, M. S., (2014). 3D QSAR studies of identified compounds as potential inhibitors for antihyperglycemic targets. *Asian Journal of Pharmaceutical and Clinical Research, 7*(3), 362–364.

Behme, M. T., Dupré, J., & McDonald, T. J., (2003). Glucagon-like peptide 1 improved glycemic control in type 1 diabetes. *BMC Endocrine Disorders,* 3:3 (https://bmcendocrdisord.biomedcentral.com/articles/10.1186/1472-6823-3-3).

Bischoff, H., (1995). The mechanism of alpha-glucosidase inhibition in the management of diabetes. *Clinical and Investigative Medicine, 18*(4), 303–311.

Blicklé, J. F., (2006). Meglitinide analogs: A review of clinical data focused on recent trials. *Diabetes & Metabolism, 32*(2), 113–120.

Chao, E. C., (2014). SGLT-2 inhibitors: A new mechanism for glycemic control. *Clinical Diabetes, 32*(1), 4–11.

Chawla, A., Chawla, P., & Dhawan, R. K., (2014). QSAR study of 2,4-dioxothiazolidine antidiabetic compounds. *Der Pharma Chemica, 6*(2), 103–110.

Chiba, S., (1997). Molecular mechanism in alpha-glucosidase and glucoamylase. *Bioscience, Biotechnology, and Biochemistry, 61*(8), 1233–1239.

Dallas, J., (2011). *Royal College of Physicians of Edinburgh*. Diabetes, doctors, and dogs: An exhibition on diabetes and endocrinology by the college library for the 43rd St. Andrew's Day Festival Symposium.

Dietrich, E., Powell, J., & Taylor, J. R., (2013). Canagliflozin: A novel treatment option for type 2 diabetes. *Drug Design, Development, and Therapy, 22*(7), 1399–1408.

Dixit, A., & Saxena, A. K., (2008). QSAR analysis of PPARγ agonists as anti-diabetic agents. *European Journal of Medicinal Chemistry, 43*(1), 73–80.

Drucker, D. J., (2007). Dipeptidyl peptidase–4 inhibition and the treatment of type 2 diabetes. *Diabetes Care, 30*(6), 1335–1343.

Dupre, J., Behme, M. T., Hramiak, I. M., McFarlane, P., Williamson, M. P., Zabel, P., & McDonald, T. J., (1995). Glucagon-like peptide I reduce postprandial glycemic excursions in IDDM. *Diabetes, 44*(6), 626–630.

Eurich, D. T., McAlister, F. A., Blackburn, D. F., Majumdar, S. R., Tsuyuki, R. T., Varney, J., & Johnson, J. A., (2007). Benefits and harms of antidiabetic agents in patients with diabetes and heart failure: Systematic review. *British Medical Journal (Clinical Research Ed.), 335*(7618), p. 497.

Ferwana, M., Firwana, B., Hasan, R., Al-Mallah, M. H., Kim, S., Montori, V. M., & Murad, M. H., (2013). Pioglitazone and risk of bladder cancer: A meta-analysis of controlled studies. *Diabetes Medicine: A Journal of the British Diabetes Association, 30*(9), 1026–1032.

Garber, A. J., (2011). Long-acting glucagon-like peptide 1 receptor agonists. *Diabetes Care, 34*(2), 279–284.

Greeley, S. A., Tucker, S. E., Naylor, R. N., Bell, G. I., & Philipson, L. H., (2010). Neonatal diabetes mellitus: A model for personalized medicine. *Trends in Endocrinology and Metabolism, 21*(8), 464–472.

Gunton, J. E., Delhanty, P. J., Takahashi, S., & Baxter, R. C., (2003). Metformin rapidly increases insulin receptor activation in human liver and signals preferentially through insulin-receptor substrate–2. *Journal of Clinical Endocrinology and Metabolism, 88*, 1323–1332.

Güthner, T., Mertschenk, B., & Schulz, B., (2006). *"Guanidine and Derivatives" in Ullmann's Encyclopedia of Industrial Chemistry*, Wiley-VCH, Weinheim.

Himsworth, H. P., (1936). Diabetes mellitus: Its differentiation into insulin-sensitive and insulin-insensitive types. *Lancet, 227*(5864), 127–130.

Hockaday, T. D. R., (1981). Should the diabetic diet be based on carbohydrate of fat restriction? In: Turner, M., & Thomas, B., (eds.), *Nutrition*, and *Diabetes, London* (pp. 23–32). Libbey Publishers.

Hollander, P., Maggs, D. G., Ruggles, J. A., Fineman, M. S., Kolterman, L., Orville, G., & Weyer, C., (2004). Effect of pramlintide on weight in overweight and obese insulin-treated type 2 diabetes patients. *Obesity, 12*(4), 661–668.

Holscher, C., (2010). Incretin analogs that have been developed to treat type 2 diabetes hold promise as a novel treatment strategy for Alzheimer's disease. *Recent Patents on CNS Drug Discovery, 5*(2), 109–117.

Jones, M. C., (2007). Therapies for diabetes: Pramlintide and exenatide. *American Family Physician, 75*(12), 1831–1835.

Mannucci, E., Monami, M., Lamanna, C., Gensini, G. F., & Marchionni, N., (2008). Pioglitazone and cardiovascular risk. A comprehensive meta-analysis of randomized clinical trials. *Diabetes, Obesity, and Metabolism, 10*(12), 1221–1238.

McIntosh, C., Demuth, H., Pospisilik, J., & Pederson, R., (2005). Dipeptidyl peptidase IV inhibitors: How do they work as new antidiabetic agents? *Regulatory Peptides, 128*(2), 159–165.

Michalik, L., Auwerx, J., Berger, J. P., Chatterjee, V. K., Glass, C. K., Gonzalez, F. J., et al., (2006). International Union of Pharmacology, LXI, peroxisome proliferator-activated receptors. *Pharmacology Reviews, 58*(4), 726–741.

Mullan, R. J., Montori, V. M., Shah, N. D., Christianson, T. J., Bryant, S. C., Guyatt, G. H., et al., (2009). The diabetes mellitus medication choice decision aid: A randomized trial. *Archives of Internal Medicine, 169*(17), 1560–1568.

Plutzky, J., (2011). The PPAR-RXR transcriptional complex in the vasculature: Energy in the balance. *Circulation Research, 108*(8), 1002–1016.

Proks, P., Reimann, F., Green, N., Gribble, F., & Ashcroft, F., (2002). Sulfonylurea stimulation of insulin secretion. *Diabetes, 51*(3), 368–376.

Rena, G., Pearson, E. R., & Sakamoto, K., (2013). Molecular mechanism of action of metformin: Old or new insights? *Diabetologia, 56*(9), 1898–1906.

Rendell, M., (2004). Advances in diabetes for the millennium: Drug therapy of type 2 diabetes. *Medscape General Medicine, 6*(3), 9–14.

Shyangdan, D. S., Royle, P. L., Clar, C., Sharma, P., & Waugh, N. R., (2010). Glucagon-like peptide analogs for type 2 diabetes mellitus: Systematic review and meta-analysis. *BMC Endocrine Disorders,* 10:20 (https://bmcendocrdisord.biomedcentral.com/articles/10.1186/1472-6823-10-20).

Thornberry, N. A., & Gallwitz, B., (2009). Mechanism of action of inhibitors of dipeptidyl-peptidase-4 (DPP-4). *Best Practice & Research in Clinical Endocrinology & Metabolism, 23*(4), 479–486.

World Health Organization, (2013). *Diabetes*. Fact sheet no 312. http://www.auburn.edu/~deruija/endo_diabetesoralagents.pdf.

CHAPTER 10

Drug Development

CORINA DUDA-SEIMAN,[1] DANIEL DUDA-SEIMAN,[2] and
MIHAI V. PUTZ[1,3]

[1]*Laboratory of Computational and Structural Physical Chemistry
for Nanosciences and QSAR, Biology-Chemistry Department, Faculty
of Chemistry, Biology, Geography at West University of Timişoara,
Pestalozzi Street No. 16, Timişoara, RO–300115, Romania*

[2]*Department of Medical Ambulatory, and Medical Emergencies,
University of Medicine and Pharmacy "Victor Babes," Avenue C. D. Loga
No. 49, RO–300020 Timisoara, Romania*

[3]*Laboratory of Renewable Energies-Photovoltaics, R&D National
Institute for Electrochemistry and Condensed Matter, Dr. A. Paunescu
Podeanu Str. No. 144, RO–300569 Timişoara, Romania,
Tel.: +40-256-592-638, Fax: +40-256-592-620,
E-mail: mv_putz@yahoo.com, mihai.putz@e-uvt.ro*

10.1   DEFINITION

Actual research in drug development is focused on the concepts of quality, high technology, and cost-effectiveness. Because of the rising costs in the pharmaceutical industry, rational drug development is based on the objective requirements of the populational state of health.

Drug development is structured as follows: (i) to establish the basic structure to be developed; (ii) to apply molecular and chemical modeling techniques in order to obtain a statistically validated model of a chemical structure; (iii) to characterize the obtained model by means of physical, chemical, and biological properties; (iv) organic synthesis; (v) physical and chemical characterization and purifying the synthesized compound; (vi) preclinical trials; (vii) clinical trials; and (viii) marketing of the obtained drug.

In order to put an agreement between clinical pharmacologists and clinical trial statisticians, drug development should be based on two interdependent models: a pharmacokinetic/pharmacodynamic model for the former and a more general statistical model for the latter (Sheiner and Steimer, 2000).

10.2 HISTORICAL ORIGINS

Morphine was firstly commercialized in Germany in 1827 by Merck, being afterward introduced in the USA, and Aspirin was introduced by Bayer in 1899 (Kinch et al., 2014).

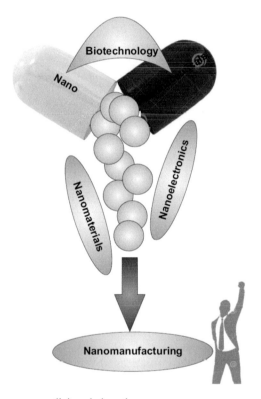

FIGURE 10.1   The nano-medicine chain-value.

A cornerstone in science and, in particular, in chemo-pharmacy, was the discovery of penicillin by Alexander Fleming. At the Alexander Fleming Museum Laboratory Museum in London, on the commemorative plaque it is written: "In 1928, at St. Mary's Hospital, London, Alexander Fleming discovered penicillin. This discovery led to the introduction of antibiotics that greatly reduced the number of deaths from infection. Howard W. Florey, at the University of Oxford working with Ernst B. Chain, Norman G. Heatley and Edward P. Abraham, successfully took penicillin from the laboratory to the clinic as a medical treatment in 1941" (Anonymous, 1999).

Pharmaceutical industry experienced several strategies in the effort of identifying and marketing of new bioactive molecules, from a random screening of biological materials to knowledge-based drug identification (Walsh, 2003).

10.3 NANO-CHEMICAL IMPLICATIONS

Nanotechnologies are assimilated to purpose technologies that whether represent the starting point for solutions of industrial requirements or a binder in order to converge other available technologies (biotechnologies, computational sciences, physical sciences, etc.). (Mangemantin and Walsh, 2012). Nanotechnology domains are (Islam and Miyazaki, 2010):

1. **Bionanotechnology**: is based upon molecular scale properties and biological nanostructures and approaches the fields of drug discovery, diagnostic, and therapy techniques at a cellular and sub-cellular level;
2. **Nanoelectronics**: provides nanoscale properties and applications of semiconductor structures with integration in computing science and communication;
3. **Nanomaterials**: approaches materials at the nanoscale dimension creating new materials with novel and improved characteristics;
4. **Nanomanufacturing and tools**: concerns the building of more intricate nanostructures. Structure of bio-colloids has got knowledge using cryo-electron microscopy. Tomography alone, or associated to microscopy is promising to highlight (macro)molecular assembly/ disassembly mechanisms, essential in nanochemistry (Frederik and Sommerdijk, 2005).

There is high interest to improve the effectiveness and safety of drugs. One of the components to achieve this desideratum is to improve drug delivery systems (DDS). A large variety of potential drug and other biolog-ical-active entities carriers has been studied: vesicles, micelles, emulsions, biodegradable nanoparticles. Okamura and his team (Okamura et al., 2008) developed albumin-based nanoparticles and phospholipid vesicles that carry recombinant fragments of platelet membrane proteins and fibrinogen dodecapeptide as a recognition site for activated IIb/IIIa glycoprotein on the surface of platelets. The authors observed that these nanoparticles have a specific recognition capacity of active bleeding injury sites. Further on, sheet-shaped carriers have a large contact area, as the targeting site, in this case being superior to the spherical-shaped carriers. So, these sheet-shaped carriers may be a viable solution as DDS for certain drugs, hemostatic reagents, wound dressings of different causes.

Nanocarriers (nanocarrier DDS) can be liposomes, dendrimers, polymeric nanoparticles, solid lipid nanoparticles, and metal nanoparticles. According to their shape, nanocarriers are nanospheres, nanocapsules, and nanoparticles within drug molecules are dispersed. Nanocarrier DDS offer several benefits: protection against drug degradation, targeting of drugs to their specific sites of action, and facilitating inclusively protein-derived drugs (Mallipeddi and Rohan, 2015).

It is very well known that in nature, one can find a large variety of bioactive molecules. Bennet et al., (2014), based on the fact that bioactive phytodrugs may be unstable and ineffective, have used poly(D,L-lactide-co-glycolide) as a carrier for apple peel ethanolic extract to enhance its photoprotective capacity. This was the case for the above-mentioned complex at a concentration of 50 μM, expressing strong and synergistic photoprotective effects in the treatment of skin damages.

Recently, it was developed a facile green method for synthesis of mono-dispersed nanocomposite materials containing silver nanoparticles (AgNPs), using a non-toxic water-soluble polymer (poly(1-vinyl-1,2,4-triazole) as stabilizing matrix, glucose, and dimethyl sulfoxide as reducing agents. AgNPs are non-toxic and heat-resistant, persist in aqueous solution over 6 months without changing size. *In vitro* tests proved that these nanoparticles express significant antimicrobial activity against Gram-positive and Gram-negative bacteria, as well as fungi (Prozorova et al., 2014).

Magnetic nanoparticles express magnetic and catalytic properties, therefore being of interest in their applications in medicine and biology. Synthesis of iron oxide magnetic nanoparticles via a reverse micelle system and modification of their surface by an organosilane agent can be exploited as an array-based bioassay system. Such a system is suitable for magnetic detection of biomolecules, with high applicability in DNA protein and microbe analyses. The advantages of magnetic particles are that they are detectable by measuring their magnetism, they are unaffected by measurement processes, samples can be stored for an undefined period of time, and are less costly than fluorescent dyes (Osaka et al., 2006).

A major challenge for researchers is to develop pharmacological active forms with increased and sustained antiretroviral activity. Anti-HIV drugs have to develop a proper distribution in quantity and quality at the specific body sites; effective drug concentrations have to be maintained at these sites for an optimal period of time. A solution for these issues is to develop DDS in terms of nanocarriers. Their entry into the target cells is possible through energy-dependant or independent mechanisms. Nanocarriers based on polymers have been proven to be effective in delivering antiretrovirals at target sites, with optimal control release of the drug. The encapsulated anti-retrovirals can be detected in peripheral blood cells in a prolonged manner. Silver nanoparticles were effective *in vitro* against a wide range of HIV-1 HIV-1 strains, acting as viral entry inhibitors by binding to the glycoprotein 120, as well as inhibiting post-entry stages of HIV-1, with the certainty of less resistance development when silver nanoparticles are used (Mallipeddi and Rohan, 2015).

An actual issue in medicine and pharmaco-chemistry is finding therapy possibilities against resistant tuberculosis (TB) strains. The preferred option is to improve available anti-TB drugs. Saifullah et al., (2012) summarized the benefits of using different nano-DDSs, enhancing anti-TB efficiency of consecrated drugs:

- Large porous particles of capreomycin: increased bioavailability, reduced side effects, avoidance of parenteral administration.
- Microparticles of rifampin-containing mannitol: pre-release (but uncontrolled release) of the drug into the lungs, not complicated to prepare.
- Microparticles of poly(D-L-lactic acid): because of the possibility of multiple drug encapsulation, a higher concentration of the drug is maintained than if administered parenterally; the drug is directly delivered to the lungs, and microparticles are rapidly phagocytized by alveolar macrophages.
- Tunable systems for controlled release into the lungs: monomers are already used in clinical practice.
- Targeted delivery of glucan particles to macrophages: the system has a good targetability, the therapeutic effect is maintained with a lower concentration than the minimum inhibitory concentration; rifampin has less side effects than if it is administered in classic fashion.
- Plant proteins (zein): possibility of multiple drug encapsulation; safety of administration.

- Gelatin nano-vehicles: The gelatin nano-vehicles are biocompatible, biodegradable, have low antigenicity; provide uniform distribution of particles, extended controlled release; have improved pharmacokinetics; could reduce dosing frequency; and non-toxic compared to the free drug.
- Chitosan-montmorillonite hydrogel: shows improved encapsulation properties, with good targetability and controlled release at lower pH.
- Nanocomposite hydrogels of poly(vinyl alcohol) and sepiolite: better drug water solubility, especially rifampin.

Modified fullerene (diadduct malonic acid-fullerene-Asn-Gly-Arg peptide [DMA-C60-NGR]) and 2-methoxyestradiol (2ME) exhibit an increased inhibition (antiproliferative) effect in the estrogen receptor-positive human breast adenocarcinoma cell line (MCF-7). 2ME induces apoptosis in some tumor cells, but not in normal cells, via several mechanisms: tubulin depolymerization, upregulation of p53 and death receptors, as well as the production of ROS (reactive oxygen species) (Shi et al., 2013).

Multi-drug resistant tumors may benefit from new nano-DDS. Wu et al., (2014) prepared a system comprising a co-loaded reversal agent and chemotherapeutic drug with shortened carbon nanotubes (CNTs) to investigate its anti-cancer capacity. CNTs were cut and purified via ultrasonication and oxidative acid treatment to optimize their length for drug-delivery vehicles, then verapamil (Ver) and doxorubicin (Dox) were co-loaded on shortened CNTs (Ver/Dox/shortened CNTs), which acted as a drug delivery system. This system was used for the treatment of multidrug-resistant leukemia K562/A02 cells, and the activity of the system upon tumor cells was assessed by a flow cytometer, 3-(4,5-dimethylthiazol-2-yl)-2,5-diphenyltetrazolium bromide assays, acridine orange/ethidium bromide staining, and Western blot analysis. Thus, the described DDS inhibits the function of P-glycoprotein by verapamil as reversal agent, facilitates an increased uptake of Dox, enhances the sensitivity of the MDR cancer cells to the chemotherapeutic agent, and induces apoptosis.

10.4 TRANS-DISCIPLINARY CONNECTIONS

Gibbons et al., (1994) state that science can be viewed in two ways: (i) the so-called old sciences (physics, chemistry, biology) are institutionalized, the specialized laboratories being disciplinary, university and government-based; (ii) science is multidisciplinary: there is a network of knowledge converging to problem-solving and social impact. Multidisciplinary chemistry and physical chemistry published papers cite in an extensive manner article belonging to the Materials Sciences macrodiscipline, as well as to Biomedical and Environmental Sciences, Chemistry, Health Sciences, and Mathematics (Porter and Youtie, 2009). Effective collaboration among pharmaceutical industry, medium size enterprises, academics, drug regulatory authorities, ethics review committees, patient organizations, and policy-makers is essential for strengthening translational research in nanomedicine (Satalkar et al., 2016).

10.5 OPEN ISSUES

Regenerative medicine is one of the most promising and challenging fields in the medicine of the future. Nanotechnology in medicine and pharmacy is applied in the synthesis of biocompatible materials supporting the growth of cells used in cell therapy. Nanomedicine will allow to develop personalized procedures by means of diagnostic steps and therapy (Boisseau and Loubaton, 2011).

In autoimmune conditions, actual available treatments, including biological therapies, are not completely specific. Using nanotechnology, new improved therapy possibilities can be developed for these conditions, more effective and with less adverse effects, with possibility of intervention both in innate immune disorders, and in acquired conditions. Nanotechnology-dependent therapies will be more targeted or personalized improving the therapeutic index, both in conditions with well-characterized autoantigens, and in autoimmune conditions without known specific autoantigens and without generation of pathogenic autoantibodies and/or autoreactive *T* cells (Gharagozloo et al., 2015).

The Nanomedicines Alliance is an organization of pharmaceutical and biotechnology companies formed in 2010 that occupies a unique place in the global pharmaceutical industry. This group of companies focuses its energy on promoting scientific development, safe use, regulatory, and legislative advancement, and expanding public knowledge of nanomedicines and nanotechnology-enabled medical devices (Malinoski, 2014).

Nanotechnology has an increasing trend in medicine with applicability form diagnosis to therapy and medical devices. A great part of available anti-cancer nano-drugs are liposomes, and polymer-based nanoformulations that lower toxicity and enhance delivery of chemotherapeutics via passive targeting. Complex formulations based on liposomes, micelles, polymeric nanoparticles, and metal nanoparticles are in current trials are attempting to reach the tumor through active targeting systems or are activated by an external source of energy only when the malign tissue or cells are specifically reached (Bregoli et al., 2016).

KEYWORDS

- **chemical modeling**
- **drug development**
- **molecular**
- **trials**

REFERENCES AND FURTHER READING

Anonymous on Behalf of the American Chemical Society and the Royal Society of Chemistry, (1999). The discovery and development of penicillin 1928–1945.

Bennet, D., Kang, S. C., Gang, J., & Kim, S., (2014). Photoprotective effects of apple peel nanoparticles. *International Journal of Nanomedicine, 9*, 93–108.

Boisseau, P., & Loubaton, B., (2011). Nanomedicine, nanotechnology in medicine (Nanomédecine et nanotechnologies pour la medicine). *Comptes Rendus Physique, 12*(7), 620–636.

Bregoli, L., Movia, D., Gavigan-Imedio, J. D., Lysaght, J., Reynolds, J., & Prina-Mello, A., (2016). Nanomedicine applied to translational oncology. *A Future Perspective on Cancer Treatment, 12*(1), 81–103.

Frederik, P. M., & Sommerdijk, N., (2005). Spatial and temporal resolution in cryo-electron microscopy: a scope for nanochemistry. *Current Opinion in Colloid & Interface Science, 10*(5/6), 245–249.

Gharagozloo, M., Majewski, S., & Foldvari, M., (2015). Therapeutic applications of nanomedicine in autoimmune diseases: From immunosuppression to tolerance induction. *Nanomedicine: Nanotechnology, Biology, and Medicine, 11*(4), 1003–1018.

Gibbons, M., Limoges, C., Nowotny, H., Schwartzman, S., Scott, P., & Trow, M., (1994). The new production of knowledge. *The Dynamics of Science and Research in Contemporary Societies*. Sage, London, England.

Islam, N., & Miyazaki, K., (2010). An empirical analysis of nanotechnology research domains. *Technovation, 30*(4), 229–237.

Kinch, M. S., Haynesworth, A., Kinch, S. L., & Hoyer, D., (2014). An overview of FDA-approved new molecular entities: 1827–2013. *Drug Discovery Today, 19*(8), 1033–1039.

Malinoski, F. J., (2014). The nanomedicines alliance: An industry perspective on nanomedicines. *Nanomedicine: Nanotechnology, Biology, and Medicine, 10*(8), 1819–1820.

Mallipeddi, R., & Rohan, L. C., (2015). Progress in antiretroviral drug delivery using nanotechnology. *International Journal of Nanomedicine, 5*, 533–547.

Mangemantin, V., & Walsh, S., (2012). The future of nanotechnologies. *Technovation, 32*(3/4), 157–160.

Okamura, Y., Utsunomiya, S., Suzuki, H., Niwa, D., Osaka, T., & Takeoka, S., (2008). Fabrication of free-standing nanoparticle-fused nanosheets and their hetero-modification using sacrificial film. *Colloids and Surfaces A: Physicochemical and Engineering Aspects, 318*(1–3), 184–190.

Osaka, T., Matsunaga, T., Nakanishi, T., Arakaki, A., Niwa, D., & Iida, H., (2006). Synthesis of magnetic nanoparticles and their application to bioassays. *Anal. Bioanal. Chem., 384*, 593–600.

Porter, A. L., & Youtie, J., (2009). How interdisciplinary is nanotechnology? *J. Nanopart. Res., 11*, 1023–1041.

Prozorova, G., Pozdnyakov, A. S., Kuznetsova, N. P., Korzhova, S. A., Emel'yanov, A. I., Ermakova, T. G., Fadeeva, T. V., & Sosedova, L. M., (2014). Green synthesis of water-soluble nontoxic polymeric nanocomposites containing silver nanoparticles. *International Journal of Nanomedicine, 9*, 1883–1889.

Saifullah, B., Hussein, M. Z. B., & Al Ali, S. H. H., (2012). Controlled-release approaches towards the chemotherapy of tuberculosis. *International Journal of Nanomedicine, 7*, 5451–5463.

Satalkar, P., Elger, B. S., Hunziker, P., & Shaw, D., (2016). Challenges of clinical translation in nanomedicine: A qualitative study. *Nanomedicine: Nanotechnology, Biology, and Medicine.* 12(4), 893–900.

Sheiner, L. B., & Steimer, J. L., (2000). Pharmacokinetic/pharmacodynamic modeling in drug development. *Annual Review of Pharmacology and Toxicology, 40*, 67–95.

Shi, J., Wang, Z., Wang, L., Wang, H., Li, L., Yu, X., Zhang, J., Ma, R., & Zhang, Z., (2013). Photodynamic therapy of a 2-methoxyestradiol tumor-targeting drug delivery system mediated by Asn-Gly-Arg in breast cancer. *International Journal of Nanomedicine, 8*, 1551–1562.

Walsh, G., (2003). In: Walsh, G., (ed.), *Biopharmaceuticals: Biochemistry and Biotechnology* (2nd edn). John Wiley & Sons Ltd., England.

Wu, P., Li, S., & Zhang, H., (2014). Design real-time reversal of tumor multidrug resistance cleverly with shortened carbon nanotubes. *Drug Design, Development, and Therapy, 8*, 2431–2438.

CHAPTER 11

Electrochemical Methods for Metal Recovery

MIRELA I. IORGA

National Institute for Research and Development in Electrochemistry and Condensed Matter, Department of Applied Electrochemistry, 300569 Timisoara, Romania

11.1 DEFINITION

The main task of metal ion wastewater treatment is to remove these ions down to the concentration required by law to discharge waters. The process is quite expensive, but it could be more cost-effective in conditions in which the metal can be recovered (particularly in the case of valuable metals). The main techniques used for metal ions removing from solutions are traditional methods or electrochemical methods. Classical methods include chemical precipitation, ion exchange, evaporation, solvent extraction, and membrane separation methods, such as reverse osmosis, ultrafiltration, and electrodialysis. Each of these methods has their advantages, but they all lack the ability to produce the metal directly, in a controlled manner, as electrochemical techniques are doing. Electrochemical methods could be recovering and/or destructive methods. Electrochemical methods for metal ions recovery from diluted solutions need serious, urgent attention, especially in recent years, due to increasing difficulties encountered in the treatment of extremely toxic contaminant waters, which often cannot be destroyed or removed by traditional methods. They refer to the favorable treatment or pretreatment of water containing heavy metals. Electrochemical recovery methods are based on metal ions discharge at the cathode, under reaction conditions provided by a suitable design of the electrochemical reactor.

11.2   HISTORICAL ORIGIN(S)

Issues arising from waste discharge into the environment have been a concern since ancient times, as shown by numerous observations and inscriptions found by archaeologists during their research.

The first mention of water supply and wastewater management seems to be in a region placed near Euphrates River and the city of Palmyra (nowadays Syria) during the Neolithic Age. At the end of the 4th millennium BC to the beginning of the 3rd millennium BC, it seems that Mesopotamian cities had networks of wastewater and stormwater drainage (Kalavrouziotis and Angelakis, 2014). Then Minoans in Crete Island developed very efficient water supply and sewerage systems in 3200–1100 BC period. The important contacts of Minoans with Egypt (in 1900–1700 BC) determined a possible influx of technology related to water, wastewater, and stormwater management. Based on the similarities of hydro-technologies developed by Mesopotamians and Egyptians, Minoans, and Indus, one can consider that some contacts between these civilizations existed.

The first mention of "dirty" water pre-treatment refers to boil water in copper cauldrons followed by cooling in clay pots, exposing to sunlight or filtration through charcoal (Rajeshwar and Ibanez, 1997). The first technical report on water treatment dated from the year 98 and belonged to Sextus Julius Frontinus, Rome water commission member.

All these hydro-technologies were permanently improved during the Classical, Hellenistic, and Roman periods as well as during several Chinese Dynasties and Empires and pre-Columbian civilizations (Kalavrouziotis and Angelakis, 2014).

Initially, until the beginning of the twentieth century, the wastewater treatment plants were built in the main industrial centers.

A major pollutant was mining waters which affect the environment both by its acidity and its metal content. In 1859 was introduced the standard treatment for metal removal from mining waters – lime or limestone precipitation (Figueroa and Wolkersdorfer, 2014).

Since sectors as mining operating, electroplating industries, metallurgical industries, battery industry, tanneries, the industry of fertilizers and pesticides, power generation facilities, electronic device manufacturing units (Chen et al., 2013) known such a massive development, the wastewaters were discharged into the environment. Those metals are not biodegradable, as organic pollutants for example, and they accumulate in live organisms. Among the most critical metals with high toxicity that are present in wastewaters, one can mention copper, zinc, nickel, cadmium, mercury, lead, chromium, etc. (Fu and Wang, 2011).

If initially, after a preliminary treatment (sedimentation, filtration), industrial effluents were discharged into the environment, nowadays the problem of environmental pollution has become a permanent concern of industry specialists and environmental quality monitors, to apply the most modern methods for pollution prevention, control, and treatment. The critical technological progress achieved in the last century determines a permanent change and improvement in water and wastewater sectors (Rajeshwar and Ibanez, 1997).

In the next period of the 1990s was found that most transition countries face serious environmental problems caused by an economy based on excessive consumption of energy and natural resources and the use of outdated and polluting technologies.

A lot of unsolved problems occur related to management principles, as processes decentralization, projects durability, sustainability, cost-effectiveness, etc. Besides, due to increasing urbanization and water sources contamination, new problems have arisen.

In recent decades, water pollution level increased a lot, especially in world regions where population and industry have developed intense and fast, without measures for water quality protection. These elements, superposed on a centralized and super planned economy, imposed the need for a severe economic restructuration, with environmental issues as a priority (Pletcher and Walsh, 2000).

The society intervenes through its control systems such as sustainable development concept, intervention tools on industrial pollution phenomenon, natural resources and waste management, environmental management, environmental impact study. Environmental externalities are social costs caused by environmental degradation due to contamination.

To remove the heavy metal ions, various technologies for wastewater treatment have been developed. Among them could be mentioned: chemical precipitation, adsorption, biosorption (Veglio and Beolchini, 1997), membrane filtration, photocatalysis (Barakat, 2011), electrochemical technologies, as electrodialysis, electrocoagulation, electro-flotation, anodic oxidation and electrochemical reduction (Chen et al., 2013).

Nowadays the electrochemical deposition processes are used for the treatment of high metal waste stream to recover the metal directly (Wendt et al., 2012). The electrochemical deposition could take place by two different processes: electroplating (applying a current to the anodic material) or cementation (due to the electropotential differences between cathodic and anodic materials) (Figueroa and Wolkersdorfer, 2014).

Many authors considered that research and development in metal recovery domain has excellent potential in the wastewater industry, due to the specificity of each treated wastewater (depending on the type and quantity of metal which will be recovered).

By studying physical, chemical, and biological processes, the technologies for metal recovery could be improved to become more profitable (Wang and Ren, 2014).

When the treatment of wastewater with metal ions content is established some basic parameters must be taken into consideration. Among them could be mentioned: the initial concentration of metal ions, pH, environmental impact, process performances, the capital investment, operational costs, but the main factors are: the technical applicability, the plant simplicity and cost-effectiveness (Barakat, 2011).

11.3   NANO-SCIENTIFIC DEVELOPMENT(S)

The main task of metal ion wastewater treatment is to remove these ions to the concentration required by law to discharge waters. In the particular case of metal ions pollution, the main metal ions pollution sources are industrial processes. These processes belong to the following areas: metallurgy (pickling, polishing), mining (ore primary output, mine water), hydroelectrometallurgy (spent electrolytes), electrochemistry (exhausted electrochemical sources – accumulators and batteries, electroplating, electrowinning, electrorefining, galvanoplastics, electroerosion, exhausted/contaminated baths), chemistry (spent catalysts, chemical reagents, dyes), leather (tanning), photographic techniques (depleted fixing solutions), etc.

Aqueous solutions with a low content of dissolute metal frequently occur in industrial production plants, processing, and finishing of metal surfaces (Grunwald, 1995; Kuhn, 1987). Besides, there are solutions from the chemical processes that use reagents, catalysts, strippers, etc. with a high content of metal ions.

It must be noticed that in every industrial unit wastewaters with metal ions content are obtained. Is must not forget that all systems and objects made by metal are subject to both erosion and especially corrosion caused by chemical and atmospheric agents (Kuhn, 1971; Pletcher and Walsh, 1990). Metal ions can get into the effluents by gas condensing (and/or dust dissolution) from burning fossil fuels (Walsh and Reade, 1994).

A brief analysis of the industrial sectors that generate metal ions solutions lead to a diversification of effluents by *nature of pollutant ions*, closely related to the industrial production areas. The main factors which must be taken into consideration are: *solution composition*, regarding contaminant ions concentration and nature, the number of metals, pH, electrolyte conductivity, etc., and flows, which determines the operation scale – from hydro-electrometallurgy (for multi-metal solution processing) to laboratory (in order to remove a particular metal ion).

In the electrochemical recovery process, the metal can be obtained as powder or flakes; plate or strip; deposited on a fuel substrate for refining; solid, reusable for soluble anodes in electrodeposition; pure metal or alloy; insoluble metal compounds, for example: oxides, hydroxides, sulfides, etc.; concentrate of metal ions; metal ions regenerated (reduced, oxidized or purified).

The metal obtained can have multiple destinations:

- direct recycling in the production process;
- recycling in other processes;
- reprocessing at the source;
- storage for refining or waste.

After Kuhn, one of the great specialists in applied electrochemistry (Kuhn, 1987) with concerns over the current role of electrochemistry in environmental protection, the concepts in applied electrochemistry for residual effluents treatment can be classified as shown in Figure 11.1. Two methods are distinguished: electrophysical and electrochemical processes (Duțu, 2005; Walsh and Reade, 1994).

In general, electrochemical methods can be divided into recovery methods and destructive methods.

Electrochemical methods have found wide application in the prevention and elimination of environmental pollution, treating liquid and gaseous effluents, the latter being most often captured in fluids (Rojanschi et al., 1997).

Compared to traditional methods, electrochemical techniques and technologies (Rajeshwar and Ibanez, 1997) provide many advantages, such as:

- compatibility with the environment:
  - the electron is used as "reagent," which can be considered a clean "reactive";
  - most often is no need for chemical additives.
- versatility:
  - as the electrochemical processes involve oxidation or reduction processes, directly or indirectly, which generate organic, inorganic or biochemical species, neutral, positive or negative;
  - solid, liquid or gaseous pollutants can be treated;

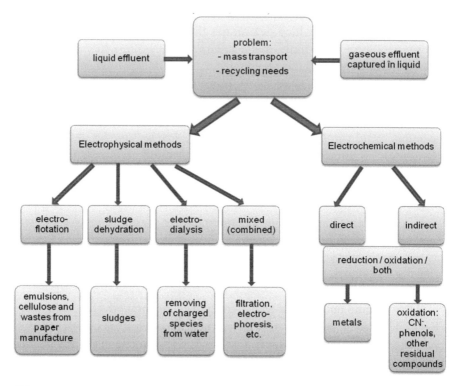

FIGURE 11.1   Concepts in the electrochemical treatment of effluents (Kuhn, 1987).

- after electrochemical processes can be obtained: precipitates, special gases, pH changes or charge neutralization–pollutants electrolysis products are often very useful;
- different types and configurations of reactors and electrodes can be used;
- the same reactor can be used with minimal changes in several processes;
- volumes of fluid from a few microliters up to million liters can be treated.
- energy efficiency:
  - since electrochemical processes typically require temperature and pressure much lower than traditional procedures;
  - applied potentials are easily controllable;
  - current leakages and voltage drops can be minimized through an appropriate design of cells and electrodes.

- operational safety:
  - since mild operating conditions characterize electrochemical processes, quantities, and nature of added chemicals are not dangerous;
  - the process can be interrupted at any time without affecting working parameters, cathode deposits quality and without endangering the operator safety.
- selectivity:
  - through a facile control of the applied potential, the reaction specificity is assured, and the occurrence of byproducts is avoided.
- automation possibility:
  - as electrical variables of electrochemical processes (current, voltage) are very suitable for automation, process control and implementation of computerized databases.
- low costs:
  - since equipment and operations are in general simple and relatively inexpensive;
  - by an appropriate design of the installations, a simple adjustment for other reactions and processes can be provided.

11.4 RECOVERING ELECTROCHEMICAL METHODS–METAL IONS ELECTRODEPOSITION

When one uses recuperative electrochemical techniques, some difficulties may occur, that once exceeded will ensure the success of these methods (Rojanschi et al., 1997).

During the recovery process, metal ion concentrations are low and decrease during the process; in this case, the process is limited by mass transport. Thus, for process efficiency, mass transfer improvement is necessary, which can be achieved through turbulence promoters, particles in a fluidized bed, agitation with gas, electrode movement, etc.

Metal ions consuming during the process will determine in time the decrease of current efficiency. This effect can be diminished by using a scheme based on an exponential scheduled dropping of working current density versus time. Besides, one can both apply voltage step control, as well as continuous potential change techniques, which results in more efficient processes (e.g., an increase of approx. 25% of the productivity and approx. 900% selectivity) (Walsh and Reade, 1994).

Low conductivity, due to metal ions low concentration, can be enhanced by support electrolyte addition.

Interference with the hydrogen release or oxygen reduction reactions must be prevented or minimized. For example, oxygen reduction becomes critical with the decrease of metal ion concentration, because its solubility at room temperature is of the order of 8 ppm; high purity hydrogen gas produced as a byproduct has great commercial value. pH adjustment can be used to facilitate the preferred reaction thermodynamically, for example, removal of cobalt ions from aqueous solutions is carried out simultaneously with the hydrogen release at low pH; however, at pH > 4 they can be efficiently removed. In some cases, the hydrogen gas production leads to mass transport improvement near the electrode and obtained metal deposits can be readily mechanically removed.

The cathode surface, the place where the metal is deposited, is changing its properties over time, leading to the need for an additional control of the process. In some cases, the electrodeposition process can be facilitated by increasing the exchange currents on the modified surface.

During the metal deposition, especially in the case of high metal content solutions, three-dimensional electrodes may be agglomerated by welding (bonding) – the solution to this problem may be the periodic removal of recovered metal.

The deposition rate and the solution composition, in some cases, can result in the production of dendritic or spongy deposits, if some substances present in the solution interfere with the electrodeposition process.

A high flow rate causes an upper limit current and low residence time, generating low recovery rates. The competition between current densities as high and low residence time may lead to different results, depending on the system. For example, in the case of Cu^{2+} ion, the higher the flow rate, the lower the recovery time, and yields are higher, while in the case of Hg^{2+} ion a higher recovery degree was observed at lower flow rates and higher residence times.

The potentiostatic control – can be used to maximize the current efficiency and minimize unwanted competing reactions – is hard to achieve in sizable cells due to the small size of the reference electrode and to the inherent difficulties of the working electrode potential measurement.

11.5 NANO-CHEMICAL APPLICATION(S)

In contrast to traditional methods (precipitation, filtration, ion exchange, reverse osmosis, physical or chemical adsorption, solidification, chemical reduction)–costly operations requiring complex technology, metal ions electrodeposition present many advantages (Pletcher and Walsh, 1990; US EPA, 2006) as follows:

- Metal is obtained in its most valuable form (pure metal) and then can be reused or recycled.
- Additional reagents are not required, so, in most cases, the treated water or the solution from the metal extraction can be recycled.
- pH can be electrochemically controlled; during water oxidation H^+ ions are produced, during reduction, OH^- ions. This control avoids the secondary or parallel reactions or, on the contrary, can initiate the production of desired substances (e.g., the production of $Cr(OH)_3$ during the reduction of Cr (VI)).
- Sludge production is minimized.
- Selective deposition of a single metal from a metal ions mixture can be achieved through a rigorous control of process parameters or by using an electrochemical reactors series, each designed for a specific metal ion.
- Alloys deposition can be achieved.
- Operating costs are reduced.
- The anodic pair reaction can be advantageously used, for example–unwanted complexing agents such as cyanides, or chelating agents or various organic compounds may be destroyed in the same cell, on the anode.
- Simple and facile operating.

11.6 REACTORS USED IN ELECTROCHEMICAL RECOVERY METHODS: REACTORS SELECTION CRITERIA

The choice of a particular type of electrochemical reactor for different applications is imposed by entirely different standards (Fahidy, 1985; Pickett, 1979). Thus, in the case of electrodeposition, the critical parameter is the deposit quality, while energy efficiency is less important since the amount of electricity used is not very high (Volgin et al., 2003; Volgin and Davidov, 2008).

In the case of electricity-intensive processes such as chloralkali industry or aluminum industry, efficiency, and productivity must be taken into account. For organic electrosynthesis, process selectivity and easy separation of reaction product must be considered (Pletcher, 1990).

In conclusion, to choose an electrochemical reactor, one must take into account more assessment criteria. These criteria are: simplicity of design, electrical connection type, terminal voltage, potential, and current distribution, the heat generating and dissipation, flow separation, convenient operation and durability in use, possibilities for process integration and versatility, minimal operating costs.

The simplicity of design: the cell design must match the process requirements as simple as; it takes into account the need for a non-expensive investment and running costs as low as possible. The choice must be attractive for users since process imposed factors determine the constructive solutions. For example, continuous operation mode requires a higher degree of difficulty due to the complexity of the design involving higher costs, but with significant technological advantages.

Electrical connection type: if a larger number of cells are required, this fact implies additional problems in electrical contacts and technical flow connecting.

Electrodes in electrochemical cells can be connected in monopolar or bipolar configuration.

Monopolar configuration implies parallel electrodes connection, and therefore the same potential difference is achieved between each anode/cathode pair, as reactor's terminals. It has the advantage of low voltages, but the disadvantage of large currents.

Bipolar configuration implies only the connection of end electrodes (first and last). Thus, the other electrodes will have an opposite polarity to the electrode in front of which they are, and each electrodes pair will have a potential corresponding to an equivalent fraction from reactor's terminals overall voltage.

Terminal voltage: terminal voltage should be reduced by decreasing of interelectrode distance and voltage drops on contact, increasing of electronic conductivity and solution electric conductivity, an appropriate selection of the counter electrode reaction, etc.

Current and potential distribution: is required to be as homogeneous as possible. Otherwise, the reactor performance can be drastically decreased by current leakage, local heating, surface reactions, unwanted conversions, etc. An appropriate potential and current distribution can be ensured by electrodes design and cell arrangement.

Heat generation and dissipation: Due to Ohmic drops (IR terms) generating Joule effect and mechanical friction forces caused by pumps, stirrers, electrodes' motion, etc. during the operation, in the electrochemical reactor extra heat is generated supplementary to that of the reaction itself. The heat may be provided in the reactor by the heat carrier fluids, and in some cases, by the electrolyte itself.

It may be dissipated by conduction through the reactor walls, by convection to the next phases, heat exchange fluids, sometimes the electrolyte itself and solvent evaporation. In the global thermal balance of a reactor should be considered caloric capacity exchanges of reactants and reaction products.

Flow separation: because occurs the possibilities of unwanted reactions caused by the reaction (cathodic or anodic) products mixed with the solution components or the reaction products of the other electrode, and that a product of an electrode reaction to undergo a reverse process to the other electrode, lead to the necessity of an interelectrode separator. Primary separation, which ensures the prevention of liquid flows mixing from two electrode compartments, is made using porous separators such as porous glass, asbestos, polymers, etc. They may fulfill the role of turbulence promoters or interelectrode insulators to prevent electrical short circuits. An efficient separation which provides the blocking of migration effects is assured by separators made by ion exchange membrane.

Advantageous operating and working resistance: the cell must allow a simple operation – product extraction and maintenance (e.g., electrodes, and separators inspection, cleaning or replacement when necessary). Depending on the process, the cell must be able to automatically operate for extended periods of time without supervision, with a high degree of resistance in operation, safety, and security.

Process integration possibilities and versatility: the reactor must be quickly and directly integrated into the overall process. In some cases, it may be necessary its occasional use, with different raw material stocks or for various reactions. If the process is under development, the reactor should be easy to scale up by increasing of dimensions, cells number and/or modules number.

Operating costs minimize: operating expenses can be kept to a minimum by:

- cell components with lower price and high running resistance;
- a lower terminal voltage obtained by:
  - electrode material and shape suitable choice;
  - counter electrode reaction optimization;
  - appropriate electrical conductivity;
  - small interelectrode distance;
  - good electric contacts, and, if possible, undivided cell.

- lack of powerful equipment (mechanisms) for electrolyte stirring or electrode movement;
- lower pressure drop in the reactor to reduce pumping costs.

The above-presented issues often determine very different requirements for the reactor's design. To obtain a suitable reactor is necessary that the designer has a vast experience and a solid preparation in this field. The electrochemical cells (reactors) cover a wide range and are made according to the specific characteristics required for the particular process.

According to various authors (Pletcher and Walsh, 1990; Rajeshwar and Ibanez, 1997; Walsh and Reade, 1994) some of reactors classification criteria are: electrolyte operating mode, temperature control, cell shape, electrodes number, electrical connection type, degree, and separation mode between anolyte and catholyte, the electrode structure, and electrode-electrolyte relative movement realization mode.

Reactors classification is usually randomly and takes into consideration one of the criteria listed above. Frequently only one of the two electrode reaction is important. If both processes are important, the reactor's design is much more complicated.

In the case of electrochemical reactors where the process of interest is metal ions recovery, they are classified according to deposited metal recovery form, metal recovery process, electrode geometry and movement, main design possibilities, reactor design, and operation mode.

11.7 REACTORS CLASSIFICATION BY METAL RECOVERING PROCEDURE

If the charge transfer from the electrode-electrolyte interface do not present any limitation and the metal ion concentration at the electrode surface is negligible, the expression for limit current is given by the following equation:

$$j_L = K_m \cdot A \cdot n \cdot F \cdot c^* \tag{1}$$

where: j_L – the limit current;

K_m – local mass transfer coefficient;

A – electrode area;

c^* – metal ions concentration in solution;

n – number of electrons transferred to an ion.

From this formula, one can notice that productivity depends on metal ions concentration. In the case of metal ions concentration higher than 10^2 ppm, recovery is usually performed in cells with plane-parallel electrodes. For concentrations lower than 10^2 ppm, a highest $K_m \cdot A$ is required.

To increase $K_m \cdot A$ one can use mass transfer improvement. This improvement can be made by the electrode-electrolyte movement which causes turbulence and increases mass transport coefficient due to diffusion layer thickness decreasing. Another approach for $K_m \cdot A$ growth appeals to the three-dimensional electrodes which provide a larger surface. This method can be combined with electrolyte recirculation and/or electrode movement.

11.8 REACTORS OPERATING PRINCIPLES

Reactors operating under simple cell regime or by recirculation provide different conversion degrees, depending on the period of time in which the electrolyte remains in the cell. The longer the time is, the metal ions depletion increases. If operating time is increased too much, the disadvantage of low productivity and decreasing of current efficiency occurs. The other operating regimes (single pass or cascade type) ensure continuous operation. No matter of operating mode, at the end of the process, once the concentration is decreasing, metal ions recovery becomes unprofitable. This is the reason why in most practical cases, a two-stage operation is preferred (Walsh and Reade, 1994; Iorga, 2009).

In the first stage, referred as "primary stage," most of the metal ions are recovered from solution as metal. In the second stage, known as "secondary stage," the depletion of the remaining metal ions in solution is achieved. Often, the secondary stage is not an electrochemical stage. It may be a chemical or physical process–eventually cementation (Agrawal and Kapoor, 1982; El-Tawil, 1988; Gheorghita et al., 2011; Mubarak, 2004), which has only the task to purify wastewater to the metal ions concentration to allow its reuse or discharge. In Figures 11.2–11.4, three versions of recovery technology treatment of wastewaters containing metal ions are shown.

An authentic electrochemical reactor for advanced removal of metal ions from solution was patented by some authors (Iorga et al., 2016a). In this case, a single reactor is made, with several compartments, each one for every electrolysis stage, circularly arranged, concentric, as presented in Figure 11.5.

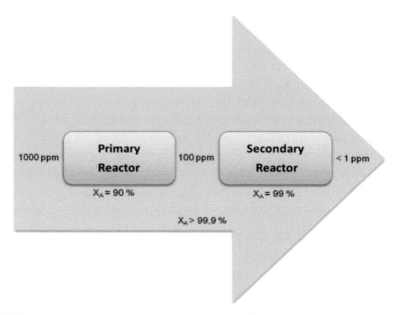

FIGURE 11.2 Recovery strategy for metal ions by flowing through a primary and a secondary reactor (one pass) (Walsh and Reade, 1994).

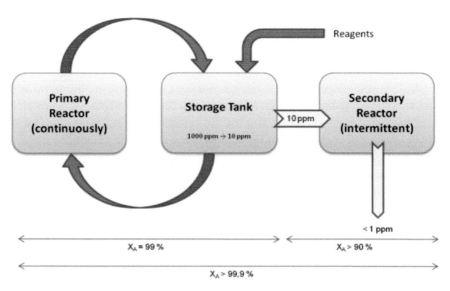

FIGURE 11.3 Recovery strategy for metal ions by recirculation cell as primary reactor and secondary treatment (possibly intermittent) (Walsh and Reade, 1994).

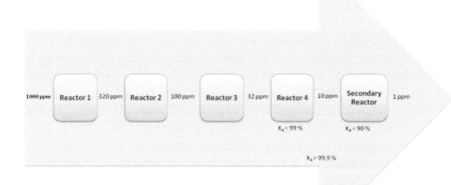

FIGURE 11.4 Recovery strategy for metal ions by cascade reactors (main stage) and secondary reactor (Walsh and Reade, 1994).

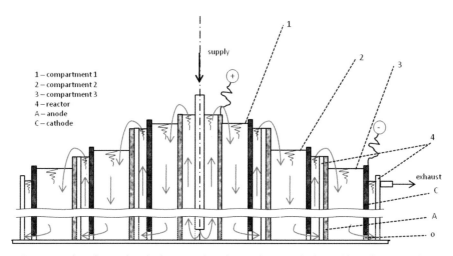

FIGURE 11.5 Electrochemical reactor for advanced removal of metal ions from solution in a three stages process (vertical profile) (Iorga et al., 2016b).

The compartments number, equal to the number of electrolysis cells is established taking into account the solution initial concentration, the imposed depletion overall degree and the intermediary depletion degree for each electrolysis cell (Iorga et al., 2016b).

11.9 MULTI-/TRANS- DISCIPLINARY CONNECTION(S)

11.9.1 THE NEED FOR METAL IONS RECOVERY FROM SOLUTIONS

Metal recovery from recyclable waste is a less energy-intensive process than getting raw material from mineral resources (taking into account the increasing limitation of natural resources). According to some authors (Damgaard et al., 2009), concerning environmental management, metal recycling involves processes with low greenhouse gas emissions–thus significantly contributing to environmental protection.

Electrochemistry studies how electricity causes chemical changes, while chemical transformations are reflected in power generation. This interaction represents the basis of a huge variety of processes, from heavy industry to battery production and biological phenomena (Walsh and Reade, 1994; Bockris, 1972). Although there are already many applications, modern research has led to a large expansion of the practical electrochemistry use. One of the most important directions in electrochemistry is environmental protection and approaches: monitoring of various polluting or environmental affecting substances, removing of any pollutants, energy production, and transport, etc.

Comparing with metal ions removing, in the case of wastewater from electroplating industry containing metal complexes, an appropriate process to destroy metal complexes and mineralize organics, is represented by electrooxidation. This process is considered to be environmentally friendly and belongs to green chemistry technologies. In this case, the metal complexes are destroyed or converted into a form that does not interfere with normal precipitation (Wang et al., 2015).

Another potential technology for metal removal from wastewaters is biosorption (Veglio and Beolchini, 1997), which is an alternative to the conventional techniques.

According to some authors (Wang and Ren, 2014), a new approach for metal recovery arose in the last years–the bioelectrochemical technology. To this purpose is used a novel technology platform named bioelectrochemical systems (BESs). The principle is to use metals as an electron acceptor in the cathodic compartment and organic waste as an electron donor in the anodic compartment. The authors reported several mechanisms, as follows: metal recovery using only abiotic cathodes or abiotic cathodes supplemented by external power sources, and metal conversion using only bio-cathodes or bio-cathodes complemented by external energy sources. This technique is considered very efficient because a flexible platform for both oxidation and reduction reaction oriented processes is offered. The BES platform has recently demonstrated excellent performances in removing and recovering metals from different wastewaters.

11.10   OPEN ISSUES

Since the globalization has taken increasingly more widespread, with more emphasis on the environmental protection and economic development inter-dependence, more strict regulations on environmental protection measures, occur (Walsh and Reade, 1994). The new environmental legislation adopted in recent years, establish a new concept of strict liability for the pollution consequences (Mirica et al., 2006).

The environmental protection issue was reconsidered at the worldwide level, and in many countries, strict laws were adopted, that boosted the scientist to find modern decontamination methods to avoid the conventional methods disadvantages.

It was necessary to establish the effects of pollutants containing metal ions present in the environment, which can act on a short or long-term, due to the accumulation of toxic substance and its derivatives, in the human body (Walsh and Reade, 1994).

Regulations in this area take into account the polluting material nature, the limits of short and long time exposure, procedures/rules for toxic substances handling and storage (ANPM, 2001).

The experience gained over time has shown that it is not enough to control the possible pollution sources, and must act in three main directions:

– pollution control;
– pollution preventing; and
– recycling and/or recovery of useful elements.

If the needs to be credible, recycling must be supported both in environmental and economic terms (Emery et al., 2002; Zhang and Forssberg, 1998).

According to principles and strategic elements to ensure sustainable development, current trends converge more and more towards wastewater treatment at the source. This kind of approach is advantageous for substances that can be recycled in the process or can be recovered to reuse in other processes or as byproducts.

The decision for this alternative is strengthened by other factors, as:

- mandatory complying with legislation that prohibits the effluent discharge into the aquatic environment, if the pollutant exceeds a particular concentration;
- implementation by local authorities of additional fees for a given pollutant content in the wastewater released by the community treatment plant;
- the necessity of a closed circuit for the fluid in the process;
- the possibility of a valuable component to be recovered.

Effluents containing metal ions meet one or more of the demands mentioned above.

ACKNOWLEDGMENT

This contribution is part of the Nucleus-Programme under the project "Deca-Nano-Graphenic Semiconductor: From Photoactive Structure to The Integrated Quantum Information Transport," PN–18–36–02–01/2018, funded by the Romanian National Authority for Scientific Research and Innovation (ANCSI).

KEYWORDS

- **electrochemistry**
- **electrodeposition**
- **environment protection**
- **metal deposition**
- **metal recovery methods**

REFERENCES AND FURTHER READING

Agrawal, R. D., & Kapoor, M. L., (1982). Theoretical considerations of the cementation of copper with iron. *Journal of the South African Institute of Mining and Metallurgy, 4,* 106–111.

Barakat, M. A., (2011). New trends in removing heavy metals from industrial wastewater. *Arab. J. Chem.*, *4*, 361–377.

Bockris, J. O. M., (1972). *Electrochemistry of Cleaner Environments*. Plenum Press, New York.

Chen, X., Huang, G., & Wang, J., (2013). Electrochemical reduction/oxidation in the treatment of heavy metal wastewater. *Journal of Metallurgical Engineering (ME), 2*(4), 161–164.

Damgaard, A., Larsen, A. W., & Christensen, T. H., (2009). Recycling of metals: Accounting of greenhouse gases and global warming contributions. *Waste Management & Research*, *27*, 773–780.

Duţu, M., (2005). Fundamental Principles and Institutions of Community Environmental Law (in Romanian), The Economic Publishing House Bucharest, Romania.

El-Tawil, Y. A., (1988). Cementation of cupric ions from copper sulfate solution using fluidized bed of zinc powder. *Materials Research and Advanced Techniques*, *79*(8), 544–546.

Emery, A., Williams, K. P., & Griffiths, A. J., (2002). A review of the UK metals recycling industry. *Waste Management & Research*, *20*, 457–467.

Fahidy, T. Z., (1985). *Principles of Electrochemical Reactor Analysis*. Elsevier, Amsterdam.

Figueroa, L., & Wolkersdorfer, C., (2014). Electrochemical recovery of metals in mining influenced water: State of the art. In: Sui, S., & Wang, (eds), *An Interdisciplinary Response to Mine Water Challenges.* China University of Mining and Technology Press, Xuzhou; pp. 627–631.

Fu, F., & Wang, Q., (2011). Removal of heavy metal ions from wastewaters: A review. *Journal of Environmental Management*, *92*(3), 407–418.

Gheorghiţă, M., Sîrbu, E., Purcaru, V., Mirica, N., Mirica, M. C., & Iorga, M., (2011). Electrochemical characterization of synthetic basic solutions with Ga and Al content. *Revista de Chimie*, *62*(5), 538–542.

Grünwald, E., (1995). *Modern Technologies of Electroplating in the Electronics and Electrotechnics Industry (Romanian translation)*, Ed. Casa Cărţii de Ştiinţă, Cluj-Napoca, Romania.

Iorga, M. I., Mirica, M. C., Mirica, N., & Balcu, I., (2016a). Reactor electrochimicşiproc edeupentruîndepărtareaionilormetalici din soluţii. Patent no. 127174, OSIM Bucuresti, Romania.

Iorga, M. I., Mirica, N., Mirica, M. C., & Buzatu, D., (2016b). Electrochemical reactor for advanced removal of metal ions from solutions. In: *22nd International Congress of Chemical and Process Engineering CHISA 2016 Prague* (pp. 5, 48).

Iorga, M., Mirica, M. C., & Buzatu, D., (2009). Monitoring autonomous station with applications in environment protection and photovoltaic systems installation. In: *Proceedings of Main Group Chemistry Conference–ZING* (pp. 10, 45). Cancun, Mexico.

Kalavrouziotis, I. K.; Angelakis, A. N. (2014), Prolegomena. In e-PROCEEDINGS of *IWA Regional Symposium on Water, Wastewater and Environment: Traditions and Culture, Greece*, (Eds.) I. K. Kalavrouziotis & A.N. Angelakis, Hellenic Open University, Patras, Greece, p. III–V.

Kuhn, A. T., (1971). *Industrial Electrochemical Processes*. Elsevier, Amsterdam.

Kuhn, A. T., (1987). *Techniques in Electrochemistry, Corrosion, and Metal Finishing: A Handbook*. Wiley, Chichester.

Mirica, N., Dragoş, A., Mirica, M. C., Iorga, M., & Macarie, C., (2006). *Electrochemical Recovery Methods of Metal Ions from Solutions (originally in Romanian)*, Mirton Publishing House, Timişoara, Romania.

Mubarak, A. A., El-Shazly, A. H., & Konsowa, A. H., (2004). Recovery of copper from industrial waste solution by cementation on reciprocating horizontal perforated zinc disc. *Desalination, 167*, 127–133.

Pickett, D. J., (1979). *Electrochemical Reactor Design.* Elsevier, Amsterdam.

Pletcher, D., (1990). The fundamentals of electrosynthesis and its scale-up. In: *Electrosynthesis from Laboratory, to Pilot, to Production.* Woodfine Printing Inc., Buffalo, NY.

Pletcher, D., & Walsh, F. C., (2000). *Industrial Electrochemistry* (2nd edn.). Springer.

Rajeshwar, K., & Ibanez, J. G., (1997). *Environmental Electrochemistry. Fundamentals and Application in Pollution Abatement.* Academic Press, Inc., San Diego.

Rojanschi, V., Bran, F., & Diaconu, G., (1997). *Environmental Protection and Engineering (originally in Romanian),* Ed. Economică, Bucureşti, Romania.

US EPA, (2006). *Solid Waste Management and Greenhouse Gases: A Life Cycle Assessment of Emissions and Sinks* (3rd edn.). United States, Environmental Protection Agency, Washington DC, USA.

Veglio, F., & Beolchini, F., (1997). Removal of metals by biosorption: A review. *Hydrometallurgy, 44*(3), 301–316.

Volgin, V. M., & Davydov, A. D., (2008). The limiting current density of copper electrodeposition on vertical electrode under the conditions of electrolyte natural convection. In: *Russian Journal of Electrochemistry, 44*(4), 459–469.

Volgin, V. M., Grigin, A. P., & Davydov, A. D., (2003). Numerical solution of the problem of the limiting current for the copper electrodeposition from a solution of cupric sulfate and sulfuric acid in conditions of natural convection. In: *Russian Journal of Electrochemistry, 39*(4), 335–349.

Walsh, F. C., & Reade, G. W., (1994). Electrochemical techniques for the treatment of dilute metal-ion solutions. In: Sequeira, C. A. C., (ed.), *Studies in Environmental Science 59, Environmental Oriented Electrochemistry.* Elsevier, Amsterdam; pp. 3–44.

Wang, H., & Ren, Z. J., (2014). Bioelectrochemical metal recovery from wastewater: A review. *In Water Research, 66*, 219–232.

Wang, J., Chen, X., Yao, J., & Huang, G., (2015). Decomplexation of electroplating wastewater in a huge electrochemical reactor with rotating mesh-disc electrodes. *International Journal of Electrochemical Science, 10*, 5726–5736.

Wendt, H., Vogt, H., Kreysa, G., Goldacker, H., Jüttner, K., Galla, U., & Schmieder, H., (2012). Electrochemistry, 2: Inorganic electrochemical processes. In: *Ullmann's Encyclopedia of Industrial Chemistry.* Wiley-VCH Verlag GmbH & Co. KgaA, Weinheim.

Zhang, S., & Forssberg, E., (1998). Mechanical recycling of electronics scrap – the current status and prospects. *Waste Management & Research, 16*, 119–128.

CHAPTER 12

Free-Wilson Model

BOGDAN BUMBĂCILĂ[1] and MIHAI V. PUTZ[1,2]

[1]*Laboratory of Computational and Structural Physical Chemistry for Nanosciences and QSAR, Biology-Chemistry Department, Faculty of Chemistry, Biology, Geography at West University of Timişoara, Pestalozzi Street No. 16, Timişoara, RO–300115, Romania*

[2]*Laboratory of Renewable Energies-Photovoltaics, R&D National Institute for Electrochemistry and Condensed Matter, Dr. A. Paunescu Podeanu Str. No. 144, RO–300569 Timişoara, Romania, Tel.: +40-256-592-638; Fax: +40-256-592-620, E-mail: mv_putz@yahoo.com, mihai.putz@e-uvt.ro*

12.1 DEFINITION

The Free-Wilson model is a simple and efficient method for the quantitative description of structure-biological activity relationships for molecules. It is the only numerical method which directly relates structural features with biological properties, in contrast to Hansch analysis, where physical-chemical properties are correlated with biological activity expressed as numerical values (Kubinyi, 1988). It is a regression technique using the presence or absence of substituents or functional groups as the only molecular descriptors in the correlations between the structure and the biological activity of the whole molecule (Mannhold et al., 1993).

12.2 HISTORICAL ORIGINS

Modern approaches for the design of bioactive species, such as drugs, pesticides are based on the quantification of the biological activity as a function of the molecular structure. The dependence of the biological activity on the partition coefficient was very early demonstrated by Hans Horst Meyer and

Charles Ernest Overton in 1899–1901. They proved that minimum alveolar concentration or MAC (the concentration of the vapors in the lungs that is needed to prevent movement in 50% of subjects in response to surgical (pain) stimulus) of a volatile substance is inversely proportional to its lipid solubility (Meyer, 1899; Overton, 1901).

The Free-Wilson approach was introduced in 1964. (Free et al., 1964) They postulated that for a series of similar compounds, analogs–differing one to another by the presence/absence of one/some substituents, the contribution of these substituents to the biological activity/potency of the whole molecule is additive and depends only by the position or type of the substituent.

12.3   NANO-CHEMICAL IMPLICATIONS

Vital to the development of the main idea of this model was the concept of the receptor site. According to this concept, biological activity depends on the recognition of a bioactive substrate (BAS) by a receptor site, followed by the binding of the BAS to the receptor site. The correlation between bioactivity and the configuration of the substrate led to the recognition of the fact that steric effects have an important contribution to the strength of this ligand-receptor binding (Voiculetz et al., 1991).

The technique was introduced in the early days of QSAR, and it uses indicator variables. Started from a series of analogs with different substituents and different associated biological activities, the aim is to generate an equation with the following expression:

$$y = a_1 x_1 + a_2 x_2 + a_3 x_3 + \ldots + \mu \tag{1}$$

where y = the biological activity of the whole ligand molecule, expressed as logarithm of critical ligand concentration, a_i is the biological activity associated with the functional group/substituent i and μ is the theoretical biological activity value of a reference or of the parent compound or of the basic molecular scaffold. x_i is a descriptor with a value of 0 or 1, depending if the substituent is absent in the molecule or present (Leach et al., 2007). In the original model, for y (the molecular biological activity), it was used the logarithmic transformation (Figure 12.1).

The Free-Wilson model can help understand and interpret substituent effects. In order to obtain a_i values, standard multiple linear regression methods are used (Leach et al., 2007).

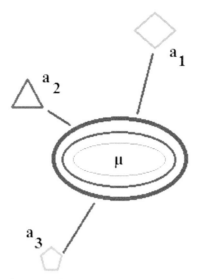

FIGURE 12.1   Conceptual presentation of the Free-Wilson theory.

In their model, the functional groups are linearly correlated to the biological activity of the series consisting of analog compounds. The basic assumption is that the contribution of a certain substituent at a specified position in the molecular scaffold to the biological activity is the same (constant) in the series. Its applicability is limited to a series of substituted compounds. Additional substituents cannot be taken into consideration. A further disadvantage is its inherent linearity. The model can, therefore, only represent an approximation. Nevertheless, this combinatorial concept has found wide application in the molecular design field for forming combinatorial compound libraries represented by analog compounds with functional groups graphed on a common core molecular structure (Schneider et al., 2008).

The advantage of the method is its simplicity. It does not need substituent parameters or other descriptors for the calculus. The biological activity associated with the substituent is the only value needed. The limitation consists in the assumption that these activity values are additive and in the one that presumes that there is no interaction between the functional groups and between the substituents and the molecular core. It is not possible, therefore, to make predictions on molecules containing other substituents that are not included in the data set. Other problems are represented by the situations when the same substituents are present in two molecules, but they are graphed in different positions (Leach et al., 2007).

Corwin Herman Hansch was also one of the first who proposed a mathematical model for determining the biological activity.

$$\log \frac{1}{C} = -a_1 (\log P)^2 + a_2 \log P + a_3 \sigma + \ldots + b \tag{2}$$

The variables a_i and b are determined by regression analysis, σ is the Hammet constant, describing the electronic parameters (electron-withdrawing/electron-donating effect) of the aryl substituents. C is the molar concentration that causes a certain biological effect. P is the partition/distribution coefficient octanol/water, and it is related to the lipophilicity of the molecule/group. Various other molecular features may enter this equation, but any term may be omitted if it is demonstrated to be irrelevant for the total biological activity (Schneider et al., 2008; Roy et al., 2015).

The equation can change and also indicator variables may be used, in order to differentiate members of the data set. Indicator variables are expressing the presence or the absence of certain molecular characteristics and therefore can have two values: 0 for the absence and 1 for the presence.

The basis of the Hansch-Fujita model is the assumption that the observed biological activity is the result of the contribution of different factors, which behave in an independent manner. Each activity contribution is represented by a structural descriptor, and the biological activity of a set of compounds is adjusted to a multilinear model (Carbó-Dorca et al., 2000).

This expression makes the following assumptions regarding the mechanism of action of a chemical compound:

1. The transport from the site of entry in the organism to the target tissue takes place across barriers. This step is expressed by the parameter logP (or π). The transport proportion through these barriers (membranes, volumes) is proportional to this parameter.
2. Its biological action is expressed following its recognition by a receptor (enzyme). This substrate-receptor interaction is translated into the biopharmacological properties of the substrate (Voiculetz et al., 1991).

In a 1985 study, Hansch et al., took a series of 29 monosubstituted sulfonamide inhibitors with the molecular formula $X-C_6H_4-SO_2-NH_2$, of human carbonic anhydrase and elaborated the following equation:

$$\log K = 1.55\sigma + 0.64\log P - 2.07I_1 - 3.28I_2 + 6.94$$
$$(R = 0.991, \ s = 0.204, \ n = 29) \tag{3}$$

where K is the enzyme inhibition constant; I_1 is 1 for substitution in position meta, and otherwise, it is 0; I_2 is 1 for substitution in position or to and 0 otherwise; R is the correlation coefficient a measure of the relative quality of a model; s is the standard deviation – a measure of the absolute quality of a model; and n is the number of studied analog compounds (Hansch et al., 1985; Supuran et al., 2004).

The structure of a molecule can be represented in several ways. One of the most common is the simplified molecular-input line-entry system or SMILES. SMILES strings are, in fact, chemical formulas. For example, acetaminophen has the molecular formula $C_8H_9NO_2$. Its SMILES string is:

$$CC(NC1=CC=C(O)C=C1)=O$$

For creating molecular models, some numerical representation is required. A wide variety of such chemical descriptors are used today: simple atom or bond counts, fingerprint descriptors (indicators for molecular substructures or fragments, continuous values (molecular weight, surface area, volume, and positive charge) (Zhang, 2016).

Some of the descriptors used are estimations based on theoretical models, like the lipophilicity. The most common representation of this characteristic is logP, where P is the partition coefficient that is measuring the ratio between the concentrations of the studied molecule in n-octanol and water. For example, polar compounds will tend to concentrate in water more; thus, they will have a lower logP value (Carbó-Dorca et al., 2000).

The Free & Wilson model relates structure to its biological potency. In some situations, fragments/substituents of a molecule can be considered as a unit that can also be substituted. An R group, a substituent can be a molecular fragment that can be attached to a molecule. The molecule can be composed by a molecular core and one or more R groups, different or not between each other. In their first model, Free, and Wilson described a common molecular core with two substituents: R_1 and R_2. R_1 could have been H or CH_3 and R_2–$N(CH_3)_2$ or $N(C_2H_5)_2$. Four analog molecules could be described by the combinations made with these substituents. (Free et al., 1964) A matrix with the numbers 0 and 1 can be created, where these 2 numbers are indicating the absence or presence of a certain substituent on the molecular scaffold. Then the linear regression formula which is relating the molecular structure with the potency/biological activity is drawn as it was previously presented.

This model makes possible the predictions for potency/biological activity of new molecules whose combinations of R_1 and R_2 groups (for example) have not been yet synthesized in the laboratory, but they are in the study for being presumptive therapeutically active. The equation model also permits to understand better how changes in the chemical structure (mutations of the R_i substituents on the molecular core) might increase or decrease the biological activity for a set of analog molecules (Zhang, 2016).

12.4 TRANS-DISCIPLINARY CONNECTIONS

An example of Free-Wilson analysis is given below. Antiadrenergic activities of 22 meta-, para-, mono-, meta- and para-disubstituted N,N-dimethyl-α-bromo-phenethylamines were assessed using Free-Wilson analysis (Figure 12.2, Tables 12.1 and 12.2).

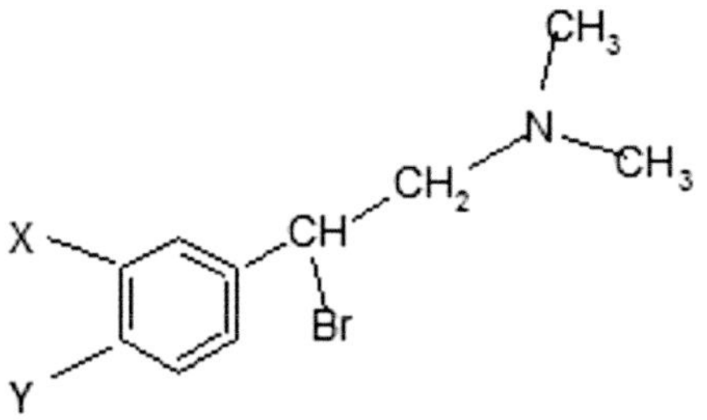

FIGURE 12.2   Meta-, mono-, and para-disubstituted N,N-dimethyl-α-brom-phenethyl-amines.

A negative coefficient indicates that the presence of that group is not favorable to activity. A positive coefficient indicates that the presence of that group is favorable to activity.

$$n = 22, r = 0.969, s = 0.194, \mu = 7.682$$

The Free-Wilson equation will be:

$$\log 1/c = -0.301[m - F] + 0.27[m - Cl] + 0.434[m - Br] + 0.579[m - I]$$
$$+ 0.454[m - CH_3] + 0.340[p - F] + 0.768[p - Cl] + 1.020 [p - Br]$$
$$+ 1.429[p - I] + 1.256[p - CH_3] + 7.821$$

where c is the concentration that causes a 50% reduction of the adrenergic effect of a certain epinephrine dose.

Few conclusions can be drawn: Biological activity increase with the lipophilicity (F to Cl to Br to I); Biological activity increase with electron-donor properties (methyl group has a larger contribution than Cl, even if both substituents have almost the same lipophilicity); meta-substituents have lower group contributions than para-substituents (Kubinyi, 1997).

TABLE 12.1 Matrix for Free-Wilson Analysis for Meta-, Para- Mono- and Meta-& Para-Disubstituted N,N-Dimethyl-α-Brom-Phenethyl-Amines (Kubinyi, 1988) *NS = Not Studied, 1=Presence, 2=Absence

Compound	Substituent position X (meta-)						Substituent position Y (para-)						log 1/C observed	log 1/C calculated
	H	F	Cl	Br	I	CH₃	H	F	Cl	Br	I	CH₃		
1	1	0	0	0	0	0	1	0	0	0	0	0	7.46	7.82
2	1	0	0	0	0	0	0	1	0	0	0	0	8.16	8.16
3	1	0	0	0	0	0	0	0	1	0	0	0	8.68	8.59
4	1	0	0	0	0	0	0	0	0	1	0	0	8.89	8.84
5	1	0	0	0	0	0	0	0	0	0	1	0	9.25	9.25
6	1	0	0	0	0	0	0	0	0	0	0	1	9.30	9.08
7	0	1	0	0	0	0	1	0	0	0	0	0	7.52	7.52
8	0	1	0	0	0	0	0	1	0	0	0	0	NS	NS
9	0	1	0	0	0	0	0	0	1	0	0	0	NS	NS
10	0	1	0	0	0	0	0	0	0	1	0	0	NS	NS
11	0	1	0	0	0	0	0	0	0	0	1	0	NS	NS
12	0	1	0	0	0	0	0	0	0	0	0	1	NS	NS
13	0	0	1	0	0	0	1	0	0	0	0	0	8.16	8.03
14	0	0	1	0	0	0	0	1	0	0	0	0	8.19	8.37
15	0	0	1	0	0	0	0	0	1	0	0	0	8.89	8.80
16	0	0	1	0	0	0	0	0	0	1	0	0	9.00	9.05
17	0	0	1	0	0	0	0	0	0	0	1	0	NS	NS
18	0	0	1	0	0	0	0	0	0	0	0	1	NS	NS
19	0	0	0	1	0	0	1	0	0	0	0	0	8.30	8.26
20	0	0	0	1	0	0	0	1	0	0	0	0	8.57	8.60

TABLE 12.1 *(Continued)*

Compound	Substituent position X (meta-)						Substituent position Y (para-)						log 1/C observed	log 1/C calculated
	H	F	Cl	Br	I	CH₃	H	F	Cl	Br	I	CH₃		
21	0	0	0	1	0	0	0	0	1	0	0	0	8.92	9.02
22	0	0	0	1	0	0	0	0	0	1	0	0	9.35	9.28
23	0	0	0	1	0	0	0	0	0	0	1	0	NS	NS
24	0	0	0	1	0	0	0	0	0	0	0	1	9.52	9.51
25	0	0	0	0	1	0	1	0	0	0	0	0	8.40	8.40
26	0	0	0	0	1	0	0	1	0	0	0	0	NS	NS
27	0	0	0	0	1	0	0	0	1	0	0	0	NS	NS
28	0	0	0	0	1	0	0	0	0	1	0	0	NS	NS
29	0	0	0	0	1	0	0	0	0	0	1	0	NS	NS
30	0	0	0	0	1	0	0	0	0	0	0	1	NS	NS
31	0	0	0	0	0	1	1	0	0	0	0	0	8.46	8.28
32	0	0	0	0	0	1	0	1	0	0	0	0	8.82	8.62
33	0	0	0	0	0	1	0	0	1	0	0	0	8.96	9.04
34	0	0	0	0	0	1	0	0	0	1	0	0	9.22	9.30
35	0	0	0	0	0	1	0	0	0	0	1	0	NS	NS
36	0	0	0	0	0	1	0	0	0	0	0	1	9.30	9.53

TABLE 12.2 Biological Activities for Different Substituents in Positions Meta- or Para- for Meta-, Para- Mono- and Meta-, Para-Disubstituted N,N-Dimethyl-α-Brom-Phenethyl-Amines (Kubinyi, 1988)

Group's biological activity in position	H	F	Cl	Br	I	CH$_3$
Meta-	0.00	–0.30	0.21	0.43	0.58	0.45
Para-	0.00	0.34	0.77	1.02	1.43	1.26

In a 2003 paper, Jha, T. and his colleagues, proved that structural variants of glutamine might antagonize enzymes involved for their possible anticancer activity by competitive inhibition of the amino acid glutamine (GLN). In the synthesis of DNA and RNA, major portions of nitrogen atoms are supplied by GLN. GLN supplies the 3rd and 9th nitrogen atoms of the purine ring, the 2nd amino group of guanine and the 3rd nitrogen atom and amino group of cytosine. It also acts as the major fuel in the tumor cell. Some cancer cells need this amino acid more in comparison with a normal cell. The only circulatory sugar D-glucose and the non-essential amino acid L-glutamine are two major substrates for cancer cells. Since all living cells, both normal and cancerous, need D-glucose for survival, L-glutamine may be the major substrate for cancer. Moreover, GLN is responsible for almost all physiological functions and cancer cases. At most of the physiological systems, tissues, and cells, as well as this amino acid, is essential for maintaining cultures of cell lines which show mutations after a certain period of time. For the QSAR study, the parent structure of 5-N-substituted-2- (substituted benzenesulphonyl)-L-glutamine was used. The anticancer activity, which is the percent of tumor weight inhibition determined against Ehrlich Ascites Carcinoma (EAC) cells in Swiss albino mice, of some substituted glutamine analogs have been considered as biological activity. The QSAR study was performed using the Fujita Ban model for the parent structure and the above-mentioned anticancer activities of glutamines. An inspection of individual contribution of substituents at positions 3' and 5' of the phenyl ring showed a general decrease of anticancer activity, on the contrary, the presence of a Br at 4'-position was correlated positively to the total activity. The anticancer activity was highly increased by a Cl at 2'-position and this substitution had the greatest contribution towards the total activity. So far the aliphatic substitutions at the 5-position was concerned, it was observed that all the substitutions were detrimental to the anticancer activity. These points should be considered in designing further glutamines (Jha et al., 2003).

These models have wide applications in pharmaceutical research. Libraries containing analog compounds are created, at first, starting from a clinical or experimental observation made over a certain molecule. The basic molecular core is maintained, and some functional groups are added on that core and/or rotated in different positions. Software programs are used for obtaining data about the presumed biological activities using the molecular characteristics of those substituents and of the core. If some molecules prove to have potential benefits (comparing them, for example, with standards or molecules already used as therapeutic agents), they can be synthesized and included in pre-clinical and clinical trials.

12.5   OPEN ISSUES

Today, most of the work in this field is concentrating on improving these models. Most of the studies are taken into consideration both the Free-Wilson model and the Hansch analysis. The problem which has to be reduced consists of the linear dependence between the variables taken into the Free-Wilson equation. A new variant of the Free-Wilson analysis is performed today: Fujita-Ban method.

   This analysis has some new aspects: the biological activity is expressed as a logarithmic value, so the dependence is parabolic instead linear, the biological activities of the hydrogen atoms are considered separately from the one of the central core and the μ value is considered to be the one of the biological activity of the unsubstituted parent molecule and not of the molecular "rest."

   So the equation for this Fujita-Ban model is: (Roy et al., 2015)

$$\log A = \sum a_i x_i + \log A_0 \tag{4}$$

where A = biological activity, a_i = a coefficient of substituent R_i which can be 0 (coefficient absent) or 1 (coefficient present), x_i = group contribution to the biological activity of the substituent i, A_0 = molecular core's biological activity (the biological activity of the unsubstituted compound) (Jha et al., 2003)

   Many applications of Free-Wilson and Hansch analysis confirm the theory that group contributions are additive for the total biological activity. Still, there are other effects causing deviations from the additive concept. Abramson et al., studied the affinities of a series of quaternary ammonium compounds to the postganglionic acetylcholine receptor. While a Free-Wilson analysis can be easily applied to that set of analogs, as long as different functional groups are considered as individual substituents, the additive concept cannot be applied anymore when these groups are dissected into smaller segments. The differences in the biological response depend on the nature of the fragments of substituents present in these larger groups (Abramson et al., 1972; Jochum et al., 1988).

KEYWORDS

- **biological activity of a substituent**
- **Hansch model**
- **SMILES**

REFERENCES AND FURTHER READING

Abramson, M. G., Yu, R. K., & Zaby, V., (1972). Ionic properties of beef brain gangliosides. *Biochimica et Biophysica Acta, 11, 280*(2), 365–372.

Carbó-Dorca, R., Robert, D., Amat, L., Gironés, X., & Besalú, E., (2000). *Molecular Quantum Similarity in QSAR and Drug Design.* Springer-Verlag, Berlin-Heidelberg.

Free, S. M. Jr., & Wilson, J. W., (1964). A mathematical contribution to structure-activity studies. *Journal of Medicinal Chemistry, 7,* 395–399.

Hansch, C., McClarin, J., Klein, T. E., & Langridge, R., (1985). A quantitative structure-activity relationship and molecular graphics study of carbonic anhydrase inhibitors. *Molecular Pharmacology, 6, 27*(5), 493–498.

Jha, T., Debnath, B., Samanta, S., & Uday De, A., (2003). QSAR study on some substituted glutamine analogs as anticancer agents. *Internet Electronic Journal of Molecular Design, 2,* 539–545.

Jochum, C., Hicks, M. G., & Sunkel, J., (1988). Physical property prediction in organic chemistry. *Proceedings of the Beilstein Workshop.* Italy, Springer-Verlag.

Kubinyi, H., (1988). Free Wilson analysis: Theory, applications, and its relationship to Hansch analysis. *Molecular Informatics, 7, 3,* 121–133.

Kubinyi, H., (1997). QSAR and 3D QSAR in drug design, Part 1. *Methodology, Drug Development, and Therapeutics, 2, 11,* 457–467.

Leach, A. R., & Gillet, V. J., (2007). *An Introduction to Chemoinformatics.* Springer, Ltd., Netherlands.

Mannhold, R., & Kubinyi, H., (1993). Methods and principles in medicinal chemistry. In: Weinheim, V. C. H., (ed.), *QSAR: Hansch Analysis and Related Approaches* (Vol. 1).

Meyer, H. H., (1899). Welcheeigenschaft der anastheticabedingtinre Narkotischewirkung? *Arch. Exp. Pathol. Pharmakol., 42*(2–4), 109–118.

Meyer, H. H., (1899). Zur Theorie der Alkoholnarkose, *Arch. Exp. Pathol. Pharmacol., 42*(2–4), 109–118.

Meyer, H. H., (1901). Zur Theorie der Alkoholnarkose. Der Einflusswechselnder Temperature auf Wirkungsstärke und Theilungscoefficient der Narcotica. *Arch. Exp. Pathol. Pharmakol., 46*(5/6), 338–346.

Overton, C. E., (1901). *Studienüber die Narkosezugleichein Beitragzurallgemeinen Pharmakologie*, Gustav Fischer, Jena, Switzerland.

Roy, K., Supratik, K., & Das, R. N., (2015). *A Primer on QSAR/QSPR Modeling: Fundamental Concepts*. Springer International Publishing, Switzerland.

Schneider, G., & Baringhaus, K. H., (2008). *Molecular Design: Concepts and Applications*. Wiley-VCH Verlag, GmbH & Co, KGaA, Weinheim, Deutschland.

Supuran, C. T., Scozzafava, A., & Conway, J., (2004). *CRC Enzyme Inhibitors Series, Carbonic Anydrase*. Its Inhibitors and Activators. CRC Press.

Voiculetz, N., Balaban, A. T., Niculescu-Duvăz, I., & Simon, Z., (1991). *Modeling of Cancer Genesis and Prevention*. CRC Press. Inc., Boca Raton, Florida.

Zhang, L., Ed. (2016). *Nonclinical Statistics for Pharmaceutical and Biotechnology Industries*, Springer International Publishing Switzerland (doi: 10.1007/978-3-319-23558-5); p. XXII+698.

CHAPTER 13

HIV Infection/AIDS/Anti-HIV and AIDS: Pharmacological Management

BOGDAN BUMBĂCILĂ,[1] CORINA DUDA-SEIMAN,[1]
DANIEL DUDA-SEIMAN,[2] and MIHAI V. PUTZ[1,3]

[1]*Laboratory of Computational and Structural Physical Chemistry for Nanosciences and QSAR, Biology-Chemistry Department, Faculty of Chemistry, Biology, Geography at West University of Timişoara, Pestalozzi Street No. 16, Timişoara, RO–300115, Romania*

[2]*Department of Medical Ambulatory, and Medical Emergencies, University of Medicine and Pharmacy "Victor Babes," Avenue C. D. Loga No. 49, RO–300020 Timisoara, Romania*

[3]*Laboratory of Renewable Energies-Photovoltaics, R&D National Institute for Electrochemistry and Condensed Matter, Dr. A. Paunescu Podeanu Str. No. 144, RO–300569 Timişoara, Romania, Tel.: +40-256-592-638; Fax: +40-256-592-620, E-mail: mv_putz@yahoo.com, mihai.putz@e-uvt.ro*

13.1 DEFINITION

Human immunodeficiency virus infection (HIV) and acquired immune deficiency syndrome (AIDS) are conditions caused by the human immunodeficiency virus, a virus which interferes with the immune system of the host causing in time opportunistic infections and development of cancers which rarely happen to people with competent immune systems. The late symptoms of the infections are severe and are referred together as "syndrome"–AIDS (Moore et al., 1999).

Today, the pharmacological management of HIV infection & AIDS includes the therapy with several antiretroviral drug classes which interfere with different steps in the HIV life-cycle. The concomitant use of multiple drugs with different mechanisms of action is called HAART (Highly Active

Anti Retroviral Therapy) and improves the life of an infected patient, reducing the incidence of the opportunistic infections and maintaining the functions of the immune system at an acceptable level (Moore et al., 1999).

There is no cure of the infection or vaccine for its prophylaxis. Still, anti-retroviral pharmacological treatment can slow the course of the disease and may lead to a near-normal life expectancy. New research is indicating that treatment should be recommended as soon as the diagnosis is made. Without treatment, the average survival time after infection is 11 years (Deeks, 2013; WHO, 2015).

13.2   HISTORICAL ORIGINS

The first antiretroviral therapeutic drug was a nucleoside-like reverse tran-scriptase inhibitor (NRTI) – zidovudine/azidothymidine. It was approved in the USA in 1987 as an anti-HIV agent, and it was previously studied as an anti-cancer drug. In the following years, more NRTIs were developed, but even in combination (cART – combination anti-retroviral therapy – the early variant of HAART), they were unable to suppress the virus for long periods of time and patients following the treatments acquired resistance to the drugs. The second class of therapeutic agents was represented by the protease inhibitors (PIs). Saquinavir and ritonavir were the first two marketed drugs, and it was shown that a combination of two NRTIs and one PI shows an impressive improvement in declining the hospitalization AIDS and death rates of those infected (Moore et al., 1999).

Later, other drug classes and other analogs of the first drugs were devel-oped, for interfering with the virus at different levels but also for lowering the adverse effects of the drugs or their administration frequency and for improving the compliance of the patients.

Today, typical combinations include 2 NRTIs and 1 NNRTI/PI/INSTI/MI.

13.3   NANO-CHEMICAL IMPLICATIONS

The life-cycle of the virus is very short–52 hours from the entry in the cell to the export of new virions able to infect other cells, from which 33 hours are taken only for the reverse transcription process (Murray et al., 2011). HIV does not have proofreading enzymes, exonucleases to correct reverse transcription errors during the reverse transcription of RNA into ADN. Its short life cycle and very high error rate are causing mutations of the virus, so HIV has a very high genetic variability (Perelson et al., 1996). Most of the mutations lead to inferior copies, with no ability to infect or reproduce, others do not modify its virulence or the response of the virus to the treatment but other mutations are leading to more active copies, more virulent, more able to skip the natural human defense mechanisms or resistant to the antiretroviral drugs (Smyth et al., 2012).

So the best choice in the HIV infection therapy is to suppress the replication as much as possible, reducing the potential mutations which lead to the superior copies (Smyth et al., 2012). This effective suppression of the replications needs a combination of drugs which puts obstacles at different levels in the viral life cycle.

HIV is entering into a cell following a few steps:

1. HIV gp120 is binding to the CD4 receptor;
2. HIV gp120 suffers a conformational change which increases its affinity for a co-receptor and exposes HIV gp41;
3. HIV gp120 binds to a co-receptor – either CCR5 or CXCR4;
4. HIV gp41 penetrates the HIV lipid membrane and the *T* cell membrane;
5. The viral core – the capsid is entering into the cell after the fuse of the viral envelope with the cell membrane.

After the entrance of the viral capsid in the cytoplasm of the cell, the core loses its structure and releases the viral RNA and some enzymes like the integrase and the reverse-transcriptase. The reverse-transcriptase with its three components (RNA-dependent DNA polymerase, ribonuclease H, DNA-dependent DNA polymerase) reverse transcripts the viral RNA into single-stranded viral DNA that forms the double-stranded viral DNA.

This DNA penetrates the nucleus membrane and with the help of the integrase enzyme is integrated into the cell's DNA, and then the transcription of infected DNA takes place. Long-chained HIV proteins are synthesized after the translation process of the infected DNA into the ribosomes of the cell, and new HIV RNA is formed, which exits the nucleus. An enzyme – HIV protease is also formed at this step, which will be helpful for cutting the long chains of the proteins in order to build smaller proteins which will assemble as new capsids and new viral envelopes.

The newly formed viral RNA and the proteins, with the help of protease enzyme, are forming new mature infectious viral particles which are released from the cell as the cell dies (aidsinfo.nih.gov, 2016).

13.4   ENTRY INHIBITORS

Entry inhibitors also known as fusion inhibitors are a class of antiretroviral drugs. They interfere with the binding, fusion, and entry of an HIV virion to a human cell.

There are several proteins involved with the HIV virion entry process:

- CD4: a protein receptor found on the surface of the Helper *T* Cells of the human immune system.
- gp120: an HIV protein that binds to the CD4 receptor.
- CXCR4 and CCR5: chemokine co-receptors found on the surface of the Helper *T* Cells and macrophages.
- gp41: an HIV protein that penetrates the host cell's membrane.

13.4.1   CCR5 CO-RECEPTOR ANTAGONISTS

HIV enters in the cell by attaching its lipid membrane gp120 (glycoprotein 120) to the CD4 receptor on the host. A conformational change in gp120 occurs, therefore, allowing it to bind also to co-receptor CCR5 expressed on the host cell. The co-attachment is triggering the expression of gp41 which creates a bridge between the viral envelope and the cell membrane. The process is called fusion–because a gap in the host cell membrane is formed, allowing the viral nucleocapsid to enter into the cell. By blocking the CCR5 co-receptor, the expression of the gp41 is not possible so the viral envelope will not establish the contact with the membrane. Some of the drugs are blocking the CCR5 co-receptor (CCR5-co-receptor antagonists), others are binding to gp41 (changing its conformation thus inactivating it–fusion inhibitors) and one drug can bind directly to gp120, blocking its activity (Gp120 antagonist–it is blocking the gp120; thus, it prevents the attachment of this viral protein with the CD4 receptor on the host cell membrane) (Table 13.1).

13.5   REVERSE TRANSCRIPTASE INHIBITORS

13.5.1   NUCLEOSIDE ANALOGS

They represent the first class of antiretroviral drugs introduced in therapy. In order to manifest their effect, their molecules have to be incorporated into the viral DNA; they are activated, therefore, in the cell, by the addition

TABLE 13.1 Entry Inhibitors

Main Pharmacological Class	Pharmacological Subclass	Drug	Mechanism(s) of action	Observations
1. Entry inhibitors	1.1. CCR5 co-receptor antagonists	1.1.1. Maraviroc	• Negative allosteric modulator of the CCR5 co-receptor; it binds to this co-receptor changing its spatial conformation, therefore, blocking the binding of gp120; (Abel et al., 2009)	• It is the only drug in this class approved today for therapy;
		1.1.2. Aplaviroc	• CCR5 receptor antagonist;	• In 2005 it was abandoned from its then evaluation in clinical trials because of its serious liver toxicity profile and poor documented efficacy; (Nichols et al., 2008)
		1.1.3. Vicriviroc	• nOn-competitive allosteric antagonist of the CCR5 co-receptor; (Schürmann et al., 2007)	• Currently in clinical trials for HIV naïve patients;
		1.1.4. Cenicriviroc	• Inhibitor of CCR2 and CCR5 co-receptors; (Klibanov et al., 2010)	• Currently in clinical studies;
	1.2. Fusion inhibitors	1.2.1. Enfuvirtide	• It binds to gp41 and changes its conformation; thus, it prevents the formation of the entry pore in the host cell membrane with the entry of the viral nucleocapsid. (Lalezari et al., 2003)	• Obtaining this drug in the lab is extremely expensive; • The molecule is unstable in aqueous solution, and it has to be administered as a subcutaneous injection twice daily; • Because the patient has to prepare the solution with the lyophilized powder before its administration and because

TABLE 13.1 (Continued)

Main Pharmacological Class	Pharmacological Subclass	Drug	Mechanism(s) of action	Observations
	1.3. Gp120 antagonists	1.3.1. Fostemsavir	• It blocks viral gp120 so the virus can not bind to the CD4 receptor on the human cell membrane; • It is temsavir's prodrug; (Nettles et al., 2012)	of its costs, enfuvirtide is maintained as an alternative when other anti-HIV agents failed; • It is an oligopeptide; (Lalezari et al., 2003) • Because gp120 is a viral protein very well conserved, this drug offers promises to those patients who developed resistance to other therapies because it's very unlikely to promote independent of CD4-binding viral particles; (Nettles et al., 2012)

with the help of cellular kinase enzymes of three phosphate groups to their deoxyribose part, to form triphosphates. The molecules compete with natural occurring nucleosides to prevent their incorporation into the viral DNA, and thus the complete reverse transcription is prevented. They are DNA chain terminators because once they are incorporated in the strain, they prevent the incorporation of other nucleosides into the DNA chain because of the absence of a 3'-OH group in their molecules, acting like competitive natural substrate inhibitors.

Usually, they selectively "inhibit" the viral reverse transcriptase (RNA-dependant DNA polymerase) but at certain doses, they lose this selectivity. This happens because human cells have mechanisms that rapidly repair the DNA damages made by the nucleoside analogs but sometimes the doses are too high, and they overcome these mechanisms (e.g., the human DNA polymerase cannot find sufficient natural substrate to continue the elongation of the DNA strain). Their secondary effects are explained so.

In time, especially after long treatments using nucleoside analogs, the viral reverse transcriptase can mutate and can recognize the "bad structures," lacking the 3'-OH group, so the enzyme will not integrate them into the elongating DNA strain.

13.5.2   NUCLEOTIDE ANALOGS

They are already phosphorylated nucleoside analogs. They also act as chain terminators.

13.5.3   NON-NUCLEOSIDE REVERSE TRANSCRIPTASE INHIBITORS

They directly inhibit the reverse transcriptase, binding to another site than the active one, changing its conformation thus blocking its activity.

13.6   INTEGRASE INHIBITORS

Integrase inhibitors or integrase strand transfer inhibitors are drugs that are blocking the integrase activity, an enzyme that inserts the viral DNA obtained from the activity of reverse transcriptase into the DNA of the host cell (Tables 13.2 and 13.3).

TABLE 13.2 Reverse Transcriptase Inhibitors

Main Pharmaco-logical Class	Pharmacological Subclass	Drug	Mechanism(s) of action	Observations
2. Reverse transcriptase inhibitors	2.1. Nucleoside Analogs	2.1.1. Zidovudine/ Azidothymidine	• Thymidine analog; • Competitive inhibitor of thymidine, it stops the viral DNA elongation when integrated by the reverse transcriptase in the DNA strain; (Yarchoan et al., 1989)	• It slows replication of HIV but does not stop it entirely; • It is mostly used to prevent HIV transmission, such as from mother to child during birth or after a needle stick injury; • For HIV infection treatment it is always used combined with other drugs because HIV can become resistant to it; (Hamilton, 2015)
		2.1.2. Didanosine	• Adenosine analog (it has hypoxanthine attached to the sugar ring); • Competitive inhibitor of adenosine;	• It was the second drug marketed for HIV therapy (FDA-approved in 1991, after FDA approved the first marketed drug, zidovudine, in 1987); • It has numerous interactions with other drugs; (Rang et al., 2007)
		2.1.3. Zalcitabine	• Pyrimidine/deoxycytidine analog;	• Less potent than other NRTIs and it has a very low half-life, so it has to be frequently administered–3 times a day; (Rang et al., 2007)
		2.1.4. Stavudine	• Thymidine analog	• It has long-term and irreversible side effects–like lipodystrophy so because of them today is less frequently used; (Wangsomboonsiri et al., 2010)

TABLE 13.2 *(Continued)*

Main Pharmaco-logical Class	Pharmacological Subclass	Mechanism(s) of action	Observations
	2.1.5. Lamivudine	• Cytidine analog	• Also used for treatment of HBV infection; • Treatment has been shown to restore zidovudine sensitivity of previously zidovudine-resistant HIV strains; (The American Society of Health-System Pharmacists, 2015)
	2.1.6. Abacavir	• Guanosine analog;	• It was approved for therapy in 1998; • It is used in patients who developed resistance to zidovudine and lamivudine; • Some genotypes can present hyper-sensitivity to abacavir, developing a hypersensitivity syndrome within the first six weeks after the treatment was initiated; The syndrome can be fatal; (Mallal et al., 2008)
	2.1.7. Emtricitabine	• Cytidine analog;	• It is also effective on hepatitis *B* virus, but it is not yet approved for the use in chronic *B* hepatitis treatment; • It was synthesized in 1996, and it was first marketed in 2003; (Lim et al., 2006)
	2.1.8. Elvucitabine	• Cytidine analog;	• Currently in phase II clinical trials;

TABLE 13.2 (Continued)

Main Pharmacological Class	Pharmacological Subclass	Drug	Mechanism(s) of action	Observations
		2.1.9. Apricitabine	• Cytidine analog;	• It appears to be active on zidovudine and lamivudine-resistant HIV strains; (Cahn et al., 2006)
		2.1.10. Amdoxovir	• Guanosine analog;	• It is currently in clinical studies; (Murphy et al., 2010)
		2.1.11. Festinavir (Censavudine)	• Thymidine analog;	• Not yet into clinical studies but it is believed to have an improved safety; (Weinberg, 2012)
	2.2. Nucleotide analogs	2.2.1. Tenofovir disoproxil	• adenosine nucleotide analog; • It prevents the formation of the 5' to 3' phosphodiester linkage necessary for DNA strain elongation because the molecule lacks an -OH group on the 3' carbon of its deoxyribose sugar; (drugbank.ca)	• Together with emtricitabine, it is now available as TRUVADA, a combination which was demonstrated in 2015 as more than 90% efficient as pre-exposure prophylaxis to HIV (Molina et al., 2015)
	2.3. Non-nucleoside inhibitors	2.3.1. Efavirenz	• The molecule binds to a distinct site on the reverse transcriptase, other than the active site (where NRTIs bind), known as the NNRTI pocket; • It is not active on HIV-2 reverse transcriptase which has a different structure of the pocket than HIV-1 reverse transcriptase; (Ren et al., 2002)	• At certain doses, it produces LSD-like hallucinogenic effects in humans; (Gatch et al., 2013)

TABLE 13.2 *(Continued)*

Main Pharmacological Class	Pharmacological Subclass	Drug	Mechanism(s) of action	Observations
		2.3.2. Nevirapine	• It allosterically binds to a site (the NNRTIs site) on the reverse transcriptase changing its conformation, therefore, inactivating it;	• Clinical studies proved that that prophylaxis of mother to child transmission during birth with single-dose nevirapine in addition to zidovudine is more effective than zidovudine alone; (Lallemant et al., 2004)
		2.3.3. Delavirdine	• It allosterically binds to a site (the NNRTIs site) on the reverse transcriptase changing its conformation, therefore, inactivating it;	• It is currently rarely used because of its short half-life (thus high frequency of administration);
		2.3.4. Atevirdine	• It directly inhibits the reverse transcriptase;	• It is currently in clinical studies; (Morse et al., 2000)
		2.3.5. Etravirine	• The molecule presents a conformational isomerism which provides a capacity of binding in several conformational possibilities with the reverse transcriptase, even if the enzyme is mutated; (Das et al., 2004)	• Apparently, it does not have HIV cross-resistance with other analogs of its class;
		2.3.6. Rilpivirine	• It is also a molecule that presents conformational isomerism, allowing a better binding to the reverse transcriptase;	• it has a higher potency, longer half-life, and reduced side-effects than all the other NNRTIs. (Goebel et al., 2006)

TABLE 13.3 Integrase Inhibitors

Main Pharmaco-logical Class	Drug	Mechanism(s) of action	Observations
3. Integrase inhibitors	3.1. Raltegravir	• It blocks the integrase enzyme;	• Initially approved for patients resistant to other HAART drugs, it was shown that the treatment with this drug led to undetectable viral loads, sooner than those taking similarly potent NNRTIs or PIs; • Today, the studies are directed to the ability of this molecule to eradicate the virus from the latent reservoirs; • It was also proved that raltegravir has anti-viral activity on herpes viruses; (Savarino, 2006)
	3.2 Elvitegravir	• It shares the molecular core with the fluoro-quinolone antibiotics; • It inhibits the integrase by binding to it;	• it is usually administered together with cobicistat, a molecule that inhibits the enzymes in the liver and wall of the gut that metabolize elvitegravir; (Stellbrink, 2007)
	3.3. Dolutegravir	• Integrase inhibitor;	• It is marketed by GlaxoSmithKline since 2013 and can be used in a very broad variety of HIV-infected patients (patients who do not have been previously treated, patients with resistance to some of the antiretroviral therapy, etc.) (Raffi et al., 2013)
	3.4. BI 224436	• It binds to another site of the integrase enzyme; thus, it has a slightly different mechanism of action than raltegravir, dolutegravir, and elvitegravir which are binding to the catalytic site of the enzyme (Levin, 2011)	• Currently in clinical trials;

TABLE 13.3 (Continued)

Main Pharmaco-logical Class	Drug	Mechanism(s) of action	Observations
	3.5. MK–2048	• It appears to be four times more potent that raltegravir–it inhibits the integrase enzyme four times longer in time than raltegravir;	• It is currently in clinical studies developed by Merck & Co.; • It is also investigated as an option of pre-exposure prophylaxis; (Mascolini, 2009)
	3.6. Carbotegravir (GSK744)	• Integrase inhibitor;	• It was found that if packed as nanoparticles it has a half-life of 21 to 50 days so, in this formulation, it can be administered as once in three months; • It is currently in clinical studies; (Borrell, 2014)

13.7   PROTEASE INHIBITORS (PIS)

Protein Inhibitors are inhibiting the viral replication by blocking the HIV proteases, enzymes that are important for the proteolysis of protein precursors that are necessary for the production of infectious mature viral particles. Because these drugs are highly specific, there is a risk, as in antibiotics, of the development of drug-resistant mutated viral particles. To reduce this risk, it is common to use several drugs together with different mechanisms of action and thus targeting different steps in the HIV replication cycle (Rang, 2007) (Table 13.4).

13.8   OPEN ISSUES: *PORTMANTEAU INHIBITORS*

Portmanteau inhibitors are molecules designed to have reverse transcriptase & integrase inhibitory activities or, more recently, integrase, and entry (CCR5 blocking) inhibitory activities.

The molecules are lab-designed, having dual-cores, corresponding to the one needed for the reverse transcriptase/entry blocking activity and to that one of the integrase inhibitory activity.

Trends in HIV/AIDS management include the development of formulations of two, three or more drugs, combined in a single tablet to facilitate the patient's adherence to the treatment. Still, physic-chemical interactions between drugs or incompatibilities between drugs and the excipients make these formulations hard to achieve.

The newest strategy in the designing process of HIV medication is the creation of a single molecule which targets the virus at different levels, thus improving the patient's compliance and reducing the mechanisms of drug resistance.

Bodiwala synthesized in 2011 portmanteau inhibitors of HIV-1 integrase and CCR5 co-receptor. They performed QSAR analysis on these structures but they've also tested their toxicity and efficacy on cultures of HIV-infected cells, and the results proved promising (Bodiwala et al., 2011).

Previously, Marchand, C. and his colleagues (2008) studied a series of 29 madurahydroxylactone (a metabolite of *Nonomuraea rubra* soil bacteria with benzo-α-naphtacenequinone structure) derivatives as dual inhibitors of HIV integrase and reverse transcriptase (particularly its ribonuclease *H* part). They proved the inhibition in vitro, by electrochemiluminescence, using recombinant HIV integrase and ribonuclease *H* and DNA substrate. The structures

TABLE 13.4 Protease Inhibitors

Main Pharma-cological Class	Drug	Mechanism(s) of action	Observations
4. Protease inhibitors	4.1. Saquinavir	• Protease inhibitor, it specifically binds to the active site of the enzyme by mimicking the tetrahedral intermediate of its substrate, and it blocks it;	• It is the first drug in this class approved for therapy;
	4.2. Ritonavir	• Protease inhibitor, it specifically binds to the active site of the enzyme by mimicking the tetrahedral intermediate of its substrate, and it blocks it;	• Today it is used as a booster for other protease inhibitors because it is a potent inhibitor of cytochrome P450–CYP3A4, a liver enzymatic system which metabolizes protease inhibitors;
	4.3. Indinavir	• Protease inhibitor, it specifically binds to the active site of the enzyme by mimicking the tetrahedral intermediate of its substrate, and it blocks it;	• It interacts with food and beverages, so it has many restrictions in this direction;
	4.4. Nelfinavir	• Protease inhibitor, it specifically binds to the active site of the enzyme by mimicking the tetrahedral intermediate of its substrate, and it blocks it;	• The oral bioavailability of this molecule is increased 2 to 5 times when taken with food. It has potential anti-cancer activity;
	4.5. Amprenavir & Fosamprenavir	• Protease inhibitor, it specifically binds to the active site of the enzyme by mimicking the tetrahedral intermediate of its substrate, and it blocks it;	• It was initially used as an antihypertensive agent;
	4.6. Lopinavir	• Protease inhibitor, it specifically binds to the active site of the enzyme by mimicking the tetrahedral intermediate of its substrate, and it blocks it;	• It is also effective against HPV;
	4.7. Atazanavir	• Protease inhibitor, it specifically binds to the active site of the enzyme by mimicking the tetrahedral intermediate of its substrate, and it blocks it;	• It is the first protease inhibitor to be approved for an once-a-day oral administration;

TABLE 13.4 (Continued)

Main Pharmacological Class	Drug	Mechanism(s) of action	Observations
	4.8. Tipranavir	• Protease inhibitor, it specifically binds to the active site of the enzyme by mimicking the tetrahedral intermediate of its substrate, and it blocks it;	• The resistance to this molecule requires too many mutations to take place, so it is an alternative for the resistance to other protease inhibitors; (Doyon et al., 2005)
	4.9. Darunavir	• Protease inhibitor, it specifically binds to the active site of the enzyme by mimicking the tetrahedral intermediate of its substrate, and it blocks it;	• Its structure allows the molecule to create the highest number of hydrogen bonds with the protease's active site than other protease inhibitors (Ghosh et al., 2007)

showed similar potencies for both enzymes, and two of these analogs had 100 times greater potencies than the others (Marchand et al., 2008).

KEYWORDS

- **entry inhibitors (fusion inhibitors)**
- **integrase inhibitors**
- **maturation inhibitors**
- **non-nucleoside analogs**
- **nucleotide analogs)**
- **protease inhibitors**
- **reverse-transcriptase inhibitors (nucleoside analogs**

REFERENCES AND FURTHER READING

Abel, S., Back, D. J., & Vourvahis, M., (2009). Maraviroc: Pharmacokinetics and drug interactions. *Antiviral Therapy, 14*(5), 607–618.

Bodiwala, H. S., Sabde, S., Gupta, P., Mukherjee, R., Kumar, R., Garg, P., Bhutani, K. K., Mitra, D., & Singh, I. P., (2011). Design and synthesis of caffeoyl-anilides as portmanteau inhibitors of HIV-1 integrase and CCR5. *Bioorganic & Medicinal Chemistry, 19*(3), 1256–1263.

Borrell, B., (2014). Long-acting shot prevents infection with HIV analog – Periodic injection keeps monkeys virus-free and could confer as long as three months of protection in humans. *Nature, 4 March (doi:10.1038/nature.2014.14819)*.

Cahn, P., Cassetti, I., Wood, R., Phanuphak, P., Shiveley, L., Bethell, R. C., & Sawyer, J., (2006). Efficacy and tolerability of 10-day monotherapy with apricitabine in antiretroviral-naive, HIV-infected patients. *AIDS, 20*(9), 1261–1268.

Das, K., Clark, A. D., Lewi, P. J., Heeres, J., De Jonge, M. R., Koymans, L. M., et al., (2004). Roles of conformational and positional adaptability in structure-based design of TMC125-R165335 (etravirine) and related non-nucleoside reverse transcriptase inhibitors that are highly potent and effective against wild-type and drug-resistant HIV-1 variants. *Journal of Medicinal Chemistry, 47*(10), 2550–2560.

Deeks, S. G., (2013). The end of AIDS: HIV infection as a chronic disease. *The Lancet, 382*(9903), 1525–1533.

Doyon, L., Tremblay, S., Bourgon, L., Wardrop, E., & Cordingley, M., (2005). Selection and characterization of HIV-1 showing reduced susceptibility to the non-peptidic protease inhibitor tipranavir. *Antiviral Research, 68*(1), 27–35.

Gatch, M. B., Kozlenkov, A., Huang, R. Q., Yang, W., Nguyen, J. D., González-Maeso, J., et al., (2013). The HIV antiretroviral drug efavirenz has LSD-like properties. *Neuropsychopharmacology, 38*(12), 2373–2384.

Ghosh, A. K., Dawson, Z. L., & Mitsuya, H., (2007). Darunavir, a conceptually new HIV-1 protease inhibitor for the treatment of drug-resistant HIV. *Bioorganic Medicinal Chemistry, 15*(24), 7576–7580.

Goebel, F., Yakovlev, A., Pozniak, A. L., Vinogradova, E., Boogaerts, G., Hoetelmans, R., De Béthune, M. P., Peeters, M., & Woodfall, B., (2006). Short-term antiviral activity of TMC278—a novel NNRTI—in treatment-naive HIV-1-infected subjects. *AIDS, 20*(13), 1721–1726.

Hamilton, R., (2015). *Tarascon Pocket Pharmacopoeia 2015 Deluxe Lab-Coat Edition* (p. 67). Jones & Bartlett Learning.

http://chemdb.niaid.nih.gov/DrugDevelopmentHIV.aspx.

http://www.drugbank.ca/drugs/DB00300.

http://www.who.int/hiv/pub/guidelines/en/-Guidelines HIV, (2015)–*World Health Organization*.

https://aidsinfo.nih.gov/contentfiles/lvguidelines/adultandadolescentgl.pdf – Guidelines for the Use of Antiretroviral Agents in HIV-1-Infected Adults and Adolescents, (2015)–US Department of Health and Human Services.

https://aidsinfo.nih.gov/education-materials/fact-sheets/19/73/the-hiv-life-cycle.

Klibanov, O. M., Williams, S. H., & Iler, C. A., (2010). Cenicriviroc, an orally active CCR5 antagonist for the potential treatment of HIV infection, *Current Opinion in Investigational Drugs, 11*(8), 940–950.

Lalezari, J. P., Eron, J. J., Carlson, M., Cohen, C., Dejesus, E., Arduino, R. C., Gallant, J. E., & Volberding, P., (2003). A phase II clinical study of the long-term safety and antiviral activity of enfuvirtide-based antiretroviral therapy. *AIDS, 17*(5), 691–698.

Lallemant, M., Jourdain, G., Le Coeur, S., Mary, J. Y., Ngo-Giang-Huong, N., Koetsawang, S., Kanshana, S., McIntosh, K., & Thaineua, V., (2004). Single-dose perinatal nevirapine plus standard zidovudine to prevent mother-to-child transmission of HIV-1 in Thailand. *The New England Journal of Medicine, 351*, 217–228.

Lamivudine. The American Society of Health-System Pharmacists. Retrieved 31 July 2015.

Levin, J., (2011). *BI 224436, a Non-Catalytic Site Integrase Inhibitor, is a Potent Inhibitor of the Replication of Treatment-Naïve and Raltegravir-Resistant Clinical Isolates of HIV-1.* Conference Reports for NATAP. 51th ICAAC Chicago, IL; September 17-20 (http://www.natap.org/2011/ICAAC/ICAAC_34.htm).

Lim, S. G., Ng, T. M., & Kung, N., (2006). A double-blind placebo-controlled study of emtricitabine in chronic hepatitis B. *Archives of Internal Medicine, 166*(1), 49–56.

Mallal, S., Phillips, E., & Carosi, G., (2008). HLA-B*5701 screening for hypersensitivity to abacavir. *New England Journal of Medicine, 358*(6), 568–579.

Marchand, C., Beutler, J. A., Wamiru, A., Budihas, S., Ilmann, U. M., Heinisch, L., et al., (2008). Madurahydroxylactone derivatives as dual inhibitors of human immunodeficiency virus type 1 integrase and RNase H. *Antimicrobial Agents and Chemotherapy, 52*(1), 361–364.

Mascolini, M., (2009). Merck offers unique perspective on second-generation integrase inhibitor. *10th International Workshop on Clinical Pharmacology of HIV Therapy.* Amsterdam, April 15–17. (http://www.natap.org/2009/PK/PK_10.htm).

Molina, J. M., Capitant, C., Spire, B., et al., (2015). On-demand preexposure prophylaxis in men at high risk for HIV-1 infection. *New England Journal of Medicine, 373*(23), 2237–2246.

Moore, R. D., & Chaisson, R. E., (1999). Natural history of HIV infection in the era of combination antiretroviral therapy. *AIDS, 13*(14), 1933–1942.

Morse, G. D., Reichman, R. C., & Fischl, M. A., (2000). Concentration-targeted phase I trials of atevirdine mesylate in patients with HIV infection: Dosage requirements and pharmacokinetic studies. *The ACTG 187 and 199 Study Teams, Antiviral Research, 45*(1), 47–58.

Murphy, R. L., Kivel, N. M., Zala, C., Ochoa, C., Tharnish, P., Mathew, J., Pascual, M. L., & Schinazi, R. F., (2010). Antiviral activity and tolerability of amdoxovir with zidovudine in a randomized, double-blind placebo-controlled study in HIV-1-infected individuals. *Antiviral Therapy, 15*(2), 185–192.

Murray, J. M., Kelleher, A. D., & Cooper, D. A., (2011). Timing of the components of the HIV life cycle in productively infected CD4+ *T* cells in a population of HIV-infected individuals. *Journal of Virology, 85*(20), 10798–10805.

Nettles, R. E., Schürmann, D., Zhu, L., Stonier, M., Huang, S. P., Chang, I., et al., (2012). Pharmacodynamics, safety, and pharmacokinetics of BMS–663068, an oral HIV-1 attachment inhibitor in HIV-1-infected subjects. *Journal of Infectious Diseases, 206*(7), 1002–1011.

Nichols, W. G., Steel, H. M., Bonny, T., Adkison, K., Curtis, L., Millard, J., Kabeya, K., & Clumeck, N., (2008). Hepatotoxicity observed in clinical trials of aplaviroc (GW873140). *Antimicrobial Agents and Chemotherapy, 52*(3), 858–865.

Perelson, A. S., Neumann, A. U., Markowitz, M., & Leonard, J. M., (1996). HIV-1 dynamics *in vivo*: Virion clearance rate, infected cell life-span, and viral generation time. *Science, 271*(5255), 1582–1586.

Raffi, F., Jaeger, H., Quiros-Roldan, E., Albrecht, H., Belonosova, E., Gatell, J. M., Baril, J. G., Domingo, P., Brennan, C., Almond, S., & Min, S., (2013). Once-daily dolutegravir versus twice-daily raltegravir in antiretroviral-naive adults with HIV-1 infection (SPRING–2 study): 96-week results from a randomized, double-blind, non-inferiority trial. *The Lancet Infectious Diseases, 13*(11), 927–935.

Rang, H. P., Dale, M. M., Ritter, J. M., & Flower, R. J., (2007). *Rang and Dale's Pharmacology* (6th edn.). Churchill Livingstone Elsevier, Philadelphia (eBook ISBN: 9780702040740; pp. 844; https://www.elsevier.com/books/rang-and-dales-pharmacology-e-book/rang/978-0-7020-4074-0).

Ren, J., Bird, L. E., Chamberlain, P. P., Stewart-Jones, G. B., Stuart, D. I., & Stammers, D. K., (2002). Structure of HIV–2 reverse transcriptase at 2.35-a resolution and the mechanism of resistance to non-nucleoside inhibitors. *Proceedings of the National Academy of Sciences of the United States of America, 99*(22), 14410–14415.

Savarino, A., (2006). A historical sketch of the discovery and development of HIV-1 integrase inhibitors. *Expert Opinion on Investigational Drugs, 15*(12), 1507–1522.

Schürmann, D., Fätkenheuer, G., Reynes, J., & Hoffmann, C., (2007). Antiviral activity, pharmacokinetics, and safety of vicriviroc, an oral CCR5 antagonist, during 14-day monotherapy in HIV-infected adults. *AIDS, 21*(10), 1293–1299.

Smyth, R. P., Davenport, M. P., & Mak, J., (2012). The origin of genetic diversity in HIV-1. *Virus Research, 169*(2), 415–429.

Steigbigel, R. T., Cooper, D. A., & Kumar, P. N., (2008). Raltegravir with optimized background therapy for resistant HIV-1 infection. *The New England Journal of Medicine, 359*(4), 339–354.

Stellbrink, H. J., (2007). Antiviral drugs in the treatment of AIDS: What is in the pipeline? *European Journal of Medical Research, 12*(9), 483–495.

Wainberg, M. A., (2012). The need for development of new HIV-1 reverse transcriptase and integrase inhibitors in the aftermath of antiviral drug resistance. *Scientifica, 1*–28.

Wangsomboonsiri, W., Mahasirimongkol, S., Chantarangsu, S., Kiertiburanakul, S., Charoenyingwattana, A., Komindr, S., et al., (2010). Association between HLA-B*4001 and lipodystrophy among HIV-infected patients from Thailand who received a stavudine-containing antiretroviral regimen. *Clinical Infectious Diseases, 50*(4), 597–604.

Yarchoan, R., Mitsuya, H., Myers, C., & Broder, S., (1989). Clinical pharmacology of 3'-azido–2,'3'-dideoxythymidine (zidovudine) and related dideoxynucleosides. *New England Journal of Medicine, 321*(11), 726–738.

CHAPTER 14

HIV-Integrase

CORINA DUDA-SEIMAN,[1] DANIEL DUDA-SEIMAN,[2] and
MIHAI V. PUTZ[1,3]

[1]*Laboratory of Computational and Structural Physical Chemistry
for Nanosciences and QSAR, Biology-Chemistry Department, Faculty
of Chemistry, Biology, Geography at West University of Timișoara,
Pestalozzi Street No.16, Timișoara, RO–300115, Romania*

[2]*Department of Medical Ambulatory, and Medical Emergencies,
University of Medicine and Pharmacy "Victor Babes,"
Avenue C. D. Loga No. 49, RO–300020 Timisoara, Romania*

[3]*Laboratory of Renewable Energies-Photovoltaics, R&D National
Institute for Electrochemistry and Condensed Matter, Dr. A. Paunescu
Podeanu Str. No. 144, RO–300569 Timișoara, Romania,
Tel.: +40-256-592-638; Fax: +40-256-592-620,
E-mail: mv_putz@yahoo.com or mihai.putz@e-uvt.ro*

14.1 DEFINITION

Human immunodeficiency viruses produce several enzymes, among them integrases which provide the integration of the viral genetic material into the DNA of the infected cell, being thus the essential part in the pre-integration complex (Liao et al., 2010).

14.2 HISTORICAL ORIGIN(S)

In the early stages of lifecycles of retroviruses, the viral RNA genome is converted into a DNA provirus. When entering into a host cell (infected cell), the intracytoplasmatic transcription stage of viral RNA is initiated in order to form a linear double-stranded DNA molecule with long terminal repeats (LTR) with the possibility of intra-chromosomial DNA integration (Bowerman et al., 1989). Viral integrase was considered an attractive and promising target for therapy: HIV viruses are not able to replicate without integration into a host chromosome (De Clercq, 2002). HIV-1 integrase (Mr 32000) is encoded at the 3'-end of the *pol* gene. Integration of HIV DNA into the host cell genome comprises certain DNA cleavage and coupling reactions; to be noticed that the viral cDNA is initially modeled in the host cell cytoplasm, and afterward, it is integrated into the nucleus. This step depends on the entirely functional HIV-integrase. (Nair and Chi, 2007). A particular aspect of HIV integration is the accumulation of increased volumes of unintegrated viral DNA in the infected cells, being considered a significant issue of the viral cytopathic effect. (Faucci, 1988).

14.3 NANO-CHEMICAL APPLICATION(S)

In human cells, viral integrase is structurally a tetramer (Chiu and Davies, 2004). HIV-1 integrase has an N-terminal domain, a catalytic core, and a C-terminal domain. The N-terminal domain is concerted opposite to the core domain sharing a similar pseudo-dyad axis in order to form almost identical dimers. This is considered the active form of the enzyme, exposing channels with positive charge capable to bind DNA (Wang et al., 2001).

Molecular dynamics and dynamic activity are concepts which give the possibility to elaborate a double simple criterion to locate clusters of homologous conformers in the manner of molecular dynamics trajectory (Gabarro-Arpa, 2000). The Hamming distance (the distance between two molecular conformations) expresses the number of bits in their sequences and describes the regions of conformational space (Laboulais et al., 2002). Cluster analysis is a process of learning by grouping sets of data maximizing intracluster similarities and minimizing intercluster similarities (Vijay et al., 2012). Analyzing the molecular dynamics of the catalytic core of HIV-1 integrase, it is shown a division of the trajectory into two different segments (2.6 ns, and 1.4 ns length, respectively) with equilibration only at the end of the first segment. (Laboulais et al., 2002). Performing two molecular dynamic simulations upon the catalytic core of HIV-1 integrase, Laboulais et al., (2001) computed the fluorescence intensity decays emitted by four tryptophan residues (the only chromophores) and found very good correlations between 5 and 30°C concerning experimental intensity decays, but only at 5°C experimental anisotropy was good. The authors concluded that this might be due to dimerization or increase of uncorrelated internal motions.

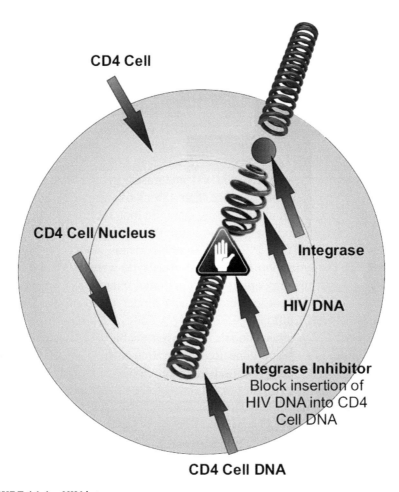

FIGURE 14.1 HIV-integrase.

The active site region of HIV-integrase contains a flexible peptide loop (Figure 14.1). Weber et al., (1998) conducted studies of molecular dynamics simulation upon this region of HIV-integrase finding that parts of the loop extend to an existing helix after 300 ps and, after completing this simulation with a lower resolution Brownian dynamics simulations, it was observed that the loop is framed into a much larger conformational space compared to the classical molecular dynamics trajectory with an aspect of grating-like movement with respect to the active site.

HIV-intergrase is characterized by a quite good stability, excepting the regions 141–148, and 190–192. Biological activity of the enzyme is multifactorial. One condition is given by the non-existence or the existence of one metal in the active site which determines the flexibility of the 146–149 region. The presence of metal near the active site stabilizes the region and the loop between 141 and 145. (Lins et al., 1999).

A leucine-zipper motif for the HIV-1 integrase protein (HIV IN-LZM) can exist as an alpha-helix: in a molecular dynamics simulation system compared to controls, and the predicted HIV IN-LZM is more stable also at high temperatures equilibration with fully stretched generated structures. (Wang et al., 1994).

Conformational dynamics of HIV-1 integrase core domain determine the enzymatic activity. A total of more than 480 ns of simulation data showed a major conformational change in the catalytic loop (gating like dynamics), and a transient intraloop structure motivating the mutational effects of certain residues on the loop (Q^{148}, P^{145}, Y^{143}). By these means, seven conformational states in the catalytic loop of the wild virus type were described. (Lee et al., 2005). The closing-opening conformational change has importance in biological activity, because mutations can hinder the catalytic loop from closing. (Lee et al., 2005).

Tetramer HIV-integrase models lead to reasonable mechanisms for *trans*-binding of viral DNA and concerted integration of both viral ends to host DNA. In the ternary complex model containing the integrase core, the viral DNA, and the inhibitor, the active site-loop of the integrase forms a potential hydrophobic binding pocket, this model being applicable to potent integrase inhibitors (elvitegravir). At the level of the active integrase binding site, only one from two ions of Mg^{2+} is interacting with a potential inhibitor. (Chen et al., 2008). Mg^{2+} ions are fundamental in the stability of the integrase-DNA complex. Also, these ions are crucial for the catalytic activity of the integrase, as well as for the binding process to a specific inhibitor. Inhibitors (e.g., raltegravir (RAL)), in the presence of Mg^{2+}, express an additive effect in the inhibitory capacity of the catalytic activity of HIV integrase because the formation of intraprotein and DNA-protein hydrogen bonds is corrupted. (Miri et al., 2014).

Experimetally, Azzi et al., (2010) immunized a mouse with a 29mer peptide (K159) formed by residues 147 to 175 of the HIV-1 integrase and obtained a monoclonal antibody (MAba4) which recognizes an epitope in the N-terminal portion of K159, but also recognizes in the same manner the epitope within the enzyme, and the entire K159 is highly antigenic. Long terminal repeated oligonucleotides, regardless of their processed or unpro-cessed state, interfere with the binding of MAba4 to the HIV-1 integrase, enforcing the idea that the α4-helix of the integrase interacts as well as with

DNA target or with MAba4 using common residues, concluding that *in vitro* MAba4 significantly decreases the enzymatic activity of HIV-1 integrase.

14.4 MULTI-/TRANS-DISCIPLINARY CONNECTION(S)

The first HIV-integrase inhibitor approved by the FDA is RAL, derived from the evolution of 5,6-dihydroxypyrimidine–4-carboxamides and *N*-methyl–4-hydroxypyrimidinone-carboxamides (Summa et al., 2008). It inhibits selectively the strand transfer activity of HIV-1 and its integration into human DNA, limiting viral replication and new cell infection. (Croxtall et al., 2008). In two combined studies (Steigbiegel et al., 2008), patients were randomized to receive RAL or placebo in a 2:1 ratio. HIV-1 RNA level decreased to a level below 50 copies/ml at week 16 of treatment in 61.8% of the patients in the RAL group compared to 34.7% of the patients in the placebo group. At week 48 of treatment, the level of less than 50 viral copies/ml was found in 62.1% of the patients receiving RAL, vs. 32.9% in the placebo group.

In a large randomized, double-blind trial, RAL was non-inferior versus efavirenz in achieving HIV-1 RNA viral levels below 50 copies/ml, when used as part of an antiretroviral therapy combination regimen, both early after treatment start and also with the maintenance of this effect after 96 weeks of treatment. In addition, the CD4 cells count improved. To notice that drug-related adverse events were less in the RAL group. (Croxtall and Scott, 2012).

Pharmacokinetics of HIV-integrase inhibitors is a very interesting issue; its understanding of providing information regarding the route of administration, dose, and efficiency. Glucuronidation is the main metabolic pathway of RAL, obtaining RAL glucuronide (RAL-GLU) via UDP-glucuronosyltransferase 1A1 (UGT1A1). RAL and RAL-GLU can be simultaneously detected and quantified using liquid chromatography-tandem mass spectrometry. This method provides knowledge about the therapeutic window of RAL in different patient categories. (Wang et al., 2011). RAL is rapidly absorbed from the gastrointestinal tract when orally administered; peak plasma concentrations are achieved after 0.5–1.3 hours. 9% of the administered dose is excreted unmodified in urine. (Gupta et al., 2015). Recently, a liquid chromatography-tandem mass spectrometry assay was proven to be sufficiently sensitive and accurate to quantify antiretroviral drugs including RAL in plasma and saliva, providing the relationship between drug concentrations in plasma and saliva. (Yamada et al., 2015).

RAL showed efficiency against multi-drug resistant HIV-1, being additive or synergistic *in vitro* with currently available antiretroviral regimens. It is important to the point that HIV resistant strains to RAL remain susceptible to other drugs. Safety of administration of RAL is comparable to placebo at all studied doses. (Grinsztejn et al., 2007).

A major problem in HIV infection treatment regimens is represented by resistance development. Rusconi et al., (2013) identified factors associated with a 24-week response in patients receiving RAL -based therapy regimens: genotypic sensitivity scores (GSS) and weighted-GSS. RAL failure was associated with the prevalence of Q148H + G140S within the integrase genotype (half of the cases that showed resistance to RAL). Each extra unit of GSS is associated with the response.

To avoid the selection of RAL resistant mutations, it is mandatory for patients to be treated with RAL and at least other two active drugs in order to maintain a viral load less than 200 copies/ml. (Malet et al., 2012).

Elvitegravir is also a first generation HIV integrase inhibitor, recommended in HIV infection treatment in combination with a pharmacokinetic booster, suitable in a single daily dose administration. In a recent study, the once-daily administration of ritonavir-boosted elvitegravir was non-inferior to the regimen of twice-daily administration of RAL in terms of efficiency and safety profile in patients with HIV-1 infection. (Desimmie et al., 2012; Molina et al., 2012).

Dolutegravir is an orally administered HIV-1 integrase strand transfer inhibitor suitable for an unboosted once-daily therapy used in combination with other retroviral agents in the HIV-1 infection treatment. (Ballantyne and Perry, 2013). It should be twice daily administered when HIV-1 integrase resistance mutations are documented. (Katlama and Murphy, 2012). Dolutegravir proved its efficiency and safety profile in all age categories, inclusively in the pediatric population: in a recent study upon an HIV-1 infected and treatment-experienced pediatric population with a median age of 15 years, a dolutegravir-based regimen was administered. After 48 weeks of treatment, dolutegravir was well tolerated with no grade 4 side effects or discontinuations due to serious adverse effects, and HIV RNA was below 50 copies/ml in 61% of the enrolled patients. (Viani et al., 2015).

14.5   OPEN ISSUES

One of the challenges for future research in HIV infections treatment via HIV-integrase inhibition is represented by patients with HIV–2 infection. Failure to RAL of HIV–2 infected patients is due to secondary changes at the integrase coding gene (I84V, Q214H) and primary changes at position 155. Treviňo et al., (2015) concluded that patients with HIV-2 infection presenting the N155H resistance mutation with failure to RAL should undergo a rescue therapy with a Dolutegravir-based regimen.

Chen et al., (2015), published recently their fundamental research, designing a series of potential HIV-1 integrase inhibitors based on 5-chloro-2-hydroxy-3-triazolyl benzoic acid scaffold. The synthesis of designed compounds uses a click chemistry reaction. The authors also conducted *in vitro* experiments to establish the HIV-integrase inhibitory capacity of the obtained compounds which are promising and candidate structures to be studied in pre-clinical trials. Bai et al., (2015) synthesized a series of N-hydroxy triazole-4-carboxamide derivatives, but with no satisfactory inhibition capacity of HIV integrase.

GSK1265744 is a new HIV integrase strand transfer inhibitor with effectiveness when delivered once daily in a low-milligram dose, not requiring a pharmacokinetic booster. Very important data derived from studies, namely that GSK1265744 demonstrated efficacy against highly resistant mutants, against clones containing RAL resistant Y143R; Q148K, N155H, G140S/ Q148H viral variants. The compound had additive or synergistic effects when used in combination with other retroviral agents, with no increase of the serious side effects. Thus, GSK1265744 is a promising potent new HIV integrase inhibitor (Yoshinaga et al., 2015).

KEYWORDS

- **DNA**
- **HIV-integrase**
- **inhibitors**

REFERENCES AND FURTHER READING

Azzi, S., Parissi, V., Maroun, R. G., Eid, P., Mauffret, O., & Fermandjian, S., (2010). *PLoS One, 5*(12), e16001, 1–9.

Bai, Y. X., Li, L. J., Wang, X. L., Zeng, C. C., & Hu, L. M., (2015). Design and synthesis of N-hydroxy triazole–4-carboxamides as HIV integrase inhibitors. *Med. Chem. Res., 24*, 423–429.

Ballantyne, A. D., & Perry, C. M., (2013). Dolutegravir: First global approval. *Drugs, 73*(14), 1627–1637.

Bowerman, B., Brown, P. O., Bishop, M., & Varmus, H. E., (1989). A nucleoprotein complex mediates the integration of retroviral DNA. *Genes & Development, 3*, 469–478.

Chen, J., Liu, C. F., Yang, C. W., Zeng, C. C., Liu, W., & Hu, L. M., (2015). Design and synthesis of 5-chloro–2-hydroxy–3-triazolyl benzoic acids as HIV integrase inhibitors. *Med. Chem. Res., 24*, 2950–2959.

Chen, X., Tsiang, M., Yu, F., Hung, M., Jones, G. S., Zeynalzadegan, A., Qi, X., Jin, H., Kim, K. U., Swaminathan, S., & Chen, J. M., (2008). Modeling, analysis, and validation of a novel HIV integrase structure provide insights into the binding modes of potent integrase inhibitors. *Journal of Molecular Biology, 380*(3), 504–519.

Chiu, T. K., & Davies, D. R., (2004). Structure and function of HIV-1 integrase. *Curr. Top Med. Chem., 4*, 965–977.

Croxtall, J. D., & Scott, L. J., (2012). Raltegravir in treatment-naïve patients with HIV-1 infection. *Drugs, 70*(5), 631–642.

Croxtall, J. D., Lyseng-Williamson, K. A., & Perry, C. M., (2008). Raltegravir. *Drugs, 68*(1), 131–138.

De Clercq, E., (2002). New anti-HIV agents and targets. *Med. Res. Rev., 22*(6), 531–565.

Desimmie, B. A., Schrijvers, R., & Debyser, Z., (2012). Elvitegravir: A once-daily alternative to raltegravir. *The Lancet Infectious Diseases, 12*(1), 3–5.

Faucci, A. S., (1988). The human immunodeficiency virus: Infectivity and mechanisms of pathogenesis. *Science, 239*(4840), 617–622.

Gabarro-Arpa, J., & Revilla, R., (2000). Clustering of a molecular dynamics trajectory with a hamming distance. *Computers & Chemistry, 24*(6), 693–698.

Grinsztejn, B., Nguyen, B. Y., Katlama, C., Gatell, J. M., Lazzarin, A., Vittecoq, D., et al., (2007). For the protocol 005 team. Safety and efficacy of the HIV-1 integrase inhibitor raltegravir (MK–0518) in treatment-experienced patients with multidrug-resistant virus: A phase II randomized controlled trial. *The Lancet, 369*, 1261–1269.

Gupta, A., Guttikar, S., Shah, P. A., Solanki, G., Shrivastav, P. S., & Sanyal, M., (2015). Selective and rapid determination of raltegravir in human plasma by liquid chromatography-tandem mass spectrometry in the negative ionization mode. *Journal of Pharmaceutical Analysis, 5*(2), 101–109.

Katlama, C., & Murphy, R., (2012). Dolutegravir for the treatment of HIV. *Expert Opinion on Investigational Drugs, 21*(4), 523–530.

Laboulais, C., Deprez, E., Leh, H., Mouscadet, J. F., Brochon, J. C., & Le Bret, M., (2001). HIV-1 integrase catalytic core. *Molecular Dynamics and Simulated Fluorescence Decays, 81*(1), 473–489.

Laboulais, C., Ouali, M., Le Bret, M., & Gabarro-Arpa, J., (2002). Hamming distance geometry of a protein conformational space: Application to the clustering of a 4-ns molecular dynamics trajectory of the HIV-1 integrase catalytic core. *PROTEINS: Structure, Function, and Genetics, 47*, 169–179.

Lee, M. C., Deng, J., Briggs, J. M., & Duan, Y., (2005). Large-scale conformational dynamics of HIV-1 integrase core domain and its catalytic loop mutants. *Biophysical Journal, 88*, 3133–3146.

Liao, C., Marchand, C., Burke, Jr., T. R., Pommier, Y., & Nicklaus, M. C., (2010). Authentic HIV-1 integrase inhibitors. *Future Med. Chem., 2*(7), 1107–1122.

Lins, R. D., Briggs, J. M., Straatsma, T. P., Carlson, H. A., Greenwald, J., Choe, S., & McCammon, J. A., (1999). Molecular dynamics studies on the HIV-1 integrase catalytic domain. *Biophysical Journal, 76*, 2999–3011.

Malet, I., Fourati, S., Morand-Joubert, L., Flandre, P., Wirden, M., Haim-Boukobza, S., et al., (2012). Risk factors for raltegravir resistance development in clinical practice. *J. Antimicrob. Chemother., 67*, 2494–2500.

Miri, L., Bouvier, G., Kettani, A., Mikou, A., Wakrim, L., Nilges, M., & Malliavin, T. E., (2014). Stabilization of the integrase-DNA complex by Mg^{2+} ions and prediction of key residues for binding HIV-1 integrase inhibitors. *Proteins, 82*, 466–478.

Molina, J. M., LaMarca, A., Andrade-Villanueva, J., Clotet, B., Clumeck, N., Liu, Y. P., et al., (2012). For the study 145 team. Efficacy and safety of once-daily elvitegravir versus twice daily raltegravir in treatment-experienced patients with HIV-1 receiving a ritonavir-boosted protease inhibitor: Randomized, double-blind, phase 3, non-inferiority study. *The Lancet Infectious Diseases, 12*, 27–35.

Nair, V., & Chi, G., (2007). HIV integrase inhibitors as therapeutic agents in AIDS. *Rev. Med. Virol., 17*, 277–295.

Rusconi, S., Vitiello, P., Adorni, F., Bruzzone, B., De Luca, A., Micheli, V., et al., (2013). On behalf of the antiretroviral resistance cohort analysis collaborative group. Factors associated with virological success with raltegravir-containing regimens and prevalence of raltegravir-resistance-associated mutations at failure in the ARCA database. *Clin. Microbiol. Infect., 19*, 936–942.

Steigbiegel, R. T., Cooper, D. A., Kumar, P. N., Eron, J. E., Schechter, M., Markowitz, M., et al., (2008). Raltegravir with optimized background therapy for resistant HIV-1 infection. *New England Journal of Medicine, 359*, 339–354.

Summa, V., Petrocchi, A., Bonelli, F., Crescenzi, B., Donghi, M., Ferrara, M., et al., (2008). Discovery of raltegravir, a potent, selective orally bioavailable HIV-integrase inhibitor for the treatment of HIV-AIDS infection. *J. Med. Chem., 51*(18), 5843–5855.

Treviño, A., Cabezas, T., Lozano, A. B., Garcia-Delgado, R., Force, L., Fernandez-Montero, J. M., De Mendoza, C., Caballero, E., & Soriano, V., (2015). Dolutegravir for the treatment of HIV–2 infection. *Journal of Clinical Virology, 64*, 12–15.

Viani, R. M., Alvero, C., Fenton, T., Acosta, E. P., Hazra, R., Townley, E., Steimers, D., Min, S., & Wiznia, A., (2015). Safety, pharmacokinetics, and efficacy of dolutegravir in treatment-experienced HIV-1 infected adolescents forty-eight-week results from IMPAACT P1093. *Pediatric Infectious Disease Journal, 34*(11), 1207–1213.

Vijay, R., Mahajan, P., & Kandwal, R., (2012). Hamming distance-based clustering algorithm. *International Journal of Information Retrieval Research, 2*(1), 11–20.

Wang, C. Y., Yang, C. F., Lai, M. C., Lee, Y. H., Lee, T. L., & Lin, T. H., (1994). Molecular dynamics simulation of a leucine zipper motif predicted for the integrase of human immunodeficiency virus type 1. *Biopolymers, 34*(8), 1027–1036.

Wang, J. Y., Ling, H., Yang, W., & Craigie, R., (2001). Structure of a two-domain fragment of HIV-1 integrase: Implications for domain organization in the intact protein. *EMBO Journal, 20*(24), 7333–7343.

Wang, L. Z., Lee, W. S. U., Thuya, W. L., Soon, G. H., Kong, L. R., Nye, P. L., et al., (2011). Simultaneous determination of raltegravir and raltegravir glucuronide in human plasma by liquid chromatography-tandem mass spectrometric method. *J. Mass. Spectrom., 46*, 202–208.

Weber, W., Demirdjian, H., Lins, R. D., Briggs, J. M., Ferreira, R., & McCammon, J. A., (1998). Brownian and essential dynamic studies of the HIV-1 integrase catalytic domain. *J. Biomol. Struct. Dyn., 16*(3), 733–745.

Yamada, E., Takagi, R., Sudo, K., & Kato, S., (2015). Determination of abacavir, tenofovir, darunavir, and raltegravir in human plasma and saliva using liquid chromatography coupled with tandem mass spectrometry. *Journal of Pharmaceutical and Biomedical Analysis, 114*, 390–397.

Yoshinaga, T., Kobayashi, M., Seki, T., Miki, S., Wakasa-Morimoto, C., Suyama-Kagitani, A., et al., (2015). Antiviral characteristics of GSK1265744, an HIV integrase inhibitor dosed orally or by long-acting injection. *Antimicrobial Agents and Chemotherapy, 59*(1), 397–406.

CHAPTER 15

Köln Model as a Toxicological Procedure for (Some) Metals

SERGIU A. CHICU,[1] MARINA A. TUDORAN,[2,3] and MIHAI *V* PUTZ[2,3]

[1]*Institute of Chemistry Timisoara of the Romanian Academy, Mihai Viteazul Boulevard 24, RO–300223 Timişoara, Romania. Permanent address: Siegstr. 4, 50859 Köln, Germany*

[2]*Laboratory of Structural and Computational Physical Chemistry for Nanosciences and QSAR, Biology–Chemistry Department, West University of Timisoara, Pestalozzi Street No. 44, Timisoara, RO–300115, Romania*

[3]*Laboratory of Renewable Energies-Photovoltaics, R&D National Institute for Electrochemistry and Condensed Matter, Dr. A. Paunescu Podeanu Str. No. 144, Timisoara, RO–300569, Romania*

15.1 DEFINITION

Metals represent the biggest part of the natural chemical elements. Corresponding to this abundance is the importance of knowing their interactions, interferences, and effects on the environment. Because these metals are not toxic "per se," the effects appear only when the internal concentration in the living organisms reach several limits ("it is the dose that makes the effect"). Investigations in this area are complex and so demands and information from apparently distinct domains such as physics, chemistry, and geochemistry, geology, biology, and not last medicine. Köln-Model represents an *in vivo* biological procedure for toxicological investigation of transitional metals Fe, Co, Ni, Cu, Pt, Ru, and Sn using the larvae metamorphosis of marine organism *Hydractinia echinata* (*H. echinata,* Fleming, 1828) as test system (*He*TS) in the polyp. The organism is a colonial marine Hydroid polyp, living on beneath rocks or on bedrock in the North-Eastern Atlantic Ocean, and most commonly, the animal is located on the shell inhabited by the Hermit Crab *Eupagurus bernhardus*. The procedure allows the toxicity quantification and interpretation of chemical derivatives singular or in mixtures, anthropogenic or naturals, which contaminates the water column and/or sediment compartments of marine environment (aquatic) of coast and profound. The study is given its direct contribution also on direct eco-physiological monitorization of these systems, where the larvae multiplication is characteristics for the evolution cycle of mollusks, barnacles, echinoderms, tunicates, etc. The experimental *He*TS fulfill all the criteria formulated by the 3R concept ("Replace," "Reduce," and "Refine") referring to using the animals in experiments: 1) do not possess spine, do not rise ethical problems due to the fact that it stays alive as larvae or polyp during the experiment; 2) do not affect the species or the marine environment as a part of plankton or as sedentary animal; and 3) is an excellent alternative to more complex, rapidly-growing animals for the study of biochemical/physiological aspects of marine pollution.

15.2 HISTORICAL ORIGIN(S)

The research debut/initiation in acute poisoning area of marine environment is represented by the projects "Einfluss von Organische Substanzen, insbesondere von Erdölbestandteilen, auf marine wirbellose Tiere" (1994) (Chicu and Berking, 1997) and "QSAR Untersuchungen a *Hydractinia echinata*" (1997) (Chicu et al., 2000). Using the metamorphosis of the marine organism *H. echinata* larvae as part of zooplankton, on the sessile polyp, the investigations were focused: 1) on determine the discrimination between effects specific for the test organisms and effects which can be found in other test systems with the same order of effectiveness; 2) on QSAR-statements about the effects of toxic substances in marine ecosystems; and 3) on the development of a procedure to calculate or estimate the toxicity of yet not tested substances. The experimental results show that the test–system developed with the occasion allows the toxicity quantification of singular derivatives or in mixture from various classes of substances: hydrocarbons, crude oil and distillation derivatives (Chicu and Berking, 1997), alcohols, and amines (Chicu et al., 2000), dyes (Chicu et al., 2014), pharmaceutical products (Chicu et al., 2016), etc. The organism *H. echinata* is characterized by particular regeneration capacity of the nervous system (Galliot et al., 2009) or of the other organs according to a mechanisms across different phyla until vertebrates (Bradshaw et al., 2015) and is intensively used in compared immunology studies (Frank et al., 2001), pluripotency (Hensel, 2013), or totipotency (Müller et al., 2004).

15.3 NANO-SCIENTIFIC DEVELOPMENT(S)

About 74 (80%) of the 92 known natural chemical elements are generically named metals, 30 of them are transitional metals characterized by the distinctive electron situated in a *d* orbital partially occupied with electrons, and form coordinative combinations. 23 of the transitional metals, e.g., Co, Cu, Ga, Au, Fe, Ni, Pt, Ag, Te, Th, and Sn with a specific gravity at least 5 times that of water (5–7 kg/dm³) are called as heavy metals (Duffus, 2002; Tchounwou et al., 2012).

The study of the combined effects of metal bioavailability and toxicity on human physiology requires, at first, the knowledge of the transfer possibilities from the natural sources and phenomena but mainly anthropogenic to the biotic environment (Boudou and Ribeyre, 1997; Moore and Luoma, 1990), as well as the knowledge of the general or specific relations between metals and environment (Di Toro et al., 2001; Santore et al., 2001). Heavy metals may enter in the human body via food, water, air, smoke (tobacco smoke to include 30 metals from which Cd, As, Pb have known toxicity), or absorption through the skin in agriculture, manufacturing, pharmaceutical, industrial, or residential settings. Due to the increased affinity towards proteins or other biological molecules, heavy metals cannot exist in these systems in their free conformation, being known their accumulation in the soft tissues (Dorne et al., 2011; Fargašová, 1998). This aspect is very important because *in vitro,* the studies with metals are not always leads to relevant results on their reactivity *in vivo* (Foulkes, 2000). The metals effect can be manifested both essential and toxic to humans (Maugh, 1973; Mertz, 1981). Thus, if Fe, Cu, Mn, Zn in traces are nutritionally essential for a healthy life, or if the anti-inflammatory effect of Diclofenac is increased by the presence of some coordinative complex of the transitional metals with Cu, Co, Ni, Mn, Fe(II), Fe(III), Pd (Konstandinidou et al., 1998), iron deficiency can cause anemia, iron overdose can cause hepatocellular necrosis (jaundice, hepatic failure), and human exposed to large amounts of cadmium in food determined a bone disease, known as itai–itai ("ouch–ouch") disease (through of cadmium in food during the latter part of World War II in Japan). Unfortunately, although is known the fact that many metals have adverse effects on the male reproductive function, information about reproductive effects of human exposure to metals is scarce and/or inconsistent (Pizent et al., 2012).

Estuaries and coastal oceans represent the marine areas intensively affected by the human activity. Here, the anthropogenic product quantity transported by the continental waters permanently modify the environment toxicity level where, for example, the Tributyltin (TBT) presence is very dangerous, the compound being well known as an extremely toxic antifouling agent (Maguire, 2000; Tchounwou et al., 2012). However, these areas represent the normal habitat of a diverse fauna: mollusks, shellfish, euryhaline (organisms able to adapt to a wide range of salinities), as well as fish species *Poecilia sphenops* and *Carcinus maenas*, or *Salmo salar* and *Anguilla anguilla* which are moving in profound waters, organisms mostly with economic importance for human (food). A study made by Hall, Jr. (Hall, Jr. and Anderson, 1995) shows that toxicity of some heavy metals such as Cd, Cr, Cu, Hg, Ni, and Zn increase with decreasing the salinity, and the alkalinity variations change the marine environment toxicity because it affects the speciation of metal ions by increasing ion-pair formation, thus decreasing free metal ion concentration (Jackson et al., 2000).

The importance given to toxicity by metals is illustrated by countless experimental studies *in vivo* and with other larvae, organisms or plants: *Crambe crambe* (marine sponge) (Becero et al., 1995), *Daphnia magna* (*D. magna*) (crustacean) (Barata et al., 1999), *Photobacterium phosphoreum* (*P. phosphoreum*) (bacterium) and *Vibrio fischeri* (*V. fischeri*) (bacterium) (Newman and McCloskey, 1996), *Bufo americanus* (*B. americanus*) (toad) and *Rana sphenocephala* (*R. spenocephala*) (frog) (James et al., 2005), *Caenorhabditis elegans* (*C. elegans*) (nematode) (Tatara et al., 1997, 1998), *Helianthus annuus* (*H. annuus*) (sunflower callus) (Enache et al., 2000, 2003), *Lemna minor* (*L. minor)* (duckweed or lesser duckweed) (Naumann et al., 2007). The scientific literature presents also the existence of some contradicting experimental results, e.g., in case of aluminum metabolisms (Venturini–Soriano and Berthon, 1998).

In the same time with the experimental test systems, there were also developed models regarding the toxicity predictions, an important aspect regarding the practical possibility of research: the Köln-Model investigates toxicity predictions based on *ESIP*'s iterations with Timisoara (Spectral–SAR) formulation provide a unified picture of computed activity in order to assess the toxicity modeling and inter-toxicity correlation maps for aquatic organisms against paradigmatic organic compounds (Chicu and Putz, 2009); Newman et al., (1997) and Tatara et al., (1998) develop a method of predicting relative metal toxicity with ion characteristics QICAR's and Todeschini et al., (1996) applies QSAR's for studying some organic compounds toxicity, as well as organo-tin's. In the flora domain, Walker et al., (2007) develop Quantitative Cationic-Activity Relationships (QCARs) for toxicity predictions of metal ions from physicochemical properties and natural occurrence levels.

The tested *Fe, Co, Ni, Cu, Pt, and Ru* derivatives have been prepared on Institute of Chemistry Timisoara of Romanian Academy. Tin (Sn) compounds were purchased from catalogs. All tested derivatives were used such as, from solutions of known concentrations, in synthetic sea water or methanol.

The organism is routinely cultured in usual biology labs so that fertilized eggs, embryos, and larvae are available daily. Larvae were induced to undergo metamorphosis into primary polyps by pulse treatment with Cs^+ (Berking, 1988a, 1998; Berking, and Walther 1994; Leitz, 1998; Walther, 1995; Walther et al., 1996) or using LWamide peptides (Frank et al., 2001; Gajewski et al., 1996; Leitz et al., 1994a, b; Müller and Leitz, 2002; Müller et al., 2004a; Plickert et al., 2003; Schmich et al., 1998). For a specific type of induction and considering the substrate nature and concentration, the larvae are metamorphosed in polyps faster or slower or stayed as they are. The process is easier to watch under the usual microscope from the lab and allows a clear and unevenly evaluation within the same experiment to what extent the test-substance presents repressive/inhibitive or stimulative influences on metamorphosis. One can states that the experimental results obtained until present with *He*TS regarding the toxicity of some singular derivatives or in the mixture, may not be obtained through direct experiments on mammalians.

The Köln-Model based on *He*TS consists of utilizing the *ESIP*'s algorithm and its corresponding iteration (while considering basically the constant toxicity in the range of ± 0.5 log u.) (Chicu et al., 2000) for 1) the evolutive toxicity C-calculation, according to the specific molecular substructures (Chicu et al., 2000; Chicu and Simu, 2009), 2) iso-toxicity (Ciso) calculation according to the specific molecular structures which form a compounds class (Chicu et al., 2008, 2011, 2014), 3) *in silico* toxicity calculation for the untested derivatives within these series, using, in all cases, directly the toxicity values experimentally measured *M* at pH of 8.2 very close to ~8.0 specific to the natural marine environment.

15.4 NANO-CHEMICAL APPLICATION(S)

15.4.1 COMPLEXES OF FE, CO, NI, CU, PT, RU, AND ORGANOMETALLIC DERIVATIVES OF SN

Metallic and organometallic complex represent two different classes of derivatives even if, in both cases, the direct bond between the metal element and the organic molecular structure is of covalent nature. The semantic border clearly differentiates these classes, although in reality, for example, a series of Pt halogenated derivatives are considered metallic complex while Sn halogenated derivatives are considered organometallic complex.

According to the experimental results from Table 15.1, the toxicities transitional elements Fe(II), Co(II), Ni(II), Cu(II), Pt(II), Ru(II) and post-transitional Sn (IV), can be appreciated/interpreted as due to the "metal" or "metal-complex" function which determines (occurrence/existence) metal/element class toxicity-isotoxicity.

The complex structures (without considering the crystallization water) consist in metallic cations (Me) coordinated with neutral or ionic molecular ligands from the inner circle (Li) of the type BAMP (N,N'–Bis(antipyril-4-methyl)-piperazine), TAMEN (N,N'-Tetra(antipyril-4-methyl)-ethylendi-amine), SCHIFF, amino-phosphonic or NH_3 and/or surrounded in the outer/exterior circle (Le) by anions monoatomic or polyatomic of the type –Cl, –NCS, etc.

By considering as model complexes of the piperazine type, the Me involvement in an interaction with the cell receptor (R) is preceded by a series of possible dissociation equilibriums, re-association, dissociation of complex in the presence of water (I):

$$\text{(I) } (Me_2Li)Le_4 \leftrightarrow (Me_2Li)^{4+} + 4Le^-$$
$$(Me_2Li)^{4+} \rightarrow 2Me^{2+} + Li^o$$

where (Le) belongs particularly to the hard electrolytes, such that their dissociation is fast. Although [Le]"[Me_2Li], an priority R↔(Le) interaction or allosteric or cooperatively (positively or negatively) influenced by (Le) cannot be considered due to the fact that, for the same type of complex will leads to different toxicity M values. Neither the R↔(Li) interaction can be considered because [Li]‹‹[Le] and always with 50% reduced than [Me] by total dissociation.

Instead, the $(Me_2Li)^{4+}$ dissociation by forming the clathrate cage between Me, (Li) and H_2O, respectively, is slower than (Le) case because the coordinative (covalent) bonds from its interior are stronger comparatively with the hydrogen bonds with the water molecules. In these conditions, concentration of [Me_2Li] is more increased than [Me] or [Li] and will be interacting highly priority with R (II) which has the biggest concentration:

$$\text{(II) } R + (Me_2Li)^{4+} \leftrightarrow (R\cdots Me_2Li)^{4+} \rightarrow (RMe_2Li)$$

The interaction mechanism thus implies the Li direct influence over the effectiveness and consists of two main stages:

1. *The fast stabilization of the new complex* (RMe$_2$Li) by:
    a) Adopting a 3D conformation of *R* corresponding to the Me$_2$Li structure.
    b) Building new coordinative connections σ by *R* reactive biostructure participation with two atomic orbitals occupied with an electron pair of the *p*-type on *d* vacant hybridized orbitals of the 2Me^{2+} atoms. In the same time, the positive charge from the donor atom can be/is compensated by doubling the σ coordinative bond with a π (d_π–p_π) bond resulted by participation of an electrons pair from the metallic orbital *d* on the vacant orbital *p*. The (d_π–p_π) bond is known in the interaction Me↔Li, if Li participates by *C* atoms or (d_π–d_π) in the interaction of *P* or As,
    c) The simultaneous formation of multiple bonds between *R* and Me$_2$Li of the van der Waals type, hydrogen bonds (important in tridimensional protein structures stabilization), fast reversible covalent bonds or Me^{2+}→π (covalent cation-π bonds formation as a specific interaction with the *R* hydrophobic area) (Mohan et al., 2013) by the electrons supplied by the aromatic rings of the enzyme structured in R. Still it is hard to assume that the (RMe$_2$Li) transition phase stabilization is also accompanied by the (Le) groups participation, strongly nucleophilic if we consider even only the steric impediments generated by –J$^-$, –SO$_4^{2-}$, etc.
2. *The enzymatic splitting* of (RMe$_2$Li) can occur differently but basically is inducted by direct involvement of hydroxyl groups or of the protein buffer system (R-ONa)/(R-OH) formed in the weak basic medium of the experiment (pH = 8.2). By neutralizing the groups presented in the reaction medium the R-protein returns on the anionic initial conformation capable of making a new transition phase, in the same time with a water molecule release and Me$_2$Li dissociate and its partial or total metabolized.

One can say that the RMe$_2$Li complex stabilization occurs by constructing all the possible bonds between the two partners through which the bidirectional or mutual exchange "information-impulse" occur at the subatomic level: towards the cell with positive or negative effect by the substrate and the other way around, the substrate receives the impulse of structural modification. From the enzymatic point of view, this mechanism represents an oxidation-reduction reaction in the presence of water. Generally speaking all transformations occur with speeds so high that can be considered as being extemporaneous.

15.4.2   COMPLEXES OF FE, CO, NI, CU, RU

The measured effectiveness increase in the Fe<Co<Ni<Cu order for BAMP complexes, and are aligned to the Pauling electronegativity scale; in case of complexes with TAMEN the order is inverse, Fe>Co. The atomic size of Fe and Co being close (atomic radius of 126 and 125, respectively), determined by the speed is the dissociation phase of (Me_2Li):

a) Faster for Co in BAMP case (otherwise determine the highest effectiveness in case of Co).

b) Faster for TAMEN in Fe complex as against the one with Co, probably from structural reasons: posses four antipyretic cycles as against the 2 of BAMP.

One can state as in the last instance, and the toxicity values M represent the stability differences of the transition compounds (Me_2Li) reflected by the dissociation-hydrolysis speeds (V) of the Li:

$$V((Co_2BAMP)^{4+} \rightarrow 2Co^{2+} + BAMP^\circ) > V((Fe_2BAMP)^{4+} \rightarrow 2Fe^{2+} + BAMP^\circ)$$
$$V((Fe_2TAMEN)^{4+} \rightarrow 2Fe^{2+} + TAMEN^\circ) > V((Co_2TAMEN)^{4+} \rightarrow 2Co^{2+}$$
$$+ TAMEN^\circ)$$

The Ni complex (Schiff) respectively Ni and Ru (amino-phosphonic) are also characterized by iso-toxicity generated by (Li) probably due to the common substructure $(C_6H_4)O^-$ and less due to the iminic, aminic, phosphonic or alcoholic groups.

15.4.3   COMPLEXES OF PT

Although the Pt electronegativity on the Pauling scale is the highest (2.28), the toxicity value M shows that there is no direct proportionality between the two sizes: effectiveness/toxicity is the direct result of a complex interaction between the cellular R and a molecular or atomic-molecular structure, while the electronegativity is referring to a characteristic linked to the singular atomic structure.

15.4.4   COMPLEXES OF SN

Sn represents an element which belongs to the IVth group together with Ge, Pb, and Bi, posses a small cationic volume with a high charge. With carbon or chlorine atoms, forms through a sp³ type hybridization 4σ polar bonds according to the different electronegativities presented on the Pauling scale: Cl (4.0) and C (2.5) as against Sn (1.9).

The most important aspect regarding the experimental results with the two test-systems *in vivo H. echinata* and *D. magna* is the possibility to evaluate, by comparison, the tendency of toxicity modification and so of R↔Sn interaction according to the studied derivatives structure. Worth mention that although not always the individual results are the same, and on some cases, the M values even coincide, their variation shows that in both test systems, respectively in the derivatives groups structurally shared, the toxicity evolution is identical, it remains in a level.

It is thus confirmed that the biochemical reaction set is, if not identical, than similar in all living organisms and from here the results which allow information transfer between two test-systems with different organisms.

Thus it appears that:

1. The toxicity of the mono-, di-, trialkyltin si phenyltin-chlor derivatives Nos. 1–16, 23–26 grows by the lipophilic effect influence paralleling the hydrocarbonate dimension; the phenomenon is absent in the case of the alkyl-trichloro derivatives Nos. 17–20, leading with the conclusion that the three chlorate atoms produce a remarkable influence by multiple interactions so canceling the previous effect (Auffinger et al., 2004).
2. In the case of the series of the mono- and di-halogenated derivatives Nos. 6–11, 14–16, where the number of the saturated carbons (Cs) vary in between 6–24 and manifest an effectivity of Cisco type. The last feature is especially significant since by starting with the hydrocarbonate structure containing 9Cs the following are observed: a) the increase/modification of hydrocarbon-chain does not play an essential role in Sn atom screening. Also, the number of 9Cs which marks the level of this screening is in the domain of ~8–12 Cs identified as the critical interface for the molecular hydrophilic-lipophilic equilibrium (expressed, as example, by the correlations of the M values with log$P$) observable in case of hydrocarbons or saturated amines (Chicu et al., 2000) and reported per se within the alkylated derivatives of the phenol (Chicu and Simu, 2009) equally for the *H. echinata* as well for test-systems *Tetrahymena pyriformis* (*T. pyriformis*) (protozoa), *Pimephales promelas* (*P. promelas*) (fish) and *D. magna*; and b) starting with the cardinal 3–4Cs the effects +I remains unaltered.

3. The phenyl substituent behaved differently by that of alkyl: a) as a consequence of an equilibrium between the volume and the sterical hindrance the Sn atomic shielding is different so that effectivity of the derivative Tetraphenyltin is higher than that of the Tetramethyltin derivative; b) the specific delocalization of π electrons allows a double valuation of the σ bond with the Sn atom (a $p\pi$–dp type bond) by involving of a single aromatic nucleus in a linear structure with one Sn atom or with one Sn atom and one Cl atom. Still, the increased effectiveness of Tri-phenyltin chloride comparatively with phenyl derivatives 22–24 is seen as a stronger and faster R↔Sn interaction, as a result of π electrons involvement from the free phenyls in bonds with the lipophilic areas of R. Comparatively, the lower effectivity of the Tetraphenyltin is due in greater respect due to the sterical hindrance exercised by the phenyl groups or by the shielding of the Sn atom.

4. Bis-(tri-n-butyltin) sulfate presents a toxicity with an order of magnitude higher than Bis-(tri-n-butyltin)oxide due mostly to the great influence exercised by the sulfonic group on the stereoelectronic equilibrium on the Sn atomic level. It is more likely that these derivatives interaction mechanism with R occurs in two successively steps with intermediate formation of Bu_3SnOH, which along with Bu_3SnCl and Bu_3SnCO_3 represent one of the most preeminent species of Sn in seawater at pH 8 (WHO, 1990).

15.5 OPEN ISSUES

Every element-metal species as an atom or as any combination, represents/behaves as a functional group in the same way as amino, carboxyl, etc.; The toxicity differences are principally due to the different nature of the metal atom. The measured toxicities M and calculated Ciso for a series of Fe, Co, Ni, Cu, Pt and Sn derivatives are determined according to The Koln-Model; they show a common characteristic of each element, i.e. class-iso-toxicity element. The Fe, Co, Ni, Cu, Pt derivatives have structures as of metallic complexes (i.e. the Me-O, -N bonds are coordinated), while the Sn derivatives have structures defined as organo-metallics (i.e. the Me-C bond is polar, whereas the carbon, having high electron density, is nucleophilic).

The metal atom toxicity from a complex is due to: a) the ligand from the inner sphere (Li), active participant in interaction with R with is structurally and geometrically-spatial molded depending on its nature and dimension, and b) do not depend on the geometry or initial spatial-structure of complex or of the nature and number of ligands from the ionic outer sphere (Le).

The difference in effectiveness between complexes of the same metal atom represents in principle, the existent differences between their hydrolysis reaction speeds. For the transitional metals Fe, Co, Ni, Cu, Pt, and Ru, the hydrolytic dissociation of $(Me_2Li)^{4+} \rightarrow 2Me^{2+} + Li^{\circ}$ can be appreciated as the speed determinant phase.

The water presence is absolutely necessary for both formation and transformation of atoms, ions, and molecules through biochemical reactions of activation-deactivation, combination-recombination and also by participating on physical processes such as dissolution, membrane processes, transport, etc. The experiments conducted with synthetic marine water exclude the organic substance presence of exogenously nature so that the effectiveness measured in case of $NH_4Fe(SO_4)_2$ $12H_2O$ and $Cu(NO_3)_2$ $3H_2O$ derivatives are due exclusively to the metallic ions Fe^{3+} and Cu^{2+} even in the presence of the $-NH_4^+$ $-(SO_4)^{2-}$ and $-(NO_3)^-$ ions excess. This affirmation is confirmed by the different toxicity of $Mg(CH_3COO)_2$ $4H_2O$ and $Zn(CH_3COO)_2$ $2H_2O$ in the presence of the same ion $-CH_3COO^-$ in excess.

The interaction $(R \cdots Me_2Li)^{4+}$ pre-stabilization is slower, but the stabilized complex (RMe_2Li) hydrolyze faster for initiation of new cycles of "pulsar" type.

The toxicity predictions *in silico* Table 15.1 (italic script) are possible by similarities and parallelism between the existent structures: if, for example, the BAMP-pyrimidinic derivatives are iso-toxic in Fe and Co case, this must be valid also in Cu or Ni case. In the case of organometallic derivatives, the iso-toxicity is due to the screening of Sn (IV) atom sp^3 hybridized combined with the electronic influences of substituents (Tables 15.1a and 15.1b).

TABLE 15.1a The Experimental Measured *M* (mol L^{-1}), by Comparison to the Calculated Average C, Differences Δ=M–C and *in silco* Values According to the Iteration of the Köln-Model, Obtained with *Hydractinia Echinata* and *Daphnia Magna* Test-Systems for Some Derivatives of Transitional Metals Fe, Co, Ni, Cu, Ru, and Pt

No	Fe-BAMP Komplex	M	Ciso	Δ	MW
1	Fe$_2$(BAMP)Cl$_4$	3.89	3.82	0.07	740.6
2	Fe$_2$(BAMP)Cl$_6$	3.53	3.82	−0.29	811.6
3	Fe$_2$(BAMP)Py$_2$Cl$_4$	4.05	3.82	0.23	896.5
4	Fe$_2$(BAMP)Py$_2$Cl$_6$	4.02	3.82	0.20	966.3
	Fe$_2$(BAMP)(ClO4)$_2$		*3.82*		
	Fe$_2$(BAMP)(NCS)$_2$		*3.82*		
	Fe$_2$(BAMP)(NCS)$_4$		*3.82*		

TABLE 15.1a *(Continued)*

No	Fe-BAMP Komplex	M	Ciso	Δ	MW
	$Fe_2(BAMP)(Ac)_4$		*3.82*		
5	$Fe_2(BAMP)(SO4)_2$	3.59	*3.82*	−0.23	790.3
	$Fe_2(BAMP)J_4$		*3.82*		
	$Fe_2(BAMP)J_6$		*3.82*		
	Fe-TAMEN Komplex				
	$Fe_2(TAMEN)Cl_4$		*5.42*		
	$Fe_2(TAMEN)Cl_6$		*5.42*		
	$Fe_2(TAMEN)Py_2Cl_4$	5.66	*5.42*	0.25	1271.03
6	$Fe_2(TAMEN)Py_2Cl_6$	5.17	*5.42*	−0.25	1342.03
7	$Fe_2(TAMEN)(ClO4)_2$		*5.42*		
	$Fe_2(TAMEN)(NCS)_2$		*5.42*		
	$Fe_2(TAMEN)(NCS)_4$		*5.42*		
	$Fe_2(TAMEN)(Ac)_2$		*5.42*		
	$Fe_2(TAMEN)(SO_4)_2$		*5.42*		
	$Fe_2(TAMEN)J_4$		*5.42*		
	$Fe_2(TAMEN)J_6$		*5.42*		
	Co-BAMP Komplex				
8	$Co_2(BAMP)Cl_4$	3.96	*4.07*	−0.11	746.46
	$Co_2(BAMP)Cl_6$		*4.07*		
9	$Co_2(BAMP)Py_2Cl_4$	4.06	*4.07*	−0.01	904.46
	$Co_2(BAMP)Py_2Cl_6$		*4.07*		
	$Co_2(BAMP)(ClO_4)_2$		*4.07*		
10	$Co_2(BAMP)(NCS)_2$	4.09	*4.07*	0.02	661.53
	$Co_2(BAMP)(NCS)_4$		*4.07*		
11	$Co_2(BAMP)(Ac)_4$	4.18	*4.07*	0.11	840.46
	$Co_2(BAMP)(SO_4)_2$		*4.07*		
	$Co_2(BAMP)J_4$		*4.07*		
11	$Co_2(BAMP)J_6$		*4.07*		
	Co-TAMEN Komplex				
12	$Co_2(TAMEN)Cl_4$	4.28	*3.88*	0.40	1120.99
	$Co_2(TAMEN)Cl_6$		*3.88*		
	$Co_2(TAMEN)Py_2Cl_4$		*3.88*		

TABLE 15.1a *(Continued)*

No	Fe-BAMP Komplex	M	Ciso	Δ	MW
	$Co_2(TAMEN)Py_2Cl_6$		*3.88*		
13	$Co_2(TAMEN)(ClO_4)_2$	3.73	3.88	−0.15	572.34
	$Co_2(TAMEN)(NCS)_2$		*3.88*		
14	$Co_2(TAMEN)(NCS)_4$	3.62	3.88	−0.26	466.07
	$Co_2(TAMEN)(Ac)_2$		*3.88*		
	$Co_2(TAMEN)(SO_4)_2$		*3.88*		
	$Co_2(TAMEN)J_4$		*3.88*		
	$Co_2(TAMEN)J_6$		*3.88*		
	Ni-BAMP Komplex				
	$Ni_2(BAMP)Cl_4$		*4.85*		
	$Ni_2(BAMP)Cl_6$		*4.85*		
	$Ni_2(BAMP)Py_2Cl_4$		*4.85*		
	$Ni_2(BAMP)Py_2Cl_6$		*4.85*		
	$Ni_2(BAMP)(ClO4)_2$		*4.85*		
	$Ni_2(BAMP)(NCS)_2$		*4.85*		
	$Ni_2(BAMP)(NCS)_4$		*4.85*		
15	$Ni_2(BAMP)(Ac)_2$	4.96*	4.85	0.10	722.02
16	$Ni_2(BAMP)(SO_4)_2$	4.75	4.85	−0.10	796.02
	$Ni_2(BAMP)J_4$		*4.85*		
	$Ni_2(BAMP)J_6$		*4.85*		
	Cu-BAMP Komplex				
17	$Cu_2(BAMP)Cl_4$	5.72*	6.05	−0.33	755.70
	$Cu_2(BAMP)Cl_6$		*6.05*		
	$Cu_2(BAMP)Py_2Cl_4$		*6.05*		
	$Cu_2(BAMP)Py2Cl_6$		*6.05*		
	$Cu_2(BAMP)(ClO4)_2$	5.92	6.05	−0.13	1011.70
	$Cu_2(BAMP)(NCS)_2$		*6.05*		
18	$Cu_2(BAMP)(NCS)_4$	6.36*	6.05	0.29	1413.70
	$Cu_2(BAMP)(Ac)_4$	6.32	6.05	0.27	849.60
	$Cu_2(BAMP)(SO_4)_2$		*6.05*		

TABLE 15.1a *(Continued)*

No	Fe-BAMP Komplex	M	Ciso	Δ	MW
19	$Cu_2(BAMP)J_4$	5.94*	6.05	−0.11	1121.30
	$Cu_2(BAMP)J_6$		*6.05*		
	CuI				
20	**BAMP**	**3.88**			486.60
	N.N'-Bis[antipyrilmethyl]piperazin				
21	**TAMEN**	4.87			861.13
	N.N'-Tetra[antipyrilmethyl]ethylendiamin				
22	Antipyrine	1.98			188.23
23	Piperazine		*2.74*		86.14
24	1.2-Diaminoethane		*2.74*		60.10
25	$NH_4Fe(SO_4)_2 \cdot 12H_2O$	3.27			482.19
	$NH_4Fe(SO_4)_2$	3.01			266.01
26	$Cu(NO_3)_2 \cdot 3H_2O$	5.35			295.60
	$Cu(NO_3)_2$	5.27			241.60
27	$Mg(CH_3COO)_2 \cdot 4H_2O$	2.26			214.40
	$Mg(CH_3COO)_2$	2.08			142.40
28	$Zn(CH_3COO)_2 \cdot 2H_2O$	4.74			219.50
	$Zn(CH_3COO)_2$	4.66			183.48
	Ni. Ru-Complex				
29	C26H20O14N2S2Na2P2]Ni2+	4.42	4.28	−0.01	906.71
30	C26H18O8N2S2Na2]Ni2+	4.27	4.28	0.14	654.71
31	C28H46O10N2P2Na4]Ru2+	4.15	4.28	−0.13	768.50

No	CAS	Pt-Complex	M	Ciso	Δ	MW
1	10025–99–7	K_2PtCl_4 PtII	4.36	4.41	−0.05	415.09
2	16921–30–5	K_2PtCl_6 PtIV	4.77	4.41	0.36	486.01
3	15663–27–1	Cisplatin cis-$(NH_3)_2PtCl_2$	4.32	4.41	−0.09	300.05
4	14913–33–8	Transplatin trans-$(NH_3)_2PtCl_2$	4.18	4.41	−0.23	300.05
5	13820–46–7	$[(NH_3)_4Pt]$ $PtCl_4$	4.44	4.41	0.03	600.09

TABLE 15.1b The Same Type of Data as in Table 15.1a, Here for the Post-Transitional Metal Sn

No	CAS	Sn compounds	H. echinata			M	D. magna		logP	MW
			M	C/Ciso	Δ		C/Ciso	Δ		
1	594–27–4	Tetramethyltin	3.40*	6.73	−3.33	3.65	5.61	−1.96	3.49	178.83
2	2176–98–9	Tetra-n-propyltin	5.72	6.73	−1.01	5.39	5.61	−0.22	7.74	291.06
3	1461–25–2	Tetra-n-butyltin	6.56	6.73	−0.17	5.37	5.61	−0.24	6.64	347.16
4	813–19–4	Hexa-n-butyl ditin	6.76	6.73	0.03				16.70	580.08
5	1066–45–1	Trimethyltin chloride	3.73	6.73	−3.00	5.62	5.61	0.01	2.06	199.27
6	994–31–0	Triethyltin chloride				6.04	5.61	0.43	3.23	241.33
7	2279–76–7	Tri-n-propyltin chloride	7.11*	6.73	0.38	6.88	5.61	1.27	4.40	283.41
8	1461–22–9	Tri-n-butyltin chloride	6.77	6.73	0.04	7.40	5.61	1.79	4.53	325.51
9	3342–67–4	Tri-n-pentyltin chloride	7.90	6.73	1.17				6.74	367.57
10	3091–32–5	Tri-cyclohexyltin chloride	6.71	6.73	−0.02				7.17	403.62
11	2587–76–0	Tri-n-octyltin chloride	5.23	6.73	−1.50				9.29/2.14*	493.82
12	753–73–1	Dimethyltin dichloride	5.76*	6.73	−0.97	3.40	5.61	−2.21	4.79	219.67
13	866–55–7	Diethyltin dichloride				4.80	5.61	−0.81	2.95	247.74
14	683–18–1	Di-n-butyltin dichloride	6.63	6.73	−0.10	5.53	5.61	−0.08	1.97/1.89*	303.84
15	74340–12–8	Di-n-heptyltin dichloride	6.52	6.73	−0.21				6.85	388.07
16	3542–36–7	Di-n-octyltin dichloride	6.75	6.73	0.02				7.63	416.05
17	993–16–8	Methyltin trichloride	4.40	4.45	−0.05	3.27	3.52	−2.34	−0.43	240.08
18	1118–46–3	n-Butyltin trichloride	3.80	4.45	−0.65	3.76	3.52	−1.85	0.41	282.17
19	59344–47–7	n-Heptyltin trichloride	4.48	4.45	0.03				4.61	324.26
20	3091–25–6	n-Octyltin trichloride	4.04	4.45	−0.41				5.00/3.72*	338.29

TABLE 15.1b *(Continued)*

No	CAS	Sn compounds	H. echinata			D. magna			logP	MW
			M	C/Ciso	Δ	C/Ciso	Δ	M		
21	56-35-9	Bis-(tri-n-butyltin)oxide	7.60*					7.61	3.84*	596.10
22	26377-04-8	Bis-(tri-n-butyltin)sulfate	8.50*						10.04	676.14
23	1124-19-2	Phenyltin trichloride	4.76	4.45	0.31				1.77	302.17
24	1135-99-5	Di-phenyltin dichloride						5.72		343.81
25	639-58-7	Tri-phenyltin chloride	7.85*					7.30		385.47
26	595-90-4	Tetraphenyltin	4.55*	4.45	0.10				4.39	427.13

*Examples of average measured values characterizing the reproducibility of the experimental measurements: 1 (3.34; 3.46); 7 (6.82; 7.40); 12 (5.70; 5.82); 21 (7.51; 7.68); 22 (8.70; 8.30); 25 (7.77; 7.92); 26 (4.51; 4.59).

KEYWORDS

- **class isotoxicity**
- ***Hydractinia echinata* test system**
- **Köln model**
- **transitional metals**

REFERENCES AND FURTHER READING

Auffinger, P., Hays, F. A., Westhof, E., & Shing, H. P., (2004). Halogen bonds in biological molecules. *Proceedings of the National Academy of Sciences of the United States of America, 101*, 16789–16794. doi: 10.1073/pnas.0407607101.

Barata, C., Baird, D. J., & Markich, S. J., (1999). Comparing metal toxicity among *daphnia magna* clones: An approach using concentration-time-response surfaces. *Archives of Environmental Contamination and Toxicology, 37*, 326–331.

Becerro, M. A., Uriz, M. J., & Turon, X., (1995). Measuring toxicity in the marine environment: Critical appraisal of three commonly used methods. *Experientia, 51*, 414–418.

Berking, S., (1988a). Ammonia, tetraethylammonium, barium, and amiloride induce metamorphosis in the marine hydroid hydractinia. *Wilhelm Roux's Archives of Developmental Biology, 197*, 1–9.

Berking, S., (1998). Hydrozoa metamorphosis and pattern formation. *Current Topics in Developmental Biology, 38*, 81–131.

Berking, S., & Walther, M., (1994). Control of metamorphosis in the hydroid hydractinia. In: Davey, K. G., Peter, R. E., & Tobe, S. S., (eds.), *Perspectives in Comparative Endocrinology* (pp. 381–388). National Research Council of Canada, Ottawa.

Boobis, A. R., & Verger, P., (2011). Human risk assessment of heavy metals: Principles and applications. *Metal Ions in Life Sciences, 8*, 27–60.

Boudou, A., & Ribeyre, F., (1997). Aquatic ecotoxicology: From the ecosystem to the cellular and molecular levels. *Environmental Health Perspectives, 105*(1), 21–35.

Bradshaw, B., Thompson, K., & Frank, U., (2015). Distinct mechanisms underlie oral vs. aboral regeneration in the cnidarian. *Hydractinia echinata. Elife, 4*, e05506.

Chicu, S. A., & Berking, S., (1997). Interference with metamorphosis induction in the marine Cnidaria *Hydractinia echinata* (Hydrozoa): A structure-activity relationship analysis of lower alcohols, aliphatic, and aromatic hydrocarbons, thiophenes, tributyl-tin, and crude oil. *Chemosphere, 34*(8), 1851–1866.

Chicu, S. A., & Putz, M. V., (2009). Köln-Timisoara activity combined models toward interspecies toxicity assessment. *International Journal of Molecular Sciences, 10*, 4474–4497.

Chicu, S. A., & Simu, G. M., (2009). *Hydractinia echinata* test system. I. Toxicity determination of some benzenic, biphenilic, and naphthalenic phenols. Comparative SAR-QSAR study. *Rev. Roum. Chim., 54*, 659–669.

Chicu, S. A., Funar-Timofei, S., & Simu, G. M., (2011). *Hydractinia echinata* test system. II. SAR toxicity study of some anilide derivatives of Naphthol-AS type. *Chemosphere, 82*, 1578–1582, doi: 10.1016/j.chemosphere.2010.11.057.

Chicu, S. A., Herrmann, K., & Berking, S., (2000). An approach to calculate the toxicity of simple organic molecules on the basis of QSAR analysis in *Hydractinia echinata* (Hydrozoa, Cnidaria). *Quantitative Structure-Activity Relationships, 19*, 227–236.

Chicu, S. A., Munteanu, M., Cîtu, I., Şoica, C., Dehelean, C., Trandafirescu, C., et al., (2014). The *Hydractinia echinata* Test-System. III: Structure-toxicity relationship study of some Azo-, Azo-anilide, and diazonium salt derivatives. *Molecules, 19*, 9798–9817. doi: 10.3390/molecules19079798.

Chicu, S. A., Schannen, L., Putz, M. V., & Simu, G. M., (2016). *Hydractinia echinata* Test-System. IV. Toxicological synergism for human pharmaceuticals in mixtures with iodoform. *Ecotoxicology Environmental Safety, 134*, 80–85. doi: 10.1016/j.ecoenv.2016.08.014.

Di Toro, D. M., Allen, H. E., Bergman, H. L., Meyer, J. S., Paquin, P. R., & Santore, R. C., (2001). Biotic ligand model of the acute toxicity of metals. 1. Technical basis. *Environmental Toxicology and Chemistry, 20*(10), 2383–2396.

Dorne, J. L., Kass, G. E., Bordajandi, L. R., Amzal, B., Bertelsen, U., Castoldi, A. F., et al., (2002). "Heavy metal": a meaningless term? *Pure and Applied Chemistry, 74*, 793–807.

Duffus, J. H., (2002). "Heavy metal": a meaningless term? *Pure and Applied Chemistry, 74*, 793–807.

Enache, M., Dearden, J. C., & Walker, J. D., (2003). QSAR analysis of metal ion toxicity data in sunflower callus cultures (Helianthus annuus "Sunspot"). *QSAR & Combinatorial Science, 22*, 234–240.

Enache, M., Palit, P., Dearden, J. C., & Lepp, N. W., (2000). Correlation of physicochemical parameters with toxicity of metal ions to plants. *Pest Management Science, 56*, 821–824.

Fargašová, A., (1998). Root growth inhibition, photosynthetic pigments production, and metal accumulation in *Sinapis alba* as the parameters for trace metal effect determination. *Bulletin of Environmental Contamination and Toxicology, 61*, 762–769.

Foulkes, E. C., (2000). Transport of toxic heavy metals across cell membranes. *Proceedings of the Society for Experimental Biology and Medicine, 223*, 234–240.

Frank, U., Leitz, T., & Mueller, W. A., (2001). The hydroid *Hydractinia echinata*: A versatile, informative cnidarian representative. *BioEssays, 23*, 963–971.

Frank, U., Leitz, T., & Müller, W. A., (2001). My favorite model organism: *Hydractinia echinata*. *BioEssays, 23*, 963–971.

Gajewski, M., Leitz, T., Schloßherr, J., & Plickert, G., (1996). LWamides from Cnidaria constitute a novel family of neuropeptides with morphogenetic activity. *Wilhelm Roux's Archives of Developmental Biology, 205*, 232–242.

Galliot, B., Quiquand, M., Ghila, L., De Rosa, R., Miljkovic-Licina, M., & Chera, S., (2009). Origins of neurogenesis, a cnidarian view. *Developmental Biology, 332*, 2–24.

Hall, Jr. L. W., & Anderson, R. D., (1995). The influence of salinity on the toxicity of various classes of chemicals to aquatic biota. *Critical Reviews in Toxicology, 25*(4), 281–346.

Hensel, K., (2013). *Wnt Signaling in Hydractinia Stem Cells*. School of natural sciences and regenerative medicine institute (REMEDI), National University of Ireland Galway, Galway, Ireland.

Jackson, B. P., Lasier, P. J., Miller, W. P., & Winger, P. W., (2000). Effects of calcium, magnesium, and sodium on alleviating cadmium toxicity to *Hyalella azteca*. *Bulletin of Environmental Contamination and Toxicology, 64*, 279–286.

James, S. M., Little, E. E., & Semlitsch, R. D., (2005). Metamorphosis of two amphibian species after chronic cadmium exposure in outdoor aquatic mesocosms. *Environmental Toxicology and Chemistry, 24*(8), 1994–2001.

Konstandinidou, M., Kourounakis, A., Yiangou, M., Hadjipetrou, L., Kovala-Demertzi, D., Hadjikakou, S., & Demertzis, M., (1998). Anti-inflammatory properties of diclofenac transition metalloelement complexes. *Journal of Inorganic Biochemistry, 70*, 63–69.

Leitz, T., (1998). Induction of metamorphosis of the marine hydrozoon *Hydractinia echinata*. *Biofouling, 12*, 173–187.

Leitz, T., Morand, K., & Mann, M., (1994a). Metamorphosin A, a novel peptide controlling development of the lower metazoan *Hydractinia echinata*. *Developmental Biology, 163*, 440–446.

Leitz, T., Morand, K., & Mann, M., (1994b). Metamorphosin A, a novel biologically active peptide from Coelenterates. *Verhandlungen der Deutschen Zoologischen Gesellschaft* [*Proceedings of the German Zoological Society*], *87*(1), 137.

Maguire, R. J., (2000). Review of the persistence, bioaccumulation, and toxicity of tributyltin in aquatic environments in relation to Canada's toxic substances management policy. *Water Quality Research Journal, 35*(4), 633–679.

Maugh, T. H. II, (1973). Trace elements: a growing appreciation of their effects on man. *Science, 181*, 253–254.

Mertz, W., (1981). The essential trace elements. *Science, 213*, 1332–1338.

Mohan, N., Suresh, C. H., Kumar, A., & Gadre, S. R., (2013). Molecular electrostatics for probing lone pair-π interactions. *Physical Chemistry Chemical Physics, 15*, 18401–18409**.**

Moore, J. N., & Luoma, S. N., (1990). Hazardous wastes from large-scale metal extraction. *Environmental Science and Technology, 24*, 1278–1285.

Müller, W. A., & Leitz, T., (2002). Metamorphosis in the Cnidaria. *Canadian Journal of Zoology, 1755–1771*.

Müller, W. A., Teo, R., & Frank, U., (2004). Totipotent migratory stem cells in a hydroid. *Developmental Biology, 275*(1), 215–224.

Naumann, B., Eberius, M., & Appenroth, K. J., (2007). Growth rate based dose-response relationships and EC-values of 10 heavy metals using the duckweed growth inhibition test (ISO, 20079) with *Lemna minor* L. clone St. *Journal of Plant Physiology, 164*, 1656–1664.

Newman, M. C., McCloskey, J. T., Wiliams, P. L., & Tatara, C. P., (1997). Using metal-ligand characteristics to predict metal toxicity: Quantitative ion character-activity relationships (QICARs). *Environmental Health Perspectives, 106*(6), 1419–1425.

Pizent, A., Tariba, B., & Živković, T., (2012). Reproductive toxicity of metals in men. *Archives of Industrial Hygiene and Toxicology, 63*(1), 35–46.

Plickert, G., Schetter, E., Verhey Van Wijk, N., Schlossherr, J., Steinbüchel, M., & Gajewski, M., (2003). The role of α–amidated neuropeptides in hydroid development-LW amides and metamorphosis in *Hydractinia echinata*. *International Journal of Developmental Biology, 47*, 439–450.

Santore, R. C., Di Toro, D. M., Paquin, P. R., Allen, H. E., & Meyer, J. S., (2001). Biotic ligand model of the acute toxicity of metals. 2. Application to the acute copper toxicity in freshwater fish and Daphnia. *Environmental Toxicology and Chemistry, 20*(10), 2397–2402.

Schmich, J., Trepel, S., & Leitz, T., (1998). The role of GLW amides in metamorphosis of *Hydractinia echinata*. *Development Genes and Evolution, 208*, 267–273.

Tatara, C. P., Newman, M. C., McCloskey, J. T., & Wiliams, P. L., (1997). Predicting relative metal toxicity with ion characteristics: *Caenorhabditis elegans* LC$_{50}$. *Aquatic Toxicology, 39*, 279–290.

Tatara, C. P., Newman, M. C., McCloskey, J. T., & Wiliams, P. L., (1998). Use of ion characteristics to predict relative toxicity of mono-, di-, and trivalent metal ions: *Caenorhabditis
elegans* LC$_{50}$. *Aquatic Toxicology, 42*, 255–269.

Tchounwou, P. B., Yedjou, C. G., Patlolla, A. K., & Sutton, D. J., (2012). Heavy metals
toxicity and the environment. *Experientia Supplementum, 101*, 133–164.

Todeschini, R., Vighi, M., Provenzani, R., Finizio, A., & Gramatica, P., (1996). Modeling
and prediction by using WHIM descriptors in QSAR studies: Toxicity of heterogeneous
chemicals on *Daphnia magna*. *Chemosphere, 32*, 1527–1545.

Venturini-Soriano, M., & Berthon, G., (1998). Aluminum speciation studies in biological fluids.
Part 4. A new investigation of aluminum-succinate complex formation under physiological
conditions, and possible implications for aluminum metabolism and toxicity. *Journal of
Inorganic Biochemistry, 71*, 135–145.

Walker, J. D., Enache, M., & Dearden, J. C., (2007). Quantitative cationic activity
relationships for predicting toxicity of metal ions from physicochemical properties and
natural occurrence levels. *QSAR & Combinatorial Science, 26*(4), 522–527.

Walther, M., (1995). *Beiträge zur Aufklärung der biochemische Prozesse bei der Metamorphoseauslösung der Planula-Larve von Hydractinia echinata FLEMING (Cnidaria, Hydrozoa)*.
Inaugural-Disseration Zur Erlangung des Doktorgrades der Mathematisch-Naturwissenschaftlichen Fakultät der Universität zu Köln [Contributions to the elucidation of biochemical
processes in metamorphosis triggering of the Planula larva of Hydractinia echinata FLEMING
(Cnidaria, Hydrozoa). Inaugural dissertation to obtain the doctoral degree of the Faculty of
Mathematics and Natural Sciences of the University of Köln].

Walther, M., Ulrich, R., Kroiher, M., & Berking, S., (1996). Metamorphosis and pattern formation in *Hydractinia echinata*, a colonial hydroid. *International Journal of Developmental
Biology, 40*(1), 313–322.

WHO, (1990). *Environmental Health Criteria 116, Tributyltin Compounds*. International
Programme on Chemicals Safety, World Health Organization, Geneva.

CHAPTER 16

Köln Model as a Toxicological Procedure for Oils and Derivatives

SERGIU A. CHICU,[1] MARINA A. TUDORAN,[2,3] and MIHAI V. PUTZ[2,3]

[1]*Institute of Chemistry Timisoara of the Romanian Academy, Mihai Viteazul Boulevard 24, RO–300223 Timişoara, Romania. Permanent address: Siegstr. 4, 50859 Köln, Germany*

[2]*Laboratory of Structural and Computational Physical Chemistry for Nanosciences and QSAR, Biology-Chemistry Department, West University of Timisoara, Pestalozzi Street No. 44, Timisoara, RO–300115, Romania*

[3]*Laboratory of Renewable Energies-Photovoltaics, R&D National Institute for Electrochemistry and Condensed Matter, Dr. A. Paunescu Podeanu Str. No. 144, Timisoara, RO–300569, Romania*

16.1   DEFINITION

The Köln model based on *Hydractinia echinata* Test System (*He*TS) is an *in vivo* biological procedure for toxicological investigation using the metamorphosis of the marine organism larvae *Hydractinia echinata* (*H. echinata*) from the lifestyle specific to plankton until the sedentary polyp (adult phase). The effectuated experiments with crude oils, components, and surfactants show that the organism is suitable also for toxicity determinations from this domain especially in case of accidents by offshore or terrestrial drilling, transport by purging procedures termed "operational discharge," or by petroleum industrial accident. The *He*TS is simple, fast, reproducible, and meets the "3R" requirements regarding the usage of the animal in laboratory experiments.

16.2   HISTORICAL ORIGIN(S)

Crude oil or raw petroleum consisting in a variable mixture of volatile components, saturated hydrocarbons (paraffinic and naphthenic) and aromatic compounds (aromatic, asphaltic) (Bestougeff, 1967) on which are added in small percentage derivatives with nitrogen, sulfur, oxygen, and metals, composition which vary depending of the formation way and place. Besides this type of oil, there is also unconventional oil with a higher content of nitrogen, sulfur, and oxygen compounds named shale oil (Cady and Seelig, 1952) as well as bituminous sand (Attanasi and Meyer, 2010) whose processing require special technologies (Cady and Seelig, 1952).

Oil spill and its components have a poisonously effect on the human organism including hematopoietic, hepatic, renal, and pulmonary abnormalities (D'Andrea et al., 2013, 2014) and immunotoxicity (Barron, 2012) for person which came in contact with it: residents, fisherman's, especially the participants on the operations of cleaning and removing of effects caused by them. A longer quantification of hydrocarbons concentrations in blood shows an accentuated persistence of derivatives with a single-ring aromatic compound (Sammarco et al., 2016). Along with this aspect exclusively medical, the petroleum industry known also tragic accidents, unworthy less remembered comparatively with the media impact of ecological catastrophes: Piper Alpha platform (Paté–Cornell, 1993), Alexander L. Kielland platform (Naess et al., 1982), Seacrest Drill Ship (Burton, 2011), etc. Also, although the hydrocarbons from oil represents an important energetic source, petroleum, and oil products represents, through accidents (Allan et al., 2012; Raloff, 1993; US EPA, 1993; Yin et al., 2015) a major pollutant of the environment (Lan et al., 2015; Lee and Anderson, 2005; Patin, 2004; Prince, 1993) affecting the marine ecosystems by immediate toxicity or lasting by persistence (Billiard et al., 2008; Carls et al., 1999, 2001; Cebrian and Uriz, 2007; Li and Boufadel, 2010; Lin et al., 2008; Neff et al., 2000; Negri et al., 2016; Odzer, 2016; Patin, 2004; Richardson et al., 2005; Rinkevich et al., 2007; Singer et al., 1990; White et al., 2012). Thus, bioaccumulation differently affects zooplankton organisms: larvae more pregnant comparatively with adults and polyaromatic hydrocarbons more toxic quantitatively present in zooplankton can be transferred through food chain to other organisms including humans (Almeda et al., 2013; Capuzzo, 1987; Howarth, 1989; Neff, 1985; Spies, 1987; Thibodeaux et al., 2011). Phytoplankton regularized the carbon quantity from the atmosphere and can play an active role through specific groups also in altering crude oil compounds in conjunction with microbial communities (McGenity et al., 2012) favoring the growth of some while inhibiting the growth of others. In this perspective, a very diverse group and microbes have been showed to have the ability to degrade petroleum hydrocarbons (Alisi et al., 2009; Atlas, 1981, 1991). Compared to the above, the problem of removing the oil spills effects and accidents leads to developing technologies based on physicochemical methods (Bouyarmane et al., 2014; Goodbody–Gringley et al., 2013; Mukherjee and Wrenn, 2009; National Research Council, 1989; Ramachandran et al., 2007; Singer et al., 1998) and on methods based on biodegradation characterized by low toxicity and special effectiveness in extreme conditions e.g., temperature (Rita de Cássia et al., 2014; Souza et al., 2014). Biosurfactants as surface-active substances synthesized by living cells are actually used in water and soil remediation or deliberate infested (Aparna et al., 2011; Khire, 2010; Malik and Ahmed, 2012; Silva et al., 2014).

*He*TS offers within the developed Koeln-Model procedure the possibility of using for the first time of metamorphosis in the investigation of toxicity evaluations for the polluted environment with crude oils and distillation derivatives (Chicu and Berking, 1997). The study allows for toxicity quantification and interpretation of derivatives singulars or in mixtures form marine environments marine (aquatic) especially form the polluted coastal environment (Becerro et al., 1995; Boudou and Ribeyre, 1997; Florea and Büsselberg, 2011; Lapota et al., 1993; Landrum et al., 1992; Plante–Cuny et al., 1993; Santore et al., 2001; Valko et al., 2005), and is making its direct contribution on inter-vertebrate monitoring (mollusks, barnacles, echinoderms, tunicates) and fishes whose progressive life cycle implies a demersal adult and a pelagic larval stage (Caffey, 1985; Christie et al., 2010; Gebauer, 2004; Grosberg and Levitan, 1992; Moser and Watson, 2006; Southward et al., 2005).

16.3 NANO-SCIENTIFIC DEVELOPMENT(S)

16.3.1 METAMORPHOSIS

In natural conditions, *H. echinata* metamorphosis is initiated by chemical factors from environment, and in the laboratory, the process is artificially influenced by larvae treatment with different substances (Berking, 1988a; Berking and Walther, 1994; Gajewski et al., 1996; Leitz, 1998; Plickert et al., 1988; Schmich et al., 1998; Walther, 1995; Walther et al., 1996). The organism is routinely cultured in usual biology labs at 18°C with a program of day/night of 14 h/10 *h* and fed with nauplii of *Artemia salina* or fish. Larvae (Galliot et al., 2009; Plickert et al., 1988; Takashima et al., 2013; Weis et al., 1985) were induced to undergo metamorphosis into primary polyps by pulse treatment with CsCl or using LWamide peptides (Frank et al., 2001; Gajewski et al., 1996; Leitz et al., 1994a, 1994b; Müller and

Leitz, 2002; Müller et al., 2004a; Plickert et al., 2003; Schmich et al., 1998). For a certain induction time and depending on the substrate nature and concentration, the larvae are transforming in polyps or stay the same. The process is easy to observe with a usual laboratory microscope and allows the clear and unequivocally evaluation on the same experiment to what extent the test-substance presents inhibitive or stimulative influences on metamorphosis.

16.3.2   WORKING ALGORITHM

All experiments were effectuated after the original project principle (Chicu and Berking, 1997): the toxicity values M (Table 16.1) are expressed in accordance with the Koeln-Model algorithm under the form $log1/MRC_{50}$ and respectively as $\mu l/ml$. All tested derivatives were used such as, from solutions of known concentrations, in synthetic seawater or methanol.

16.3.3   RESULTS

The experimental results of the study are presented in Table 16.1.

The hydrocarbons interaction with a biologic system occurs at cellular level – the base unit of organism – whose membrane allows through proteins with amphipathic character the spontaneous and selective diffusion of ions, molecules, and water from areas with high concentration in areas with low concentrations. This way the cell homeostasis can be affected and therefore all normal physicochemical and biological processes; the cell survival and therefore the organism survival is due to their capacity to adapt on environmental conditions.

Because the electrons repartition is equal within the C–H bonds, hydrocarbons are nonpolar and hydrophobic. Their interaction with the cell occurs however in presence of non-covalent interactions (Anslyn and Dougherty, 2004), e.g., complex formation between saturated hydrocarbons and benzene (Ran and Wong, 2006; Wheeler, 2012), between hydrocarbons and water as aqueous-π electron interactions (Suzuki et al., 1992; Zhang et al., 2013) or between aromatic cycles which can form hydrogen bonds like interactions with functional groups (Sinnokrot et al., 2002).

The toxic action exercised by oils and derivatives is first of all due to hydrophobic effect of hydrocarbons on biological systems *H. echinata* and *D. magna*. In general, aliphatic hydrocarbons have: 1) stronger interactions

TABLE 16.1 Toxicity *M* Values as mol/L or µl/ml of Two Crude Oils, Some Distillation Derivatives, Surfactants, and Different Components with *H. echinata* and *D. magna* Test-Systems

No	CAS		Cs	Car	*H. echinata* M		*D. magna*		logP	MW	d
					mol/L	µl/ml	mol/l	µl/ml	(M/C*)	g/mol	g/cm³
1		Kuwait crude oil				0.45					
2		Statfjord crude oil				0.19					
3		Jet fuel				0.90					
4		Heating oil				0.40					
5		Diesel oil				0.033					
6		Gas oil				0.23					
7	109–66–0	Pentane	5		2.78	0.19	3.87	0.015	3.34*	72.15	0.63
8	287–92–3	Ciclopentane	5		2.80	0.15	3.82	0.014	2.70*	70.13	0.74
9	291–64–5	Cycloheptane	7		4.06	0.0105			3.91*	98.19	0.811
10	493–02–7	trans-Decalin	10		5.32	0.00074			4.79*	138.25	0.87
11	71–43–2	Benzene		6	1.65	2.00	3.39	0.036	2.13*	78.11	0.874
12	108–88–3	Toluene	1	6	1.86	1.54	3.90	0.014	2.73	97.14	0.87
13	108–38–3	Xilene	2	6	2.18	0.81	4.20	0.0077	3.20	106.17	0.868
14	100–41–4	Ethylbenzene	2	6	2.64	0.28	4.70	0.0024	3.15	106.17	0.87
15	98–82–8	Isopropylbenzene	3	6	3.49	0.045	5.30	0.00069	3.66	120.19	
16	91–20–3	Naphthalene		10	3.18	0.074	3.89	0.014	3.30	128.17	1.14
17	90–12–0	1-Methylnaphthalene	1	10	4.62	0.0033	5.00	0.0014	3.87	142.2	1.02
18	91–57–6	2-Methylnaphthalene	1	10	3.54	0.04	4.89	0.0018	3.86	142.2	1.006
19	573–98–8	di-Methylnaphthalene	2	10	4.00	0.015			4.31	156.22	1.013
20	92–52–4	Biphenyl		12	4.05	0.013	3.70	0.029	4.01	154.21	1.04
21	85–01–8	Phenanthrene		14	4.72	0.003	5.19	0.0011	4.47	178.23	1.063
22	120–12–7	Anthracene		14	5.42	0.0005	4.77	0.0024	4.45	178.24	1.25
23	779–02–2	9-Methylanthracene	1	14	5.60	0.00045	5.64	0.00045	5.07	192.26	1.066

TABLE 16.1   *(Continued)*

No	CAS		Cs	Car	H. echinata M		D. magna		logP	MW	d
					mol/L	µl/ml	mol/l	µl/ml	(M/C*)	g/mol	g/cm^3
24	129–00–0	Pyrene		16	3.68	0.033	5.05	0.0014	4.88	202.26	1.27
25	109–86–4	2-Methoxyethanol	3		1.51	2.44			–0.77	76.09	0.965
26	110–80–5	2-Ethoxyethanol	4		1.48	3.21			–0.32	90.12	0.93
27	2807–30–9	2-Propoxyethanol	5		1.98	1.19			0.08*	104.15	0.913
28	122–99–6	2-Phenoxyethanol	2	6	2.03	1.25			1.16	134.18	1.005
29	9016–45–9	Tergitol NP–40	89	6	2.66	4.13				1982.47	1.05
30		SAT 980138			2.99						
31		SAT 981176			2.47						
32		SAT 990188			2.97						
33	108–95–2	Phenol		6	2.89	0.11	3.32	0.042	1.46	94.11	1.07
34	90–05–1	2-Methoxy-phenol	1	6	2.77	0.19			1.32	124.13	1.12
35	104–40–5	4-n-Nonylphenol	9	6	4.95	0.0026			4.69	220.35	0.937
36	98–11–3	Acid benzene sulfonic		6	2.65	0.27			0.47*	158.18	1.32
37	88–44–8	Acid-4-toluidin-3-S	1	6	2.71	0.24			0.55	187.22	1.49
38	91–22–5	Quinoline		9	2.94	0.13			2.03	129.16	1.10
39	91–63–4	2-Methylquinoline	1	9	2.98	0.14			2.59	143.19	1.06
40	110–02–1	Thiophene		4	3.11	0.06			1.81	84.14	1.051
41	554–14–3	2-methylthiophene	1	4	3.03	0.09			2.33	98.17	1.014

Reproductibility: 2-Methoxyethanol 1.47;1.54; Phenol 3.15/2.66/2.85; Quinoline 3.07/2.79/2.96; 2-Methylquinoline 2.73/2.98.

Abbreviations: Cs saturated carbon; Car aromated carbon; S sulfonic; *C calculated P

SAT 980138 (alcoholethoxylat) (Henkel)

SAT 981176 (alchylbenzene-sulfonat) (Henkel)

SAT 990188 (nonionic tensid) (Henkel)

than aromatic compounds; 2) the carbon chain ramification by stearic impediments will reduce the hydrophobic effect of that molecule; and 3) the linear carbon chain determines the most intense hydrophobic interaction. The toxicity of saturated or aromatic hydrocarbons increases with increasing the number of carbon atoms and in general with log *P*; isopropylbenzene is the most toxic due to the push-electronic effect exercised by the isopropyl group, a *p*–*π* type interaction (Egli et al., 2007; Jain et al., 2009; Mohan et al., 2013) being possible also with the *π*-electrons of the biologic system (Arendorf, 2011; Hunter and Sanders, 1990).

The toxicity dependence of the C-number from a saturated catena is in accordance with the results obtained in case of amines and alcohols (Chicu et al., 2000) or with alkylate derivatives of phenols (Chicu and Simu, 2009) being confirmed also by obtained results with test-systems *Tetrahymena pyriformis* (*T. pyriformis*), *Daphnia magna* (*D. magna*), *Pimephales promelas* (*P. promelas*), and *Vibrio fischeri* (*V. fischeri*) (Chicu et al., 2000; Chicu and Simu, 2009). This way one can prove that "*if the organisms can be different within their anatomy and physiology the diversity is significantly reduced on the cell level and on its biochemistry*" (Dimitrov et al., 2004).

The study regarding the substance mixture (Chicu et al., 2016) shows: 1) their toxicity is decreasing as the components number increase; and 2) the mixture toxicity is influenced by the component with the highest percentage participation, independent of its singular effectiveness. The lower toxicity of Kuwait crude oil is due to its principal components: 23.3% and parrafins 20.9% (Ferek et al., 1992) comparative with Statfjord crude oil characterized by a low concentration of normal paraffin. Also, Jet fuel which consists in a high number of components has a lower toxic effect than Diesel oil consist in about 75% saturated hydrocarbons (primarily paraffin including n-, iso-, and cycloparaffin) (Nos. 7–10), and 25% aromatic hydrocarbons (including naphthalene and alkylbenzene) (Nos. 11–18) has the strongest toxic effect, followed by Gas oil which contains mostly hydrocarbons with condensates nuclei (Nos. 16–24).

The tenside toxicity with dispersed function (Nos. 29 and SAT 30–32 from Henkel–Germany) is compatible with the one of singular chemical derivatives (inferiors terms) from the ethers (Nos. 25–28), of the benzene minor substituted (Nos. 33–37) or some derivatives with *N* and *S* (Nos. 38–41). Due to the fact that dispersants consist normally of one or more surfactants, their using in case of accidents with oil spilling, drilling or transport involves problems regarding the toxic effect not only on the marine environment (DeLeo et al., 2016; Kleindienst et al., 2015), but also in the terrestrial environment through the residual waters or by direct spill by excessive industrial and domestic consumer, the half-lives of aerobic degradation being dependent on many factors such as their structure and concentration, the microorganisms presence, etc. (Tezel, 2009).

One can also observe that: 1) surfactants toxicity as singular derivatives of anionic type alkyl-sulphonate (Nos. 36, 37), nonionic-ethers (Nos. 25–29) and even quaternary ammonium salts of benzoic acid is closed (Chicu et al., 2008), but much smaller than the one of hydrocarbons; and 2) the molecular mass do not have always (Chicu et al., 2014) a prime order: Tergitol NP40 posses a special molecular mass but an M value practical similar with the compounds with the same SAT type action or with the singular derivatives (Nos. 33, 34, 36–41). The molecules behavior with large dimensions can be due to bioconcentration and hydrophobic effects (Opperhuizen et al., 1985, 1987) but first of all the interaction of functional group with the cellular receptor (Chicu et al., 2014).

The Koeln-Model and *He*TS offer the possibility of using for the first time of the metamorphosis in clear, fast, and cheap investigations of toxicity evaluation in general, as well as in the problem of marine environment moni-toring in special. We found that crude oil of the North Sea is more harmful than oil from Kuwait, and Diesel as other derivatives Gas oil > Heating oil > Jet fuel. Systematic evaluation of hydrocarbons revealed that the distil-lation derivatives of oil are more toxic than brut oil and the benzene and inferior alkyl-benzene are less toxic than corresponding superior-saturated derivatives or polycyclic aromatic hydrocarbons. The biological model is important because: 1) can evaluate the degree in which the environment can be affected in case of all kind of accidents with crude oil or products: by offshore or terrestrial drilling, by transport, by allowing purging procedures termed "operational discharge," or by petroleum industrial accident, 2) the investigations are accessible and applicable for any biologic laboratory which has just an optical microscope, 3) the procedure is reproducible, 4) *He*TS meets the "3R" criteria and represents besides the mentioned test-systems a simple format more flexible and faster.

In the same time, the investigation possibilities and the research direc-tion related to the *H. echinata* organism are multiple because also allows studying some regenerative mechanisms (Bradshaw et al., 2015; Galliot et al., 2009), comparative immunology studies (Frank et al., 2001), plury-potence (Hensel, 2013) or totipotence (Müller et al., 2004). One can state that the experimental results obtained until present with *He*TS will not been possible by direct experiments not only on mammalians but also with other organisms.

KEYWORDS

- *Hydractinia echinata* **test system**
- **hydrophobic effect**
- **Köln-Model**
- **noncovalent interactions**

REFERENCES AND FURTHER READING

Alisi, C., Musella, R., Tasso, F., Ubaldi, C., Manzo, S., Cremisini, C., & Sprocati, A. R., (2009). Bioremediation of diesel oil in a co-contaminated soil by bioaugmentation with a microbial formula tailored with native strains selected for heavy metals resistance. *Science of the Total Environment, 407*, 3024–3032.

Allan, S. E., Smith, B. W., & Anderson, K. A., (2012). Impact of the deepwater horizon oil spill on bioavailable polycyclic aromatic hydrocarbons in Gulf of Mexico coastal waters. *Environmental Science & Technology, 46*(4), 2033–2039.

Almeda, R., Wambaugh, Z., Chai, C., Wang, Z., Liu, Z., & Buskey, E. J., (2013). Effects of crude oil exposure on bioaccumulation of polycyclic aromatic hydrocarbons and survival of adult and larval stages of gelatinous zooplankton. *PLoS One, 8*(10), c74476.

Anslyn, E. V., & Dougherty, D. A., (2004). *Modern Physical Organic Chemistry.* Sausalito, CA: University Science, California.

Aparna, A., Srinikethan, G., & Hedge, S., (2011). Effect of addition of biosurfactant produced by Pseudomonas ssp. on biodegradation of crude oil. *International Proceedings of Chemical, Biological & Environmental Engineering, Proceedings of the 2ⁿᵈ International Proceedings of Chemical* (pp. 71–75), Singapore.

Arendorf, J. R. T., (2011). *A Study of Some Non-Covalent Functional Group π Interactions.* Doctoral thesis, UCL (University College London).

Atlas, R., (1981). Microbial-degradation of petroleum-hydrocarbons, an environmental perspective. *Microbiology Reviews, 45*, 180–209.

Atlas, R., (1991). Microbial hydrocarbon degradation-bioremediation of oil spills. *Journal of Chemical Technology and Biotechnology, 52*, 149–156.

Attanasi, E. D., & Meyer, R. F., (2010). *Natural Bitumen and Extra-Heavy Oil* (pp. 123–140). Survey of energy resources (PDF) (22ⁿᵈ edn.). World Energy Council.

Barron, M. G., (2012). Ecological impacts of the deepwater horizon oil spill: Implications for immunotoxicity. *Toxicologic Pathology, 40*(2), 315–320.

Becerro, M. A., Uriz, M. J., & Turon, X., (1995). Measuring toxicity in marine environment: Critical appraisal of three commonly used methods. *Experientia, 51*, 414–418.

Berking, S., (1988a). Ammonia, tetraethylammonium, barium, and amiloride induce metamorphosis in the marine hydroid Hydractinia. *Wilhelm Roux's Archives of Developmental Biology, 197*, 1–9.

Berking, S., & Walther, M., (1994). Control of metamorphosis in the hydroid hydractinia. In: Davey, K. G., Peter, R. E., & Tobe, S. S., (eds.), *Perspectives in Comparative Endocrinology* (pp. 381–388). National Research Council of Canada, Ottawa.

Bestougeff, M. A., (1967). Petroleum hydrocarbons. In: Nagy, B., & Colombo, U., (eds.), *Fundamental Aspects of Petroleum Geochemistry* (pp. 77–108). Amsterdam, Elsevier.

Billiard, S. M., Meyer, J. N., Wassenberg, D. M., Hodson, P. V., & Di Giulio, R. T., (2008). Non-additive effects of PAHs on early vertebrate development: Mechanisms and implications for risk assessment. *Toxicological Sciences, 105*, 5–23.

Boudou, A., & Ribeyre, F., (1997). Aquatic ecotoxicology: From the ecosystem to the cellular and molecular levels. *Environmental Health Perspectives, 105*(1), 21–35.

Bouyarmane, H., Saoiabi, S., Laghzizil, A., Saoiabi, A., Rami, A., & El Karbane, M., (2014). Natural phosphate and its derivative porous hydroxyapatite for the removal of toxic organic chemicals. *Desalination and Water Treatment, 52*(37–39), 7265–7269.

Bradshaw, B., Thompson, K., & Frank, U., (2015). Distinct mechanisms underlie oral vs. aboral regeneration in the cnidarian. *eLife, 4*, e05506.

Burton, S., (2011). *The Seacrest Dill Ship. Thai Wreck Diver*. 31 July 2011. [Online]. Available: http://www.thaiwreckdiver.com/seacrest_drill_ship.htm (accessed 26 February 2019).

Cady, W. E., & Seelig, H. S., (1952). Composition of Shale Oil. *Industrial & Engineering Chemistry* 44(11), 2636–2641.

Caffey, H. M., (1985). Spatial and temporal variation in settlement and recruitment of intertidal barnacles. *Ecological Monographs, 55*, 313–332.

Capuzzo, J. M., (1987). Biological effects of petroleum hydrocarbons: Assessments from experimental results. In: Boesch, D. F., & Rabalais, N. N., (eds.), *Long-Term Environmental Effects of Offshore Oil and Gas Development* (pp. 343–410). Elsevier Applied Science, New York, Chapter 8.

Carls, M. G., Babcock, M. M., Harris, P. M., Irvine, G. V., Cusick, J. A., & Rice, S. D., (2001). Persistence of oiling in mussel beds after the Exxon Valdez oil spill. *Marine Environmental Research, 51*(2), 167–190.

Carls, M. G., Rice, S. D., & Hose, J. E., (1999). Sensitivity of fish embryos to weathered crude oil: Part I. Low-level exposure during incubation causes malformations, genetic damage, and mortality in larval Pacific herring (*Clupea pallasi*). *Environmental Toxicology and Chemistry, 18*(3), 481–493.

Cebrian, E., & Uriz, M. J., (2007). Contrasting effects of heavy metals and hydrocarbons on larval settlement and juvenile survival in sponges. *Aquatic Toxicology, 81*, 137–143.

Chicu, S. A., & Berking, S., (1997). Interference with metamorphosis induction in the marine Cnidaria Hydractinia echinata (Hydrozoa): A structure-activity relationship analysis of lower alcohols, aliphatic, and aromatic hydrocarbons, thiophenes, tributyltin, and crude oil. *Chemosphere, 34*(8), 1851–1866.

Chicu, S. A., & Simu, G. M., (2009). *Hydractinia echinata* test-system. I. Toxicity determination of some benzenic, biphenilic, and naphthalenic phenols. Comparative SAR-QSAR study. *Revue Roumaine de Chimie, 54*(8), 659–669.

Chicu, S. A., Grozav, M., Kurunczi, L., & Crisan, M., (2008). SAR for amine salts of carboxylic acids to *Hydractinia echinata*. *Revista de Chimie, 5*, 82–587.

Chicu, S. A., Herrmann, K., & Berking, S., (2000). An approach to calculate the toxicity of simple organic molecules on the basis of QSAR analysis in *Hydractinia echinata* (Hydrozoa, Cnidaria). *Quantitative Structure-Activity Relationship, 19*, 227–236.

Chicu, S. A., Munteanu, M., Cîtu, I., Şoica, C., Dehelean, C., Trandafirescu, C., et al., (2014). The *Hydractinia echinata* Test-System. III: Structure-toxicity relationship study of some azo-, azo-anilide, and diazonium salt derivatives. *Molecules, 19*(7), 9798–9817.

Chicu, S. A., Schannen, L., Putz, M. V., & Simu, G. M., (2016). *Hydractinia echinata* test-system. IV. Toxicological synergism for human pharmaceuticals in mixtures with iodoform. *Ecotox. Environ. Saf., 134*, 80–85, doi: 10.1016/j.ecoenv.2016.08.014.

Christie, M. R., Tissot, B. N., Albins, M. A., Beets, J. P., Jia, Y., Ortiz, D. L., Thompson, S. E., & Hixon, M. A., (2010). Larval connectivity in an effective network of marine protected areas. *PLoS One, 5*(12), e15715.

D'Andrea, M. A., & Reddy, G. K., (2014). Health risks associated with crude oil spill exposure. *American Journal of Medicine, 127*(9), 886, 9–13.

D'Andrea, M. A., Singh, O., & Reddy, G. K., (2013). Health consequences of involuntary exposure to benzene following a flaring incident at British Petroleum refinery in Texas City. *American Journal of Disaster Medicine, 8*(3), 169–179.

DeLeo, D. M., Ruiz, R. D. V., Baums, I. B., & Cordes, E. E., (2016). Response of deep-water corals to oil and chemical dispersant exposure. *Deep Sea Research Part II: Topical Studies in Oceanography, 129*, 137–147.

Dimitrov, S., Koleva, Y., Schultz, T. W., Walker, J. D., & Mekenyan, O., (2004). Interspecies quantitative structure-activity relationship model for aldehydes: Aquatic toxicology. *Environmental Toxicology and Chemistry, 23*(2), 463–470.

Egli, M., & Sarkhel, S., (2007). Lone pair-aromatic interactions: To stabilize or not to stabilize. *Accounts of Chemical Research, 40*, 197–205.

Ferek, R. J., Hobbs, P. V., Herring, J. A., & Laursen, K. K., (1992). Chemical composition of emissions from the Kuwait oil fires. *Journal of Geophysical Research, 97*(14), 483–14, 489.

Florea, A. M., & Büsselberg, D., (2011). Metals and breast cancer: Risk factors or healing agents? Hindawi Publishing Corporation. *Journal of Toxicology*, p. 8. Article ID 159619.

Frank, U., Leitz, T., & Mueller, W. A., (2001). The hydroid *Hydractinia echinata*: A versatile, informative cnidarian representative. *Bioessays, 23*, 963–971.

Gajewski, M., Leitz, T., Schloßherr, J., & Plickert, G., (1996). LW amides from Cnidaria constitute a novel family of neuropeptides with morphogenetic activity. *Wilhelm Roux's Archives of Developmental Biology, 205*, 232–242.

Galliot, B., Quiquand, M., Ghila, L., De Rosa, R., Miljkovic-Licina, M., & Chera, S., (2009). Origins of neurogenesis, a cnidarian view. *Developmental Biology, 332*, 2–24.

Gebauer, P., Paschke, K., & Anger, K., (2004). Stimulation of metamorphosis in an estuarine crab, *Chasmagnathus granulata* (Dana, 1851): Temporal window of cue receptivity. *Journal of Experimental Marine Biology and Ecology, 311*, 25–36.

Goodbody, G. G., Wetzel, D. L., Gillon, D., Pulster, E., Miller, A., & Ritchie, K. B., (2013). Toxicity of deepwater horizon source oil and the chemical dispersant, Corexit® 9500, to coral larvae. *PLoS One, 8*(1), e45574.

Grosberg, R. K., & Levitan, D. R., (1992). For adults only? Supply-side ecology and the history of larval biology. *Trends in Ecology & Evolution, 7*, 130–133.

Hensel, K., (2013). *Wnt Signaling in Hydractinia Stem Cells*. School of natural sciences and regenerative medicine institute (REMEDI), National University of Ireland Galway, Galway, Ireland, Thesis submission September 2013.

Howarth, R. W., (1989). Determining the ecological effects of oil pollution in marine ecosystems. In: Levin, S. A., Harwell, M. A., Kelly, J. R., & Kimball, K. D., (eds.), *Problems in Ecotoxicology* (pp. 69–97). Springer-Verlag, New York, NY.

Hunter, C. A., & Sanders, J. K. M., (1990). The nature of π–π interactions. *Journal of the American Chemical Society, 112*(14), 5525–5534.

Jain, A., Ramanathan, V., & Sankararamakrishnan, R., (2009). Lone pair-π interactions between water oxygens and aromatic residues: Quantum chemical studies based on high-resolution protein structures and model compounds. *Protein Science, 18*, 595–605.

Khire, J. M., (2010). Bacterial biosurfactants, and their role in microbial enhanced oil recovery (MEOR). *Advances in Experimental Medicine and Biology, 672*, 146–157.

Kleindienst, S., Seidel, M., Ziervogel, K., Grim, S., Loftis, K., Harrison, S., et al., (2015). Chemical dispersants can suppress the activity of natural oil-degrading microorganisms. *Proceedings of the National Academy of Sciences of the United States of America*, pp. 1–6.

Lan, D., Liang, B., Bao, C., Ma, M., Xu, Y., & Yu, C., (2015). Marine oil spill risk mapping for accidental pollution and its application in a coastal city. *Marine Pollution Bulletin, 96*(1/2), 220–225.

Landrum, P. F., Fadie, B. J., & Faust, W. R., (1992). Variation in the bioavailability of polycyclic aromatic hydrocarbons to the amphipod *Diporeia* (sp.) with sediment aging. *Environmental Toxicology and Chemistry, 11*, 1197–1208.

Lapota, D., Rosenberger, D. E., Platter, R. M. F., & Seligman, P. F., (1993). Growth and survival of *Mytilus edulis* larvae exposed in low levels of dibutyltin and tributyltin. *Marine Biolog., 115*, 413–419.

Lee, R. F., & Anderson, J. W., (2005). Significance of cytochrome P450 system responses and levels of bile fluorescent aromatic compounds in marine wildlife following oil spills. *Marine Pollution Bulletin, 50*(7), 705–723.

Leitz, T., (1998). Induction of metamorphosis of the marine hydrozoon *Hydractinia echinata*. *Biofouling, 12*, 173–187.

Leitz, T., Morand, K., & Mann, M., (1994a). Metamorphosin A, a novel peptide controlling development of the lower metazoan *Hydractinia echinata*. *Developmental Biology, 163*, 440–446.

Leitz, T., Morand, K., & Mann, M., (1994b). Metamorphosin A, a novel biologically active peptide from Coelenterates. *Verhandlungen der Deutschen Zoologischen Gesellschaft [Proceedings of the German Zoological Society], 87*(1), 137.

Li, H., & Boufadel, M. C., (2010). Long-term persistence of oil from the Exxon Valdez spill in two-layer beaches. *Nature Geoscience, 3*, 96–99.

Lin, C. Y., & Tjeerdema, R. S., (2008). Crude oil, oil, gasoline, and petrol. In: Jorgensen, S. E., & Fath, B. D., (eds.), *Encyclopedia of Ecology* (Vol. 1, pp. 797–805). Ecotox. Oxford, UK.

Malik, Z. A., & Ahmed, S., (2012). Degradation of petroleum hydrocarbons by oil field isolated bacterial consortium. *African Journal of Biotechnology, 11*, 650–658.

McGenity, T. J., Folwell, B. D., McKew, B. A., & Sanni, G. O., (2012). Marine crude-oil biodegradation: A central role for interspecies interactions. *Aquatic Biosystems, 8*(10), 1–19.

Mohan, N., Suresh, C. H., Kumar, A., & Shridhar, R. G., (2013). Molecular electrostatics for probing lone pair-π interactions. *Physical Chemistry Chemical Physics, 15*, 18401–18409.

Moser, H. G., & Watson, W., (2006). Ichthyoplankton. In: Allen, L. G., Pondella, D. J., & Horn, M. H., (eds.), *Ecology of Marine Fishes* (pp. 269–319). California and adjacent waters University of California Press.

Mukherjee, B., & Wrenn, B. A., (2009). Influence of dynamic mixing energy on dispersant performance: Role of mixing systems. *Environmental Engineering Science, 26*, 1725–1737.

Müller, W. A., & Leitz, T., (2002). Metamorphosis in the Cnidaria. *Canadian Journal of Zoology, 80*, 1755–1771.

Müller, W. A., Teo, R., & Frank, U., (2004). Totipotent migratory stem cells in a hydroid. *Developmental Biology, 275*(1), 215–224.

Naess, A. A., Haagensen, P. J., Lian, B., Moan, T., & Simonsen, T., (1982). *Offshore Technology Conference,* 3–6. May, Houston, Texas.

National Research Council (NRC), (1989). *Using Oil Spill Dispersants on the Sea* (p. 335). National Academy Press, Washington, DC.

Neff, J. M., (1985). Polycyclic aromatic hydrocarbons. In: Rand, G. M., & Petrocelli, S. R., (eds.), *Fundamentals of Aquatic Toxicology* (pp. 416–454). Hemisphere Publishing Corporation, New York.

Neff, J. M., Ostazeski, S., Gardiner, W., & Stejskal, I., (2000). Effects of weathering on the toxicity of three offshore Australian crude oils and a diesel fuel to marine animals. *Environmental Toxicology and Chemistry, 19*, 1809–1821.

Negri, A. P., Brinkman, D. L., Flores, F., Botté, E. S., Jones, R. J., & Webster, N. S., (2016). Acute ecotoxicology of natural oil and gas condensate to coral reef larvae. *Scientific Reports, 6.* Article number: 21153.

Odzer, N., (2016). *The Effect of the Polycyclic Aromatic Hydrocarbon Naphthalene on Porites Divaricate.* AAAS 2016 Annual Meeting. 11 - 15 February, Washington, DC.

Opperhuizen, A., Damen, H. W. J., Asyee, G. M., & Van der Steen, W. M. D., (1987). Uptake and elimination by fish of polydimethylsiloxanes (silicones) after dietary and aqueous exposure. *Toxicological & Environmental Chemistry, 13*, 265–285.

Opperhuizen, A., Van der Velde, E. W., Gobas, F. A. P. C., Liem, D. A. K., & Van der Steen, J. M. D., (1985). Relationship between bioconcentration in fish and steric factors of hydrophobic chemicals. *Chemosphere, 14*, 1871–1896.

Paté–Cornell, M. E., (1993). Learning from the piper alpha accident: A postmortem analysis of technical and organizational factors. *Risk Analysis, 13*(2), 215–232.

Patin, S. A., (2004). Environmental impact of crude oil spills. In: Cleveland, C. J., (ed.), *Encyclopedia of Energy* (pp. 737–748). Elsevier, New York.

Plante, C. M. R., Salen-Picard, C., Grenz, C., Plante, R., Alliot, E., & Barranguet, C., (1993). Experimental field study of the effects of crude oil, drill cuttings, and natural biodeposits om microphyto-and macrozoobenthic communities in a Mediterranean area. *Marine Biology, 117,* 355–366.

Plickert, G., Kroiher, M., & Munck, A., (1988). Cell proliferation and early differentiation during embryonic development and metamorphosis of *Hydractinia echinata. Development, 103*, 795–803.

Plickert, G., Schetter, E., Verhey Van Wijk, N., Schlossherr, J., Steinbüchel, M., & Gajewski, M., (2003). The role of amidated neuropeptides in hydroid development–LW amides and metamorphosis in *Hydractinia echinata. International Journal of Developmental Biology, 47*, 439–450.

Prince, R., (1993). Petroleum spill bioremediation in marine environments. *Critical Reviews in Microbiology, 19*, 217–242.

Raloff, J., (1993). Valdez spill leaves lasting impacts. *Science News, 143*(7), 103, 104.

Ramachandran, S. D., Hodson, P. V., Khan, C. W., & Lee, K., (2004). Oil dispersant increases PAH uptake by fish exposed to crude oil. *Ecotoxicology and Environmental Safety, 59*(3), 300–308.

Ran, J., & Wong, M. W., (2006). Saturated hydrocarbon-benzene complexes: Theoretical study of cooperative CH/pi interactions. *Journal of Physical Chemistry A., 110*(31), 9702–9709.

Richardson, A. J., Sims, D. W., Smith, T., Walne, A. W., & Hawkins, S. J., (2005). Long-term oceanographic and ecological research in the Western English Channel. *Advances in Marine Biology, 47*, 1–105.

Rinkevich, B., (2007). Short and long term toxicity of crude oil and oil dispersants to two representative coral species. *Environmental Science & Technology, 41*, 5571–5574.

Rita de Cássia, F. S., Silva, A. D. G., Rufino, R. D., Luna, J. M., Santos, V. A., & Sarubbo, L. A., (2014). Applications of biosurfactants in the petroleum industry and the remediation of oil spills. *International Journal of Molecular Sciences, 15*(7), 12523–12542.

Sammarco, P. W., Kolian, S. R., Warby, R. A., Bouldin, J. L., Subra, W. A., & Porter, S. A., (2016). Concentrations in human blood of petroleum hydrocarbons associated with the BP/Deepwater Horizon oil spill, Gulf of Mexico. *Archives of Toxicology, 90*(4), 829–837.

Santore, R. C., Di Toro, D. M., Paquin, P. R., Allen, H. E., & Meyer, J. S., (2001). Biotic ligand model of the acute toxicity of metals. 2. Application to the acute copper toxicity in freshwater fish and *Daphnia*. *Environmental Toxicology and Chemistry, 20*(10), 2397–2402.

Schmich, J., Trepel, S., & Leitz, T., (1998). The role of GLWamides in metamorphosis of *Hydractinia echinata*. *Development Genes and Evolution, 208*, 267–273.

Silva, E. J., Rocha e Silva, N. M. P., Rufino, R. D., Luna, J. M., Silva, R. O., & Sarubbo, L. A., (2014). Characterization of a biosurfactant produced by Pseudomonas cepacia CCT6659 in the presence of industrial wastes and its application in the biodegradation of hydrophobic compounds in soil. *Colloids and Surfaces B: Biointerfaces, 117*, 36–41.

Singer, M. M., Smalheer, D. L., Tjeerdema, R. S., & Martin, M., (1990). Toxicity of an oil dispersant to the early life stages of four California marine species. *Environmental Toxicology and Chemistry, 9*(11), 1387–1395.

Singer, M., George, S., Lee, I., Jacobson, S., Weetman, L., Blondina, G., Tjeerdema, R., Aurand, D., & Sowby, M., (1998). Effects of dispersant treatment on the acute aquatic toxicity of petroleum hydrocarbons. *Archives of Environmental Contamination and Toxicology, 34*(2), 177–187.

Sinnokrot, M. O., Valeev, E. F., & Sherrill, C. D., (2002). Estimates of the ab initio limit for $\pi-\pi$ interactions: The benzene dimer. *Journal of the American Chemical Society, 124*(36), 10887–10893.

Southward, A. J., Langmead, O., Hardman-Mountford, N. J., Aiken, J., Boalch, G. T., Dando, P. R., et al., (2006). Saturated hydrocarbon−benzene complexes: Theoretical study of cooperative CH/π interactions. *Journal of Physical Chemistry A., 110*(31), 9702–9709.

Souza, E. C., Vessoni, P. T. C., & Souza, O. R. P., (2014). Biosurfactant-enhanced hydrocarbon bioremediation: An overview. *International Biodeterioration & Biodegradation, 89*, 88–94.

Spies, R. B., (1987). The biological effects of petroleum hydrocarbons in the sea: Assessments from the field and microcosms. In: Boesch, D. F., & Rabalais, N. N., (eds.), *Long-Term Environmental Effects of Offshore Oil and Gas Development* (pp. 411–467). Elsevier Applied Science, New York.

Suzuki, S., Green, P. G., Bumgarner, R. E., Dasgupta, S., Goddard, W. A., & Blake, G. A., (1992). Benzene forms hydrogen bonds with water. *Science, 257*(5072), 942–945.

Takashima, S., Gold, D., & Hartenstein, H., (2013). Stem cells and lineages of the intestine: A developmental and evolutionary perspective. *Development Genes and Evolution, 223*(0), 10.1007/s00427–012–0422–8.

Tezel, U., (2009). *Fate and Effect of Quaternary Ammonium Compounds in Biological Systems*. Ph. D. Thesis, Georgia Institute of Technology, Atlanta, GA, http://smartech.gatech.edu/handle/1853/28229.

Thibodeaux, L. J., Valsaraj, K. T., John, V. T., Papadopoulos, K. D., Pratt, L. R., & Pesika, N. S., (2011). Marine oil fate: Knowledge gaps, basic research, and development needs, a perspective based on the deepwater horizon spill. *Environmental Engineering Science, 28*(2), 87–93.

U.S. Environmental Protection Agency, (1993), *Understanding Oil Spills and Oil Spill Response* (pp. 5–105). Publication Number 9200, Washington, D.C.

Valko, M., Morris, H., & Cronin, M. T., (2005). Metals, toxicity, and oxidative stress. *Current Medicinal Chemistry, 12*(10), 1161–1208.

Walther, M., (1995). *Beiträge zur Aufklärung der biochemische Prozesse bei der Metamorphoseauslösung der Planula-Larve von Hydractinia echinata* FLEMING (Cnidaria, Hydrozoa). Inaugural-Disseration Zur Erlangung des Doktorgrades der Mathematisch-Naturwissenschaftlichen Fakultät der Universität zu Köln [Contributions to the elucidation of biochemical processes in metamorphosis triggering of the Planula larva of Hydractinia echinata FLEMING (Cnidaria, Hydrozoa). Inaugural dissertation To obtain the doctoral degree of the Faculty of Mathematics and Natural Sciences of the University of Köln.].

Walther, M., Ulrich, R., Kroiher, M., & Berking, S., (1996). Metamorphosis and pattern formation in *Hydractinia echinata*, a colonial hydroid. *International Journal of Developmental Biology, 40*(1), 313–322.

Weis, V. M., Keene, D. R., & Buss, L. W., (1985). Biology of hydractiniid hydroids 4. Ultrastructure of the planula of *Hydractinia echinata*. *Biological Bulletin, 168*(3), 403–418.

Wheeler, S. E., (2012). Understanding substituent effects in noncovalent interactions involving aromatic rings. *Accounts of Chemical Research, 46,* 1029–1038.

White, H. K., Hsing, P. Y., Cho, W., Shank, T. M., Cordes, E. E., Quattrini, A. M., et al., (2012). Impact of the deep-water horizon oil spill on a deep-water coral community in the Gulf of Mexico. *Proceedings of the National Academy of Science of the United States of America, 109,* 20303–20308.

Yin, F., John, G. F., Hayworth, J. S., & Clement, T. P., (2015). Long-term monitoring data to describe the fate of polycyclic aromatic hydrocarbons in deep-water horizon oil submerged off Alabama's beaches. *Science of the Total Environment, 508,* 46–56.

Zhang, M., Zhao, J., Liu, J., Zhou, L., & Bu, Y., (2013). Coexistence of solvated electron and benzene-centered valence anion in the negatively charged benzene-water clusters. *Journal of Chemical Physics, 138*(1), 014310.

CHAPTER 17

Logistic Enzyme Kinetics

MIHAI V. PUTZ,[1,2] ANA-MARIA PUTZ,[3] CORINA DUDA-SEIMAN,[1] and DANIEL DUDA-SEIMAN[4]

[1]*Laboratory of Computational and Structural Physical Chemistry for Nanosciences and QSAR, Biology-Chemistry Department, Faculty of Chemistry, Biology, Geography at West University of Timișoara, Pestalozzi Street No. 16, Timișoara, RO–300115, Romania*

[2]*Laboratory of Renewable Energies-Photovoltaics, R&D National Institute for Electrochemistry and Condensed Matter, Dr. A. Paunescu Podeanu Str. No. 144, RO–300569 Timișoara, Romania, Tel.: +40-256-592-638; Fax: +40-256-592-620, E-mail: mv_putz@yahoo.com, mihai.putz@e-uvt.ro*

[3]*Institute of Chemistry Timisoara of the Romanian Academy, 24 Mihai Viteazul Bld., Timisoara 300223, Romania*

[4]*Department of Medical Ambulatory, and Medical Emergencies, University of Medicine and Pharmacy "Victor Babes," Avenue C. D. Loga No. 49, RO–300020 Timisoara, Romania*

17.1 DEFINITION

Enzyme kinetics concerns the rate of an enzyme-catalyzed reaction: a substrate S is converted into a product P, under a strict dependence on the concentration of the enzyme E (Schnell and Maini, 2003). Talking about the mechanism of the reaction, in the reaction mixture an intermediate enzyme–substrate is obtained ES with undergoing reaction rates k_1 and k_{-1}. The next step of the enzymatic reaction is the irreversible obtaining by breaking down of the ES complex into the product P and the enzyme E. (Michaelis and Menten, 1913)

$$E + S \underset{k_{-1}}{\overset{k_1}{\leftrightarrow}} ES \overset{k_2}{\to} E + P \tag{1}$$

It is an open issue to undergo complete analytical aspects of the classical Michaelis-Menton reaction, modifying it under logistic form with constant in vitro reaction parameters (as in practical laboratory work).

17.2   NANO-CHEMICAL IMPLICATIONS

In order to have a logistic approach of enzyme kinetics, one approach consists in probabilistic calculations starting from the law of mass action with the ability to characterize *in vitro* enzymatic reactions of type (1) (Putz et al., 2007):

$$1 = P_{REACT}([S]_{bind}) + P_{UNREACT}([S]_{bind}) \tag{2}$$

Here $P_{REACT}([S]_{bind})$ represents the probability that the reaction (1) takes place under a certain concentration of substrate binding to the enzyme $[S]_{bind}$. Following situations are possible: $P_{REACT}([S]_{bind}) = 0$ when the enzymatic reaction does not take place, or it stops because of several reasons (either the substrate does not exist, or it is consumed); $P_{REACT}([S]_{bind}) = 1$ when the enzymatic reaction takes place. It is important to emphasize that the probability that the enzymatic reaction (Eq. 1) takes place is ranged between these two limits. $P_{UNREACT}([S]_{bind})$ defines the probability (in the same manner) that the enzymatic catalysis does or does not take place.

Enzymatic kinetics derives from the chemical bonding capacity of enzymes on the implied substrate, providing data upon the quantitative treatment of enzymatic catalysis (Voet and Voet, 1995). In this endeavor, the binding substrate concentration $[S]_{bind}$ can be assimilated to the instantaneous substrate concentration $[S](t)$. Another assumption that has to be made is that the enzymatic reaction takes place under constant association–dissociation rates, so that the probability of reaction is provided by the rate of substrate consumption.

$$v(t) = -\frac{d}{dt}[S](t) \tag{3}$$

to saturation (Putz et al., 2007):

$$P_{REACT}([S](t)) = \frac{v(t)}{V_{max}} = -\frac{1}{V_{max}}\frac{d}{dt}[S](t) \tag{4}$$

The expression (Putz et al., 2007):

$$P_{UNREACT}([S](t))^{MM} = \frac{K_M}{[S](t) + K_M} \tag{5}$$

corresponds to all probability requirements, providing the instantaneous version of the consecrated Michaelis-Menten reaction and being considered a common factor to define the efficiency of the Michaelis-Menten reaction written as follows:

$$v = \frac{d}{dt}[P] = \frac{V_{max}[S]}{[S] + K_M} \tag{6}$$

The expression (5) is only a certain and particular aspect for a probabilistic enzymatic kinetic model conservation law (2). In order to cover all the probabilistic aspects, Eq. (5) may be formulated as follows (Putz et al., 2007):

$$P_{UNREACT}([S](t))^* = e^{-\frac{[S](t)}{K_M}} \tag{7}$$

From this point, the term (6) is returned for the situation in which the bound substrate reaches the zero value (consumption), while the [S](t) registers a first order expansion (Putz et al., 2007):

$$P_{UNREACT}([S](t))^* = \frac{1}{e^{\frac{[S](t)}{K_M}}} \overset{[S](t) \to 0}{\cong} \frac{1}{1 + \frac{[S](t)}{K_M}} = P_{UNREACT}([S](t))^{MM} \tag{8}$$

Equation (8) gives the physical and chemical significance for the Michaelis-Menten term (5) and for the reaction kinetics in the case of fast enzymatic reactions and fast consumption of [S](t). Using Eq. (7), the reaction limits are enlarged, and the kinetic equation is obtained in the form of a logistic expression (Putz et al., 2007):

$$-\frac{1}{V_{max}} \frac{d}{dt}[S](t) = 1 - e^{-\frac{[S](t)}{K_M}} \tag{9}$$

This equation provides an initial velocity of reaction (v_0^*), higher, but in a uniform manner, than that calculated by (5). The exception is constituted when [S_0]→0.

The main acceptance in deriving the original Michaelis-Menten kinetics is that there is an equivalence between the *ES* complex synthesis rate and consumption rate until the substrate is almost exhausted. It has been shown (Laidler, 1955) that the *quasi-steady-state approximation* (QSSA) is superposable to the well known physiological pattern in which the substrate exceeds in a large manner the enzyme:

$$[S_0] >> [E_0] \tag{10}$$

In order (1) to take place, it is necessary that the following relation to be accomplished (Putz et al., 2007):

$$P_{REACT}([S]_{bind}) \rightarrow 1 \Leftrightarrow P_{UNREACT}([S]_{bind}) \rightarrow 0 \qquad (11)$$

The first term (the left side) of the probabilistic relation (11) is valid for QSSA. Best QSSA is obtained when $P_{UNREACT}([S](t))$ is close to zero and, respectively, $P_{REACT}([S](t))$ reaches a value close to one. The logistic probability $P_{UNREACT}$ is lower than the Michaelis-Menten in all conditions. So, QSSA is better when the logistic approach is taken into consideration.

17.3   TRANS-DISCIPLINARY CONNECTIONS

In the case of metabolic proteins, enzymes, it has been shown that their activities may be influenced and regulated by other molecules than substrates, of small shape, identified as activators or inhibitors (Cantor and Schimmel, 1980; Copeland, 2000). In the theory of allosteric regulation, it is established that the binding of a substrate (ligand) at a specific site is influenced by the binding of another different ligand identified as an inhibitor at a different or allosteric site on the protein/enzyme. This principle is applicable at the cellular level resulting in cellular regulation where organism needs are translated into metabolic processes via different protein/enzymatic pathways. Protein synthesis determined by gene expression is only a part of this process of RNA transcription which has to provide a perfect balance in the genome maintenance by repeated signalization started by other cells or bio-inspired nano-implants (Curran et al., 2005). The complex substrate-enzyme plays the role of molecular messengers.

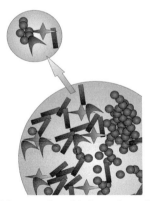

FIGURE 17.1   The sketch of the transferred information within the Ligand - Receptor, as the pattern for Substrate-Enzyme, interaction.

KEYWORDS

- **enzyme kinetics**
- **logistic**
- **Michaelis-Menton reaction**

REFERENCES AND FURTHER READING

Cantor, C. R., & Schimmel, P. R., (1980). *Biophysical Chemistry, Part III: The Behavior of Biological Macromolecules*. W. H. Freeman and Company, San Francisco.

Copeland, R. A., (2000). *Enzymes: A Practical Introduction to Structure, Mechanism, and Data Analysis*. Wiley-VCH, New York.

Curran, J. M., Gallagher, J. A., & Hunt, J. A., (2005). The inflammatory potential of biphasic calcium phosphate granules in osteoblasts/macrophage co-culture. *Biomaterials, 26*, 5313–5320.

Laidler, K. J., (1955). Theory of transient phase in kinetics, with special reference to enzyme systems. *Can. J. Chem., 33,* 1614–1624.

Michaelis, L., & Menten, M. L., (1913). Die kinetik der invertinwirkung. *Biochem. Z., 49*, 333–369.

Putz, M. V., Lacrama, A. M., & Ostafe, V., (2007). Introducing logistic enzyme kinetics. *J. Optoelectron. Adv. Mater., 9*(9), 2910–2916.

Schnell, S., & Maini, P. K. A., (2003). Century of enzyme kinetics: Reliability of the k_m and v_{max} estimates. *Comm. Theor. Biol., 8*, 169–187.

Voet, D., & Voet, J. G., (1995). *Biochemistry* (2nd edn.). John Wiley & Sons Inc., New York, Chapter 13.

CHAPTER 18

Minimal Steric Difference

BOGDAN BUMBĂCILĂ[1] and MIHAI V. PUTZ[1,2]

[1]*Laboratory of Computational and Structural Physical Chemistry for Nanosciences and QSAR, Biology-Chemistry Department, Faculty of Chemistry, Biology, Geography at West University of Timișoara, Pestalozzi Street No.16, Timișoara, RO–300115, Romania*

[2]*Laboratory of Renewable Energies-Photovoltaics, R&D National Institute for Electrochemistry and Condensed Matter, Dr. A. PaunescuPodeanu Str. No. 144, RO–300569 Timișoara, Romania, Tel.: +40-256-592-638; Fax: +40-256-592-620, E-mail: mv_putz@yahoo.com, mihai.putz@e-uvt.ro*

18.1 DEFINITION

Minimal steric difference is a steric mapping method for QSAR studies. It usually helps for determining the fitting spatial characteristics of the binding site for a ligand molecule on its specific receptor. The physical-chemical and steric properties of the receptor but also of the molecule are very important for this mapping technique. It is the initial method of currently-used minimal topological difference (MTD) method (Simon et al., 1994).

18.2 HISTORICAL ORIGINS

The MSD method was introduced by Zeno Simon in the early 70s as a very original method for determining of quantitative structure-activity relationships (QSAR) between chemical structures and their biological activity (Simon et al., 1973).

   The steric conformation of the responsible sites from the receptor for the biological activity of a ligand is usually theoretically assessed using the model of the natural substrate molecule (a molecule with assumed maximum biological effect). This natural ligand molecule may have several isoenergetic 3D conformations, and the exact conformation which best fits the receptor's site is not known (Simon, 1976).

The method is used to parameterize the similarities or differences of the molecular shape between a studied ligand and the natural ligand for a specific receptor. Initially, only the 2D representation of the ligand molecule was considered and most of the analyzed series of analog molecules with cyclic fragments. So, the Minimal Steric Difference between them was calculated. Later, the molecules were considered as 3D objects, some of them having the same molecular weights but different spatial conformations. So the Minimal Topological Difference had to be calculated for better precision (Ariëns, 979).

18.3   NANO-CHEMICAL IMPLICATIONS

The importance of the stereochemistry in the biological activity assessment of a molecule was recognized more than one hundred years ago, by Emil Fischer, who described the interaction between the drug molecule and its macromolecular receptor as a "steric fit" guided by the "key-lock" principle. A year later, this steric fit proved to be quite flexible – the "lock" (receptor) can be deformed by the key (drug), or the "key" can change its conformation for a better fit in the "lock" (Voiculetz et al., 1993).

The steric fit depends on the shape of both of the receptor and the ligand/ effector molecule. Steric parameters are describing the shape of the molecule. The MSD concept is that the linkage affinity of a series of compounds is a linearly decreasing function of the sum of non-superposable volumes of those molecules and the volume of the receptor's site (usually a concavity) (Balaban et al., 1980).

We should also introduce the term "steric misfit"–the difference between the steric hindrance of the standard compound and one of the molecules in the study. It is a measure of the similarity between two molecules. Low values of steric misfit between two molecules correspond to high similarity between them (Todeschini et al., 2009). Sometimes a steric misfit between the van der Waals volumes of the drug molecule and the receptor can generate local differences of energy, which may lead to conformational changes (for example changes of the bond angles) of the molecules implied with a steric fit effect (Voiculetz et al., 1993). MSD is, in fact, the parameter which approximates the steric misfit.

The first step for determining MSD/MTD is defining a standard molecule, which best binds the receptor. This standard ligand may be the most potent compound from a series, potency of its biological activity being suggested with a Free-Wilson analysis of the series or a known natural substrate. The second step is to get a hypermolecule from the superposing all the structural diagrams of the molecules in the series. The structural diagrams are, in fact, shapes of the molecules. These shapes are oriented in convenient positions. Eventual differences between bond lengths and angles and the Hydrogen atoms in the molecules are not taken into consideration (Ariëns, 1979).

The shape of the receptor's binding site should be hypothesized. The minimal steric difference MSD_i of a molecule M_i, considering a standard molecule M_0, is equal to the number of unsuperposable atoms when M_i and M_0 are overlapped, atom-by-atom (with neglect of the Hydrogen atoms). The relative orientation of M_i taking M_0 into consideration and the chosen conformation for M_i (if the molecule has more than one low-energy conformations) has to be as to allow a maximal superposition of the two molecules for obtaining a minimal steric difference between them (Voiculetz et al., 1993).

The biological activity of the M_i molecule expressed as A_i should depend on MSD respecting the following equation (Balaban et al., 1980):

$$A_i = \alpha - \beta MSD_i \qquad (1)$$

MSD_i represents the number of non-overlapping and non-hydrogen atoms when M_i and M_0 are superposed. Hydrogen atoms are neglected for the ease of the superposition because they have small van der Waals volumes and covalent lengths. Thus, atoms X in/and groups $-XH_3$, $-XH_2$, $-XH$, etc. are considered equivalent.

So the Minimal Steric Difference between a given molecule and its natural substrate is the non-overlapping volume of those lowest energy molecular conformations which permits a maximum overlap between the two molecules (Simon, 1976):

Because there are differences between van der Waals volumes of atoms in different periods, an adjustment has to be made:

- Second-period atoms (C, N, O) can be attributed a weight coefficient 1;
- Third-period atoms (S, Cl) should be then attributed a weight coefficient of 1.5;
- Higher period atoms (Br, I) should be attributed to a weight coefficient of 2.

If a heavier atom is superposed on a lighter atom (from a smaller period), par example a Br atom over a C atom of a $-CH_3$ group the difference of the weight coefficients should be added to the MSD value. If a heavier atom is superposed on a lighter atom (from a smaller period), but the lighter atom is engaged in a ramification, par example a I atom over a C atom from a –iPr group, the result added to the MSD value is obtained by extracting the weight of the higher period atom from the sum of weights of the ramification (3 C atoms, trifurcation, in our example) (Balaban et al., 1980).

As it was previously said, differences between bond lengths or bond angles are also neglected. Atoms bonded in one molecule should not be overlapped with non-bonded atoms in the other molecule, because van der Waals contact distances (3–4 Å) are much larger than covalent distances (1.2–2 Å). If the molecule has several low-energy conformations, the one who fits best to the receptor site should be chosen. For enzymatic studies, certain atoms from the studied molecule M_i (the atoms implied in bonds that are assumed to be cleaved) should be superposed on a certain group of atoms from the standard molecule M_0 (atoms implied in bonds that are known to be cleaved). Groups that are considered to have strong interactions with the receptor from the M_i should be superposed with groups known to highly interact with the binding site (for example electrically charged groups) of the M_0 (Balaban et al., 1980).

The relative orientation of the superposition is given by the pharmacophore features (molecular characteristics necessary for the recognition of M_i at the binding site on the macromolecular receptor) or by the common nucleus/ring presented by both M_0 and M_i.

We can summarize all the above said with a formula for the calculation of MSD of one molecule:

$$MSD_i = \sum_{j=1}^{m} d_j \left| w_{ij} x_{ij} - v_j \right| \tag{2}$$

where d_j is the relevance (1) or irrelevance (0) of atom j for the steric fit, x_{ij} indicates if atom j is occupied (1) or unoccupied (0) in molecule i and v_j indicates if the atom is present in the standard (1) or absent (0); $w = 0$ (for hydrogen atoms), 1, 1.5 or 2 for second/third/higher period elements.

For example, if the standard is considered R-cyclopentyl (S), the MSD values for R-phenyl (B) and R-cyclohexyl (A) are both 1. The MSD value of R-i-propyl would be 2, and that of R-t-butyl would be 3. The structures of these molecular species are shown in Figure 18.1 (Ariëns, 1979).

This calculus permits the summarization of all the differences between arrangements of atoms in the studied molecule and the standard one (Ariëns, 1979).

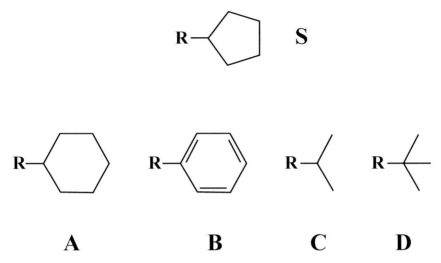

FIGURE 18.1 Structures of R-cyclopentyl (S), R-cyclohexyl (A), R-phenyl (B), R-i-propyl (C) and R-t-butyl (D).

A conclusion can be drawn from the MSD calculus: any relevant atom in a molecule which is also contained in the standard molecule is bringing the same incremental value for the receptor binding, and any relevant atom from the molecule which is not presented by the standard molecule involves a negative effect for the receptor binding.

The MSD value can be used as a variable in the Hansch equation which expresses the correlation between the biological activity and the relevant parameters:

$$BA = a + b(\log P) + c\,(\log P)^2 + \ldots + nMSD \qquad (3)$$

In QSAR studies, the molecule considered as Standard and the values of the relevance of atoms for the steric fit (d_j) in the hypermolecule are the most important. All the analogs in the molecule series have their own MSD values and corresponding statistical measures of "fitness" to the receptor, measures expressed as different values of the biological activity in the Hansch equation. A map of the drug-receptor interaction or a "negative" copy of the receptor can be realized with these MSD values of the analog molecules in the studied series (Ariëns, 1979).

In the following figure, superpositions, and MSD values for oxytocin derivatives with amino acid substitutions in position 3 are presented. A decrease of the biological activity (represented by A value) can be expressed as a decrease of the oxytocic activity with oxytocin being the standard. For

MSD, the standard was isoleucine (Ile). Atoms marked with * are unsuperposable upon Ile side chain (Voiculetz et al., 1993).

M_0 is considered the molecule with the best affinity for the receptor; thus, it is considered a "negative" of the receptor's cavity (Minailiuc et al., 2001) (Figures 18.2 and 18.3).

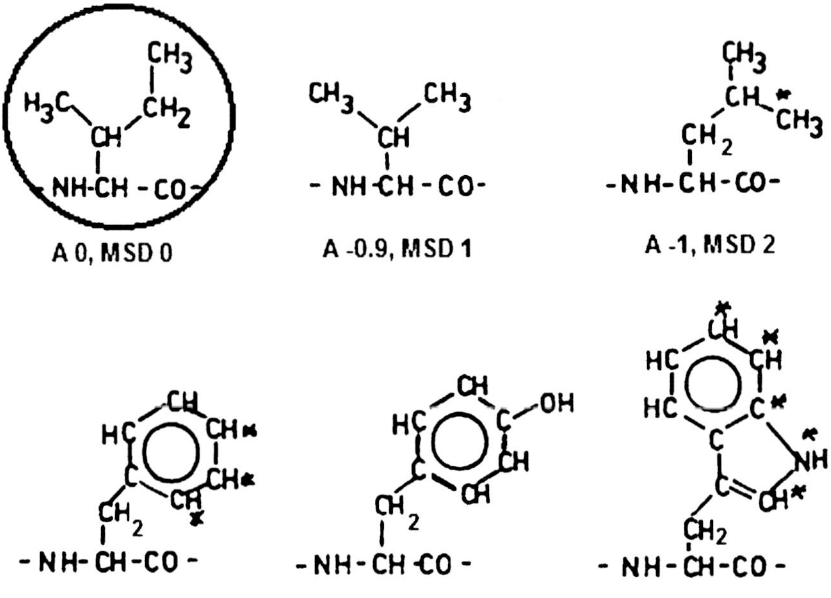

A 0, MSD 0 — A -0.9, MSD 1 — A -1, MSD 2

A -1.25, MSD 5 — A -3.65, MSD 6 — A -4.05, MSD 8

FIGURE 18.2 MSD values and biological activities for oxytocin derivatives with other amino acid substitutions in position 3, where Ile is highlighted (Voiculetz et al., 1993).

The result of the superposition of the M_i molecules over the M_0 one is the hypermolecule, an estimation of the binding space on the receptor which is more efficiently described in the MTD improved variant of MSD (Simon, 1984).

The affinity for a receptor of a ligand molecule depends not only on its steric fit at the binding site but also on its other properties, like solvation, electronic reactivity characteristics (if the biological activity implies not only the binding in the cavity but also a chemical reaction between the ligand and the macromolecule) and other intermolecular forces which can appear if both the molecule and its receptor are presenting certain surface functional groups (Balaban et al., 1980).

18.4 TRANS-DISCIPLINARY CONNECTIONS

MSD is very accurate for correlations implying amino acid replacements in biologically active oligopeptides. The natural oligopeptide is considered the standard M_0.

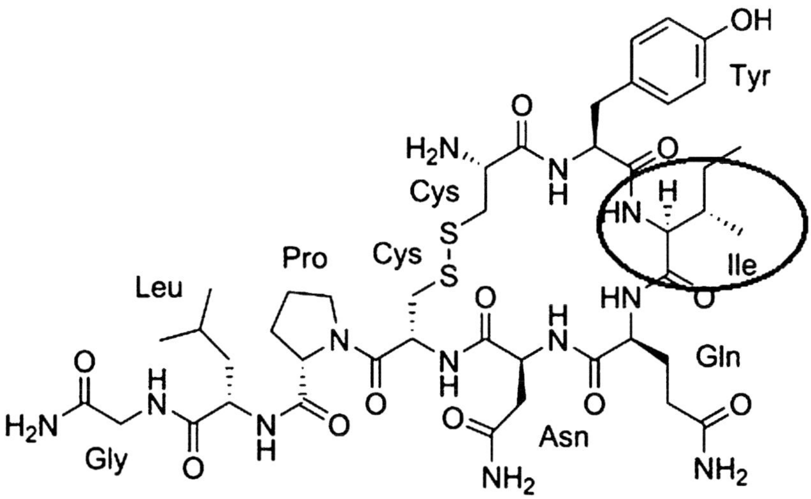

FIGURE 18.3 Oxytocin structure. Ile rest is highlighted.

For example, the decrease of oxytocic activity was correlated with MSD for a series of 42 compounds. The difference between the precise compound's activity and the standard activity ($\Delta A_{i, i = 1-42}$) was evaluated using oxytocin as a standard (Figure 18.4).

(1)Cys-(2)Tyr-(3)Ile-(4)Gln-(5)Asn-(6)Cys-(7)Pro-(8)Leu-(9)Gly

FIGURE 18.4 Oxytocin primary structure.

In Table 18.1, MSD values for amino acid-amino acid superpositions were calculated by Balaban et al., (1980) as follows: between L-amino acids – above the diagonal, between the same D- and L-amino acids – on the diagonal and between D- and L-amino acids – below the diagonal. In Table 18.2, the same researchers had calculated MSD values and the variation of the biological activity of 42 oxytocine derivatives obtained by different amino acid replacements.

TABLE 18.1 MSD Values for Aminoacid-Aminoacid Replacements in Peptides/Proteins (Balaban et al., 1980)

	L-Arg	L-Lys	L-Glu	L-Asp	L-Gln	L-Asn	L-Thr	L-Ser	L-Hys	L-Trp	L-Tyr	L-Phe	L-Pro	L-Met	L-Cys	L-Ile-Leu	L-Val	L-Ala	Gly	
Gly	7	5	5	4	5	4	3	2	6	10	8	7	3	4.5	2.5	4	3	1	0	Gly
L-Ala	6	4	4	3	4	3	2	1	5	9	7	6	2	3.5	1.5	3	3	2	1	D-Ala
L-Val	6	4	4	3	4	3	0	1	5	9	7	6	2	3.5	1.5	1	6	4	3	D-Val
L-Leu	5	3	3	0	3	0	3	2	2	6	4	3	3	2.5	1.5	2	5	5	4	D-Leu
L-Ile	5	3	3	2	3	2	1	2	4	8	6	5	3	2.5	1.5	4	5	5	4	D-Ile
L-Cys	4.5	2.5	2.5	1.5	2.5	1.5	1.5	0.5	3.5	7.5	5.5	4.5	1.5	3	5	4.5	5.5	3.5	2.5	D-Cys
L-Met	3.5	1.5	2.5	2.5	2.5	2.5	3.5	2.5	2.5	6.5	4.5	3.5	3.5	3	4	3.5	4.5	5.5	4.5	D-Met
L-Pro	6	4	4	3	4	3	2	1	3	7	5	4	6	7.5	5.5	7	6	4	3	D-Pro
L-Phe	2	2	4	3	4	3	6	5	1	4	1	8	10	7.5	9.5	9	10	8	7	D-Phe
L-Tyr	1	3	5	4	5	4	7	6	2	3	10	9	11	8.5	10.5	10	11	9	8	D-Tyr
L-Trp	5	5	7	6	7	6	9	8	4	12	14	13	13	7.5	10.5	10	11	11	10	D-Trp
L-Hys	3	1	3	2	3	2	5	4	2	6	10	9	9	3.5	6.5	6	7	7	6	D-Hys
L-Ser	5	3	3	2	3	2	1	4	6	12	10	9	5	4.5	4.5	4	5	3	2	D-Ser
L-Thr	6	4	4	3	4	3	6	5	7	11	11	10	6	5.5	5.5	5	6	4	3	D-Thr
L-Asn	5	3	3	0	3	4	5	4	4	8	10	9	7	2.5	5.5	4	5	4	4	D-Asn
L-Gln	4	2	0	3	6	5	6	5	6	13	10	9	8	5.5	4.5	5	6	6	5	D-Gln
L-Asp	5	3	3	4	5	4	5	4	4	8	10	9	7	2.5	5.5	4	5	5	4	D-Asp
L-Glu	4	2	6	5	6	5	6	5	6	13	10	9	8	5.5	4.5	5	6	6	5	D-Glu
L-Lys	2	4	5	3	5	3	5	5	3	9	7	6	6	5.5	5.5	3	5	6	5	D-Lys
L-Arg	4	6	7	5	7	5	7	7	5	7	7	6	7	7.5	7.5	5	7	8	7	D-Arg

TABLE 18.2 MSD Values and ΔA (Decrease in Activity) in Oxytocin Derivatives Obtained by Amino Acid Replacements (Balaban et al., 1980)

Derivative	Replacement	MSD	ΔA	Derivative	Replacement	MSD	ΔA
1	Phe2	1	1.15	22	Phe2Arg8	6	2.35
2	Ser2	6	5	23	Ser2Hys3	7	5
3	Phe3	5	1.25	24	Ser2Lys8	9	5
4	Tyr3	6	3.65	25	Hys2Phe3	7	5
5	Trp3	8	4.05	26	Phe3Lys8	8	1.95
6	Leu3	2	1	27	Phe3Arg8	10	0.8
7	Val3	1	0.9	28	Phe2Hys8	7	2.4
8	Ser4	3	0.3	29	Tyr3Lys8	9	4.6
9	Ala4	1	1.1	30	Trp3Lys8	11	5
10	Asn4	3	0.6	31	Ser4Ile8	5	0.5
11	Ser5	2	2.8	32	Asn4Gln5	7	3.05
12	Ala5	3	5	33	Phe2Phe3Lys8	9	3.2
13	Gln5	3	2.65	34	Phe2Phe3Arg8	11	3.35
14	Val5	3	5	35	Phe2Tyr3Lys8	10	5
15	Ile8	2	0.2	36	Ser2Hys3Lys8	10	5
16	Val8	3	0.35	37	Hys2Ser3Lys8	7	5
17	Lys8	3	0.75	38	Hys2Phe3Lys8	10	5
18	Arg8	5	0.75	39	Phe3Asn4Lys8	11	2.2
19	Phe2Phe3	6	2.1	40	Phe3Ala4Lys8	9	2.95
20	Phe2Tyr3	7	5	41	Phe3Ser4Lys8	11	2.7
21	Phe2Lys8	4	2.65	42	Phe3Ser5Lys8	10	5

The biological activity was determined with the following equation:

$$A = 0.993 + 0.484\Delta\pi + 0.646\Delta AR + 0.761\Delta CT - 0.685\Delta EC$$
$$+ 0.182\Delta HB + 0.958\Delta\mu + 0.067 \, MSD \qquad (4)$$

where π is Tanford hydrophobicity, AR is the Aromatic Character, EC is the electric charge at pH 7, CT is the charge transfer character, HB is the hydrogen bonding character and μ is the orientation of the dipole moment.

According to the obtained values, it was found that replacements in positions 3, 4 and 8 are not greatly influencing the biological activity to the corresponding space on the receptor to the side chains of the amino acids in these three positions has a "low rigidity" and accepts these replacements (Balaban et al., 1980).

18.5   OPEN ISSUES

The hypermolecule created by superpositions presents "vertices," and these vertices correspond to the positions of superposable atoms and the edges to the bonds between these atoms in the "i" studied molecules. The hypermolecule spatially resembles to the cavity of the receptor where the ligands are binding. The vertices can be categorized in "cavity vertices" with beneficial effect for interaction, "wall vertices" with detrimental effect for interaction (they are protruding the wall of the receptor) and "irrelevant vertices" with no effect on the interaction (probably exterior to the cavity–they are not able to promote a biological effect) (Draber et al., 2000).

Simon improved in 1984 the MSD method and called the improvement the MTD (Minimal Topological Difference) method, also a measure of the steric misfit of the ligand molecules regarding the shape of the receptor's cavity, where the superposition procedure is repeatable for atoms occupying similar positions in similar structures. Finally, a "map" of the receptor is obtained using this hypermolecule's network with vertices. Later, computer software was created for a better overlapping of the molecular volumes for avoiding miscalculations (Voiculetz et al., 1993).

The challenge, today, consists of the ongoing improvement of these programs for a more accurate "mapping" process of the receptive macromolecule.

KEYWORDS

- **hypermolecule**
- **molecular superposition**
- **steric fit**
- **steric misfit**

REFERENCES AND FURTHER READING

Ariëns, E. J., (1979). *Drug Design in Medicinal Chemistry: A Series of Monographs* (Vol. II-VIII). Academic Press, New York.

Balaban, A., Chiriac, A., Moţoc, I., & Simon, Z., (1980). *Steric Fit in Quantitative Structure-Activity Relationships*. Springer Verlag, New York.

Draber, F., & Fujita, T., (2000). *Rational Approaches to Structure, Activity, and Ecotoxicology of Agrochemicals* (pp. 60–64), CRC Press, Inc.

Minailiuc, O. M., & Diudea, M. V., (2001). *QSPR/QSAR Studies by Molecular Descriptors* (pp. 363–388). Nova Science, Huntington, New York.

Simon, Z., (1976). *Quantum Biochemistry and Specific Interactions*. Abacus Press, Tunbridge Wells, Kent, England.

Simon, Z., (1984). *Minimum Steric Difference: The MTD Method for QSAR Studies*. Research Studies Press.

Simon, Z., (1987). Stereochemical and informational aspects in QSAR and MTD. *Rev. Roum. Chim., 32,* 103–1107.

Simon, Z., & Szabadai, Z., (1973). Minimal steric difference-parameter and the importance of steric fit for quantitative structure–activity correlations. *Studia Biophys. (Berlin), 39*, 123–132.

Simon, Z., Chiriac, A., Holban, S., Ciubotariu, D., & Mihalas, G. I., (1994). *Minimum Steric Difference* (p. 173). The MTD method for QSAR studies, Research Studies Press Ltd., Letchworth.

Todeschini, R., & Consonni, V., (2009). *Molecular Descriptors for Chemoinformatics*. WILEY-VCH Verlag GmbH & Co. KGaA, Weinheim.

Voiculetz, N., Moţoc, I., & Simon, Z., (1993). *Specific Interactions and Biological Recognition Processes* (pp. 49–51). CRC Press, Inc, USA.

CHAPTER 19

Minimal Topological Difference

CORINA DUDA-SEIMAN,[1] DANIEL DUDA-SEIMAN,[2] and
MIHAI V. PUTZ[1,3]

[1]*Laboratory of Computational and Structural Physical Chemistry
for Nanosciences and QSAR, Biology-Chemistry Department, Faculty
of Chemistry, Biology, Geography at West University of Timișoara,
Pestalozzi Street No.16, Timișoara, RO–300115, Romania*

[2]*Department of Medical Ambulatory, and Medical Emergencies,
University of Medicine and Pharmacy "Victor Babes,"
Avenue C. D. Loga No. 49, RO–300020 Timisoara, Romania*

[3]*Laboratory of Renewable Energies-Photovoltaics, R&D National
Institute for Electrochemistry and Condensed Matter, Dr. A. Paunescu
Podeanu Str. No. 144, RO–300569 Timișoara, Romania,
Tel.: +40-256-592-638; Fax: +40-256-592-620,
E-mail: mv_putz@yahoo.com, mihai.putz@e-uvt.ro*

19.1 DEFINITION

First ideas underlying the QSAR studies (quantitative relationship chemical structure-biological activity) occurred after 1890 with the observations of Richet, Meyer, and Overton on the relationship between water/lipid solubility and toxicity or narcosis, the theory of Emil Fischer on the importance of the stereo-configuration of the substrate in enzyme reactions, and the antibacterial chemotherapy studies made by Ehrlich, which led him to formulate the concept of the receptor (Kubinyi, 1993). There is an approach in which the molecular modeling techniques relate to mapping techniques [e.g., the minimal topological difference method (MTD)], using equations of the type 1; but their application leads to values likely to be generalized to other types of biological activities, or for compounds from other *L* series of effectors (Simon et al., 1985).

$$A_x = a + b_1 t_x + b_2 t_x^2 + b_3 \sigma_{1,x} + b_4 \sigma_{D,x} + b_5 S \qquad (1)$$

Steric effects depend on the biological receptor form (R) and of the bioactive molecule (ligand –L); interaction and complex formation L–R is in strict dependence with the bioactive ligand L shape complementarity in relation to the biological receptor site for the structure R. In order to assess the degree of quantitative steric L–R mismatch, in QSAR techniques there was introduced the minimal steric difference (MSD). MTD was extended (Simion et al., 1984; Ciubotariu et al., 2000) to consider both compounds are exhibiting multiple conformations, and those whose activities cannot be determined with certainty.

19.2   HISTORICAL ORIGINS

MTD was developed as a molecular modeling technique by the QSAR group from Timisoara (Romania): Prof. Dr. Zeno Simon, Prof. Dr. Adrian Chiriac, Prof. Dr. Dan Ciubotariu.

19.3   NANO-CHEMICAL IMPLICATIONS

For a set of molecules L_i, $i = 1,2., N$ of organic compounds that share a common biological property P, known, and measurable. It is assumed that the known values of the biological activities A_i, $i = 1., n; n < N$, are determined under the same conditions. To broaden the scope of the method, there will be considered also the cases where a number k of molecules can have the values A_i, $i = 1., k$ that are lower than a limit A_{max}, p molecules that can have values A_i, $i = 1., p$ above than a limit A_{min}, or for m molecules $A_i \in [A_1 - A_2]$. Obviously, $N = n + k + m$. In these circumstances the question of establishing a relationship QSAR using MSD as a measure of steric disparity of L_i ligands in relation to the receptor site. Minimum steric differences (MTD$_i$ values) are obtained by comparing the shape of each molecule L_i with a hypothetical form of the receptor's cavity, the H hypermolecule. The ultimate goal is to achieve optimized shapes of the receptor, which can provide information on the structure of the receptor, on the functional and steric requirements imposed for the L_i structures, and on the mechanism of interaction while the L-R complex formation. Consequently, the method involves the development of algorithms to build the H hypermolecule (describing the stereochemistry of R, based on information provided by L_i)

and to optimize the topography of *H* (and, implicitly of R) (Simion et al., 1984; Ciubotariu et al., 2000).

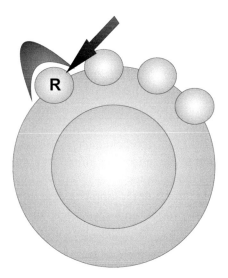

FIGURE 19.1   The sketch of the "Ligand" substitution/addition on the "Receptor".

H, the hypermolecule building, is achieved by superposition, atom by atom, of all the L_i molecules from the studied. Usually, the most active molecule in the series is considered as standard, *S*, the other molecules being superimposed over S_r, taking into account a number of structural rules. Being the most active compound in the studied series, *S* can be considered as the "best model" for the receptor, actually a negative copy (complementary) of the receptor.

The relative orientation of the L_i molecules is such as to achieve a superposition of the pharmacophore groups or of those atomic groups similar electrically charged, capable of strong interactions with the receptor, of common molecular fragments, etc.

In the superposition process, hydrogen atoms are neglected, being considered "included" in the atoms in the second period and the following; thus, it is considered the molecular graph without hydrogen atoms, similar to topological quantitative models. If necessary, it is taken into account the structural 3D aspect.

Atom by atom superposition of the other (N–1) L_i molecules is realized so that the overlapping degree to be maximized. To achieve this goal, slight differences that may occur between lengths of bonding and valence angles of different molecules are neglected, but, it is important not to overlap pair of atoms covalently bonded in one molecule (the distance between atoms is of 1, 2–2 Å) over pairs of non-bonded atoms from other molecules (van der Waals contact distance between atoms is of 3–4 Å). For molecules with conformational flexibility, all low-energy conformers are considered.

From the above, it is noted that the result of the superposition process, the H hypermolecule, is actually a topological network (2D with possible 3D elements) composed from a number of h vertices. These correspond roughly to the positions of atoms in the L_i molecules, while the edges of the network can be treated as chemical bonds between atoms. In this way, the H hypermolecule reflects overall the stereochemistry of bioactive molecules which interact with the receptor site, encoding indirectly information regarding the structure of the biological receptor.

The hypermolecule allows describing the stereochemistry of L_i bioactive molecules and of the site of the biological receptor R. For this purpose, for each conformation k of the L_i molecule, it is used a number of h structural variables x_{ijk}, $i = 1, ..., N; j = 1, ..., h; k = 1, ..., C_i$, with following significance: $x_{ijk} = 1$ if the j vertex in H is occupied by L_i in k conformation, $x_{ijk} = 0$ if it is not occupied.

The stereochemistry of the site of the biological receptor is described with the aid of a ternary parameter ε_j, attached to each vertex in $\mathbf{H}$, as follows:

$\varepsilon_j = -1$ corresponds to vertices located in the cavity of the receptor;

$\varepsilon_j = +1$ is the attachment to vertices in the receptor's walls;

$\varepsilon_j = 0$ is referring to the vertices outside of the receptor, in the steric irrelevant zone.

The three assignments of vertices–cavity (−1), outside (0), or in the walls of the receptor (+1)–encodes a positive, negative or non-relevant effect within the considered series of molecules upon biological activity; the effect is due to the occupancy of the j vertices in $\mathbf{H}$.

The sequences of the h values ε_j represent the investigated space of the receptor (Hansch, 1993), or the image, the map, or the chart of the receptor (Hyde, 1988), actually an approximate model of a part of the receptor in an active biological confirmation.

The MTD_i of the L_i molecule in relation to the receptor chart $S^0 = \{\varepsilon_1, \varepsilon_2., \varepsilon_h\}$ is defined as the sum between the number of vertices belonging to the unoccupied receptor cavity by L_i and the number of in the receptor's wall occupied by L_i. MTD values are calculated as follows:

$$MTD_{ik} = s + \sum_{j=1}^{h} \varepsilon_j x_{ijk} \tag{2}$$

In (2), s is the total number of vertices in the cavity, ε_j represents the assignments of vertices in **H**, and x_{ijk} show that the j vertex in H is occupied by L_i in the k conformation ($x_{ijk} = 1$) or is not occupied ($x_{ijk} = 0$).

The general form of the correlation analysis model is:

$$\hat{A}_i = a_0 + a_1 S_{1i} + a_2 S_{2i} \cdots - bMTD_i \tag{3}$$

S is the parameters that measure other effects, possibly involved in the observed biological response.

The optimization algorithm involves the following steps:

1. Selecting an initial chart $S^0 = \{ \varepsilon_1^0, \varepsilon_2^0., \varepsilon_h^0 \}$ for the H hypermolecule. This is done automatically, taking into account the relative occupancy degree of each vertex in the most active molecules and in the less active ones.
2. Calculating MTD_i values in relation to the adopted S^0 in step 1. In cases of multiple conformations for molecule L_i, one will choose the k conformation that will match to the minimal MTD_{ik} value, i.e., $MTD_i = MTD_{ik}$.
3. Obtaining the equation (2) using the usual methods of correlation analysis. Optimization criterion is the correlation coefficient, r.
4. At each step, p, of the optimization procedure, in the previous S^{p-1} chart, the assignments ε_j of each vertex are changed, for each being calculated the correlation coefficient, r (ε_j). In this way, one you will obtain 2h images of the receptor, respectively 2 h values of the correlation coefficient, r.
5. The S^p is selected corresponding to the maximum correlation coefficient obtained in step p–1 $(r_p = r_{p-1}^{max})$ and continues the optimization procedure (steps 2–4) until the correlation coefficient obtained in step p is lower than that obtained in step p–1 $(r_p < r_{p-1})$. The optimized receptor chart (S*) is that corresponding to the step p–1, i.e.

$S_{p-1} = S^* = \{\varepsilon_1^*, \varepsilon_2^*., \varepsilon_h^*); \varepsilon_1^*, \varepsilon_2^*., \varepsilon_h^*$ values represent final assignments of vertices in **H**. Equation (2) enables the calculation of MTD_{ik} values or any number of conformations of a M_i molecule. The method, in its original form (Simon et al., 1984), makes no distinction between the size of atoms or atomic groups in considered molecules, occupying the same vertex in **H**. The method was extended (Hansch, 1993), allowing an exhaustive investigation of the space of the receptor.

19.4   OPEN ISSUES

The MTD method allows obtaining rapidly information regarding interactions between ligands L and involved receptors R. Its use can be very effective if the interpretation of results is made with caution (especially the significance of the assigned vertices) and in an interaction model of L–R judiciously chosen.

KEYWORDS

- **ligand**
- **minimal topological difference (MTD)**
- **QSAR**
- **receptor**

REFERENCES AND FURTHER READING

Ciubotariu, D., Medeleanu, M., & Gogonea, V., (2000). The Van Der Waals molecular descriptors and the new variant of the MTD method, Chapter 11. In: Diudea, M., (ed.), *Molecular Descriptors for QSPR/QSAR Studies*. USA, Nova Science.

Hansch, C., (1993). Quantitative structure-activity relationships and the unnamed science. *Acc. Chem. Res.,* 26(4), 147–153.

Hyde, R. M., & Livingstone, D. J., (1988). Perspectives in QSAR: Computer chemistry and pattern recognition. *Journal of Computer-Aided Molecular Design, 2*(2), 145–155.

Kubinyi, H., (1993). 3D QSAR in drug design. *Theory, Methods, and Applications*. ESCOM, Science Publishers, B. V., Leiden.

Simon, Z., Chiriac, A., Holban, S., Ciubotariu, D., & Mihalas, I., (1984). *Minimum Steric Difference*. The MTD Method for QSAR Studies, Research Studies Press, Letchworth, UK.

Simon, Z., Ciubotariu, D., & Balaban, A. T., (1985). Reactivity and stereochemical parameters in QSAR for carcinogenic polycyclic hydrocarbon derivatives. In: Seydel, J. K., (ed.), *QSAR, and Strategies in the Design of Bioactive Compounds* (pp. 370–373). VCH Verlagsgesellschaft, Weinheim.

CHAPTER 20

Macromolecular Crowding

ADRIANA ISVORAN,[1,2] LAURA PITULICE,[1,2] EUDALD VILASECA,[3]
ISABEL PASTOR,[4] SERGIO MADURGA,[3] and FRANCESC MAS[3]

[1]*Department of Biology-Chemistry, West University of Timisoara,
16 Pestalozzi, Timisoara, Romania, E-mail: adriana.isvoran@e-wut.com*

[2]*Advanced Environmental Research Laboratory, 4 Oituz, Timisoara,
Romania*

[3]*Department of Materials Science & Physical Chemistry and Research
Institute of Theoretical and Computational Chemistry (IQTCUB) of
Barcelona University, C/Martí i Franquès, 1, 08028-Barcelona, Spain,
E-mail: eudald.vilaseca@ub.edu, s.madurga@ub.edu*

[4]*Small Biophysics Lab of Condensed Matter Department, Barcelona
University, C/Martí i Franquès, 1, 08028-Barcelona, Spain, and Carlos III
Health Institute, Madrid, Spain*

20.1 DEFINITION

Macromolecular crowding (MC) refers to a high concentration of macromolecules that is present in a limited space such as the intracellular environment. A substantial proportion of the considered volume is physically occupied, reducing the effective volume available for all present molecules. This volume exclusion is a relevant factor in analyzing soft interactions, viscosity, perturbed diffusion and physical interactions between the present molecules and it affects both the properties of molecules and the physicochemical processes taking place in such environment. Crowding especially influences diffusion processes and biochemical reactions with great implications in biochemistry and cell biology.

20.2 HISTORICAL ORIGIN

The study of some physicochemical processes in concentrated environments started in the late '50s using proteins and some polysaccharides. It generated the theory of excluded volumes in the 1960s (Ogston and Phelps, 1961; Laurent, 1964). More studies followed considering the effects of excluded volume on the behavior of macromolecular complexes. The concept of MC appeared at the beginning of the 1980s, along with the necessity to study properly the biochemical processes taking place in living cells (Minton, 1981, 1983).

20.3 NANO-SCIENTIFIC DEVELOPMENT

In living cells, the total concentration of macromolecules is high (60% proteins) occupying about 30% of the cellular volume (Luby-Phelps, 2000). Such an environment is denoted as crowded (Figure 20.1). The average distance between macromolecules in a crowded media can be much smaller than their own size, and the volume they occupy is unavailable to other molecules, conducting to excluded volume effects. Also, the living cell is a highly structured and compartmented media, and thus it enhances the effects of MC.

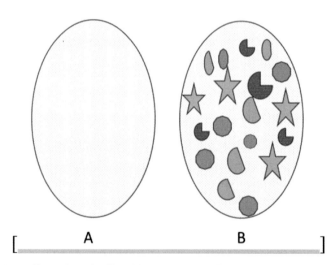

FIGURE 20.1   Illustration of a diluted homogenous media (A) by comparison to a crowded media (B).

As a consequence, the random distribution of the particles decreases and leads to the decrease of the medium entropy with corresponding implications on their biological and biochemical processes (Ralston, 2005). It was extensively believed that the major mechanism responsible for the consequences of crowded environments on biological molecules and their processes may be described in terms of excluded volume effects. Quite recently, literature data have shown that other factors are also important: viscosity, partition, soft interactions, the effects of crowders on the solvent properties, physical interactions between crowders and other molecules (Kuznetsova et al., 2015). Furthermore, the crowding agents are non-equivalent, and the crowding they produce can be grouped into two different categories (Ma and Nussionov, 2013): uniform crowding, generated by synthetic particles with a narrow size distribution and structured crowding, corresponding to cellular environments.

Both experimental and computational studies are performed considering the effects of MC on biological molecules and their processes. The experimental studies are performed *in vitro* using concentrated solutions of crowding agents (e.g., polymers) as environments for the analyzed systems. Computational studies are based on *on-* and *off-lattice* simulation algorithms for biological processes in which the molecules of interest and the species of crowding agents are accordingly considered. These studies equally revealed that MC might affect macromolecular interactions and proved its impact in disturbing many biological molecules and biological processes: protein folding, stability, and association (van der Berg et al., 1999, 2000; Minton, 2000; Ellis and Minton, 2006), genome structure and function (Zimmerman and Harisson, 1987; Zimmerman, 1993), diffusion rates (Saxton, 1987, 1989, 1990, 1992, 1993, 1994; Dix and Verkman, 2008), reaction rates and equilibrium constants of biochemical reactions (Zimmerman and Minton, 1993; Minton, 2001; Zhou et al., 2008). The effects produced by MC on biological processes are summarized in Figure 20.2.

With regard to the diffusion process, in diluted solutions, it is normal and characterized by the mean squared displacement ($< r^2(t) >$) increasing linearly with time (t)

$$< r^2(t) >= 2dDt \qquad (1)$$

where d is the topological dimension of the media, D is the diffusion coefficient. This equation is known as Einstein-Smolukovski equation (Havlin and Ben-Avraham, 2002). This equation stands valid if the diffusing particles are independent, their individual displacements between two successive collisions are statistically independent and if displacements also correspond to a typically mean free path distributed symmetrically in positive and negative directions.

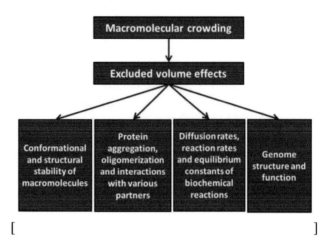

[   ]

FIGURE 20.2 Schematic representation of the potential effects of macromolecular crowding on the biological and biochemical processes.

MC affects quantitatively and qualitatively the diffusion processes. The diffusion of a kind of molecule is hindered by the other molecules that are present in the diffusion environment and may result in anomalous diffusion (Saxton, 1987, 1989, 1990, 1992, 1993, 1994; Dix and Verkman, 2008). For anomalous diffusion, at least one of the validity conditions of the Einstein-Smolukovski equation is not fulfilled, and the mean squared displacement depends on time as follows:

$$<r^2(t)> = 2d\Gamma t^{\alpha} \tag{2}$$

where Γ is the generalized diffusion coefficient and α is the anomalous diffusion exponent with $0 < \alpha < 1$ for subdiffusion and $\alpha > 1$ for superdiffusion (Havlin and Ben-Avraham, 2002; Banks and Fradin, 2005; Pastor et al., 2010). The quantitative effects of MC on diffusion are also denoted by reduced diffusion coefficients and they dependence on time. The corresponding time scale laws have exponents influenced by size and mobility of the crowding molecules (Vilaseca et al., 2011a). Both simulation and experimental studies have confirmed the presence of anomalous diffusion in crowded media and its dependency on size and conformation of the traced particle and total concentration of the crowded solution.

Protein folding, stability, and association processes have been studied for many years in the diluted solution, but the effects of the cytosolic environment on these processes have only been considered for a few decades. The crowding agents do not bind covalently to the protein, but they change the properties of the environment of the protein. As protein misfolding and/or aggregation have been associated with several neurodegenerative diseases (Apetri and Surewicz, 2002; Winklhofer et al., 2008), it becomes important to study the protein's behavior in their natural environment and take into consideration the effects of the environment.

Minton (2000) has shown that MC has positive effects on protein folding and aggregation. It stabilizes the compact native protein structure compared to the expanded, denatured one and enhances the formation of protein complexes. Under pathological circumstances, MC may conduct to the formation of nonfunctional aggregates of proteins (Minton, 2000; Yadav, 2013; Kuznetsova et al., 2015). MC also enhances the stability of proteins (Pozdnyakova and Wittung-Stafshede, 2010). The effects of crowding agents seem to be very protein-specific. These effects underlie that the simple excluded volume effects added to non-specific steric interactions are not the only ones to be considered and some protein-crowders specific interactions are also important (Kuznetzova et al., 2015).

The biological activities of enzymes are usually determined by minor conformational transitions. As MC affects the conformational state of the protein, it is expected that it also modulates the catalytic activities of enzymes. The enzyme activity may be enhanced in crowded environments because of the structural changes induced by the crowding agents (Jiang and Guo, 2007; Akabayov et al., 2013). However, there are also enzymes that do not change their activity in the presence of crowding agents (Asaad and Engberts, 2003; Vopel and Makhatadze, 2012). It is considered that enzyme activity is affected by MC if catalysis is accompanied by changes in size, or shape of the enzyme (Kuznetsova et al., 2015). For oligomeric enzymes, the formation of oligomers is advantaged by the crowded media through enhancement of molecule association.

Dedicated literature also contains experimental and simulation studies that have explored the effects of MC on enzyme catalysis (Laurent, 1971; Minton and Wilf, 1981; Gellerich et al., 1998; Wenner and Bloomfield, 1999; Berry, 2002; Derham and Harding, 2006; Olsen, 2006; Olsen et al., 2007; Sasaki et al., 2007; Moran-Zorzano et al., 2007; Pastor et al., 2010, 2011, 2014; Vilaseca et al., 2011a, 2011b; Pitulice at al., 2013, 2014; Balcells et al., 2014, 2015). For enzymatic reactions taking place in crowded environments, Kopelman introduced the concept of *fractal-like kinetics* and considered that the rate coefficient for diffusion-controlled reactions depends on time (Kopelman, 1988)

$$k(t) = k_0 t^{-h} \tag{3}$$

where k_0 is a constant and $0 \leq h \leq 1$ is called the fractal parameter, and it is a measure of the topological dimensionality of the system. In homogenous media, $h = 0$ and $k = k_0$ as in classical kinetics. Another equation was proposed by Pitulice et al., (2014), who improved the one proposed earlier by Schnell and Turner (2004)

$$k(t) = \frac{k_0}{\left(1 + \dfrac{t}{\tau}\right)^h} \; ; (0 \leq h \leq 1) \tag{4}$$

where k_0 and h have the same meaning as in Eq. (2) and τ is a positive constant. This equation fits better the simulation data concerning enzymatic reactions following a Michaelis Menten mechanism.

Published data from simulations and experiments have predicted that the rate of enzymatic reactions either can decrease or increase when MC increases, depending on the mechanism involved by the enzymatic reaction. In other words, for a reaction following the Michaelis Menten mechanism, both the Michaelis Menten (K_m) and catalytic (k_{cat}) constants, related to maximum velocity (v_{max}) can either decrease or increase as an effect of the excluded volume or other possible causes, as for example diffusion control, conformational changes, inhibition by product, mixed activation-diffusion control, etc. (Pastor et al., 2014; Balcells et al., 2015). On the other hand, the kinetics of some oligomeric enzymes can be affected by the relative size ratio of crowders versus enzymes (Homcaudhuri et al., 2006; Balcells et al., 2014, 2015; Poggi and Slade, 2015). MC also affects the behavior of small-molecule substrates because they may interact with crowding agents (Aumiller et al., 2014).

20.4 NANO-CHEMICAL APPLICATION(S)

MC effects on biological molecules have important applications for *in vitro* biotechnological and nanotechnological processes: drug delivery systems, dispersion of nanomaterials, assembly of nanoparticles, interfacial reactions, electrochemical DNA sensors, and gene expression in synthetic cellular nanosystems (Myoshi et al., 2013; Tan et al., 2013).

20.5 MULTI-/TRANS-DISCIPLINARY CONNECTION(S)

A crowded milieu incorporates heterogeneous viscosity, buffer-like cavities, excluded volumes, weak interactions and associations that are based on the chemical structure and size of the crowding agents and on those of molecular probes. It means that understanding MC and its effects brings together many fields of science, especially physics, chemistry, biochemistry, biology, and computer sciences. Applications of MC effects also involve engineering and technology, all these illustrating the multi-trans-disciplinary connections of this concept.

20.6   OPEN ISSUES

There are few contradictory data reported in specific literature: some studies reveal clear effects of MC on the structural and dynamic properties of biological molecules as others do not. These contradictory data may arise from the different type of the crowding agents used in experimental studies. Here it needs to be taken into account that uniform MC agents may affect protein behavior differently than solutions of a variety of individual crowders. Also, the results may be distinct when the macromolecules diffuse in a real cellular media.

Due to its huge biological importance, judiciously organized studies and more sophisticated models and theories for understanding crowding effects are needed. Moreover, one should consider that the complexity of the intracellular milieu ranges far outside the MC concept and thus understanding the specific physical properties of the intracellular environment is also crucial.

KEYWORDS

- **biochemical reactions**
- **crowding**
- **diffusion**
- **volume exclusion**

REFERENCES AND FURTHER READING

Akabayov, S. R., Akabayov, B., Richardson, C. C., & Wagner, G., (2013). Molecular crowding enhanced ATP-ase activity of the RNA helicase EIF4A correlates with compaction of its quaternary structure and association with EIF4G. *Journal of the American Chemical Society, 135,* 10040–10047.

Apetri, A. C., & Surewicz, W. K., (2002). Kinetic intermediate in the folding of human prion protein. *Biological Chemistry, 277*, 44589–44592.

Asaad, N., & Engberts, J. B., (2003). Cytosol-mimetic chemistry: Kinetics of the trypsin-catalyzed hydrolysis of p-nitrophenyl acetate upon addition of polyethylene glycol and N-tert-butylacetoacetamide. *Journal of the American Chemical Society, 125*, 6874–6875.

Aumiller, W. M. Jr., Davis, B. W., Hatzakis, E., & Keating, C. D., (2014). Interactions of macromolecular crowding agents and cosolutes with small-molecule substrates: Effect on horseradish peroxidase activity with two different substrates. *The Journal of Physical Chemistry B., 118*, 10624–10632.

Balcells, C., Pastor, I., Pitulice, L., Hernández, C., Via, M., Garcés, J. L., Madurga, S., Vilaseca, E., Isvoran, A., Cascante, M., & Mas, F., (2015). Macromolecular crowding upon *in vivo-like* enzyme-kinetics: Effect of enzyme-obstacle size. *New Frontiers in Chemistry, 24*, 3–16.

Balcells, C., Pastor, I., Vilaseca, E., Madurga, S., Cascante, M., & Mas, F., (2014). Macromolecular crowding effect upon *in* vitro enzyme kinetics: Mixed activation-diffusion control on the oxidation of NADH by pyruvate catalyzed by lactate dehydrogenase. *The Journal of Physical Chemistry B., 118,* 4062–4068.

Banks, D. S., & Fradin, C., (2005). Anomalous diffusion of proteins due to molecular crowding. *Biophysical Journal, 89*, 2960–2971.

Berry, H., (2002). Monte Carlo simulations of enzyme reactions in two dimensions: Fractal kinetics and spatial segregation. *Biophysical Journal, 83*(4), 1891–1901.

Derham, B. K., & Harding, J. J., (2006). The effect of the presence of globular proteins and elongated polymers on enzyme activity. *Biochimica et Biophysica Acta, 1764*(6), 1000–1006.

Dix, J. A., & Verkman, A. S., (2008). Crowding effects on diffusion in solutions and cells. *Annual Review of Biophysics, 37*, 247–263.

Ellis, R. J., & Minton, A. P., (2006). Protein aggregation in crowded environments. *Biological Chemistry, 387*(5), 485–497.

Gellerich, F. N., Laterveer, F. D., Korzeniewski, B., Zierz, S., & Nicolay, K., (1998). Dextran strongly increases the Michaelis constants of oxidative phosphorylation and of mitochondrial creatine kinase in heart mitochondria. *European Journal of Biochemistry, 254*, 172–180.

Havlin, S., & Ben-Avraham, D., (2002). Diffusion in disordered media. *Advances in Physics, 51*, 187–292.

Homchaudhuri, L., Sarma, N., & Swanjminathan, R., (2006). Effect of crowding by dextrans and ficolls on the rate of alkaline phosphatase-catalyzed hydrolysis: A size-dependent investigation. *Biopolymers, 83*, 477–486.

Jiang, M., & Guo, Z., (2007). Effects of macromolecular crowding on the intrinsic catalytic efficiency and structure of enterobactin-specific isochorismate synthase. *Journal of the American Chemical Society, 129*(4), 730–731.

Kopelman, R., (1988). Fractal reaction kinetics. *Science, 241*(4873), 1620–1626.

Kuznetsova, I. M., Turoverov, K. K., & Uversky, V. N., (2014). What macromolecular crowding can do to a protein. *International Journal of Molecular Sciences, 15*, 23090–23140.

Kuznetsova, I. M., Zaslavsky, B. Y., Breydo, L., Turoverov, K. K., & Uversky, V. N., (2015). Beyond the excluded volume effects: Mechanistic complexity of the crowded milieu. *Molecules, 20*(1), 1377–1409.

Laurent, T. C., (1964). The interaction between polysaccharides and other macromolecules. The exclusion of molecules from hyaluronic acid gels and solutions. *Biochemical Journal, 93*, 106–112.

Laurent, T. C., (1971). Enzyme reactions in polymer media. *European Journal of Biochemistry, 21*, 498–506.

Luby-Phelps, K., (2000). Cytoarchitecture and physical properties of cytoplasm: Volume, viscosity, diffusion, intracellular surface area. *International Review of Cytology, 192*, 189–221.

Ma, B., & Nussinov, R., (2013). Structured crowding and its effects on enzyme catalysis. *Topics in Current Chemistry, 337*, 123–137.

Minton, A. P., (1981). Excluded volume as a determinant of macromolecular structure and reactivity. *Biopolymers, 20*, 2093–2120.

Minton, A. P., (1983). The effect of volume occupancy upon the thermodynamic activity of proteins: Some biochemical consequences. *Molecular and Cellular Biochemistry, 55*, 119–140.

Minton, A. P., (2000). Implications of macromolecular crowding for protein assembly. *Current Opinion in Structural Biology, 10*(1), 34–39.

Minton, A. P., (2001). The influence of macromolecular crowding and macromolecular confinement on biochemical reactions in physiological media. *Journal of Biological Chemistry, 276*(14), 10577–10580.

Minton, A. P., & Wilf, J., (1981). Effect of macromolecular crowding upon the structure and function of an enzyme: Glyceraldehyde-3-phosphate dehydrogenase. *Biochemistry, 20*, 4821–4826.

Moran-Zorzano, M. T., Viale, A., Muñoz, F., Alonso-Casajas, N., Eydaltm, G., et al., (2007). Escherichia coli AspP activity is enhanced by macromolecular crowding and by both glucose–1: 6-bisphosphate and nucleotide-sugars. *FEBS Letters, 581*, 1035–1040.

Myoshi, D., Fujimoto, T., & Sugimoto, N., (2013). Molecular crowding and hydration regulating of G-quadruplex formation. *Topics in Current Chemistry, 330*, 87–110.

Ogston, A. G., & Phelps, C. F., (1961). The partition of solutes between buffer solutions and solutions containing hyaluronic acid. *Biochemical Journal, 78*(4), 827–833.

Olsen, S. N., (2006). Applications of isothermal titration calorimetry to measure enzyme kinetics and activity in complex solutions. *Thermochimica Acta, 448*, 12–18.

Olsen, S. N., Ramlev, H., & Westh, P., (2007). Effects of osmolytes on hexokinase kinetics combined with macromolecular crowding test of the osmolyte compatibility hypothesis towards crowded systems. *Comparative Biochemistry and Physiology Part A: Molecular & Integrative Physiology, 148*, 339–345.

Pastor, I., Pitulice, L., Balcells, C., Vilaseca, E., Madurga, S., Isvoran, A., Cascante, M., & Mas, F., (2014). Effect of crowding by dextrans in enzymatic reactions. *Biophysical Chemistry, 185*, 8–13.

Pastor, I., Vilaseca, E., Madurga, S., Garcés, J. L., Cascante, M., & Mas, F., (2010). Diffusion of alpha-chymotrypsin in solution-crowded media: A fluorescence after photobleaching recovery study. *The Journal of Physical Chemistry B., 114*, 4028–4034. Erratum, *ibid, 114*, 12182.

Pastor, I., Vilaseca, E., Madurga, S., Garcés, J. L., Cascante, M., & Mas, F., (2011). Effect of crowding by dextrans on the hydrolysis of N-succinyl-L-phenyl-Ala-p-nitronilide catalyzed by alpha-chymotrypsin. *The Journal of Physical Chemistry B., 115*, 1115–1121.

Pitulice, L., Pastor, I., Vilaseca, E., Madurga, S., Isvoran, A., Cascante, M., & Mas, F., (2013). Influence of macromolecular crowding on the oxidation of ABTS by hydrogen peroxide catalyzed by HRP. *Journal of Biocatalysis & Biotransformation, 2*, 1.

Pitulice, L., Vilaseca, E., Pastor, I., Madurga, S., Garces, J. L., Isvoran, A., & Mas, F., (2014). Monte Carlo simulations of enzymatic reactions in crowded media. Effect of the enzyme-obstacle relative size. *Mathematical Biosciences, 251*, 72–82.

Poggi, C. G., & Slade, K. M., (2015). Macromolecular crowding and the steady-state kinetics of malate dehydrogenase. *Biochemistry, 54*(2), 260–267.

Pozdnyakova, I., & Wittung-Stafshede, P., (2010). Non-linear effects of macromolecular crowding on enzymatic activity of multi-copper oxidase. *Biochimica et Biophysica Acta, 1804*, 740–744.

Ralston, G. B., (2005). Effects of "crowding" in protein solutions. *Journal of Chemical Education, 67*, 857–860.

Sasaki, Y., Miyoshi, D., & Sugimoto, N., (2007). Regulation of DNA nucleases by molecular crowding. *Nucleic Acids Research, 35*, 4086–4093.

Saxton, M. J., (1987). Lateral diffusion in an archipelago. The effect of mobile obstacles. *Biophysical Journal, 52*(6), 989–997.

Saxton, M. J., (1989). Lateral diffusion in an archipelago. Distance dependence of the diffusion coefficient. *Biophysical Journal, 56*(3), 615–622.

Saxton, M. J., (1990). Lateral diffusion in a mixture of mobile and immobile particles. A Monte Carlo study. *Biophysical Journal, 58*(5), 1303–1306.

Saxton, M. J., (1992). Lateral diffusion and aggregation. A Monte Carlo study. *Biophysical Journal, 61*(1), 119–128.

Saxton, M. J., (1993). Lateral diffusion in an archipelago. Dependence on tracer size. *Biophysical Journal, 64*(4), 1053–1062.

Saxton, M. J., (1994). Anomalous diffusion due to obstacles: A Monte Carlo study. *Biophysical Journal, 66*, 394–401.

Schnell, S., & Turner, T. E., (2004). Reaction kinetics in intracellular environments with macromolecular crowding: Simulations and rate laws. *Progress in Biophysics & Molecular Biology, 85*, 235–260.

Tan, C., Saurabh, S., Bruchez, M. P., Schwartz, R., & LeDuc, P., (2013). Molecular crowding shapes gene expression in synthetic cellular nanosystems. *Nature Nanotechnology, 8*, 602–608.

Van Den Berg, B., Ellis, R. J., & Dobson, C. M., (1999). Effects of macromolecular crowding on protein folding and aggregation. *EMBO Journal, 18*(24), 6927–6933.

Van Den Berg, B., Wain, R., Dobson, C. M., & Ellis, R. J., (2000). Macromolecular crowding perturbs protein refolding kinetics: Implications for folding inside the cell. *EMBO Journal, 19*(15), 3870–3875.

Vilaseca, E., Isvoran, A., Madurga, S., Pastor, I., Garces, J. L., & Mas, F., (2011a). New insights into diffusion in 3D crowded media by Monte Carlo simulations: Effect of size, mobility, and spatial distribution of obstacles. *Physical Chemistry Chemical Physics, 13*(16), 7396–7407.

Vilaseca, E., Pastor, I., Isvoran, A., Madurga, S., Garces, J. L., & Mas, F., (2011b). Diffusion in macromolecular crowded media: Monte Carlo simulation of obstructed diffusion vs. FRAP experiments. *Theoretical Chemistry Accounts, 128*(4–6), 795–805.

Vopel, T., & Makhatadze, G. I., (2012). Enzyme activity in the crowded milieu. *PLoS One, 7*, e39418.

Wenner, J. R., & Bloomfield, V. A., (1999). Crowding effects on EcoRV kinetics and binding. *Biophysical Journal, 77*, 3234–3241.

Winklhofer, K. F., Jörg, T., & Haass, C., (2008). The two faces of protein misfolding: Gain- and loss-of-function in neurodegenerative diseases. *EMBO Journal, 27*(2), 336–349.

Yadav, J. K., (2013). Macromolecular crowding enhances catalytic efficiency and stability of α-amylase. *ISRN Biotechnology, ID 737805*, p. 7. doi: 10.5402/2013/737805.

Zhou, H. X., Rivas, G., & Minton, A. P., (2008). Macromolecular crowding and confinement: Biochemical, biophysical, and potential physiological consequences. *Annual Review of Biophysics, 37*, 375–397.

Zimmerman, S. B., (1993). Macromolecular crowding effects on macromolecular interactions: Some implications for genome structure and function. *Biochimica et Biophysica Acta, 1216*(2), 175–185.

Zimmerman, S. B., & Harrison, B., (1987). Macromolecular crowding increases binding of DNA polymerase to DNA: An adaptive effect. *Proceedings of the National Academy of Sciences of the United States, 84*(7), 1871–1875.

Zimmerman, S. B., & Minton, A. P., (1993). Macromolecular crowding: Biochemical, biophysical, and physiological consequences. *Annual Review of Biophysics and Biomolecular Structure, 22*, 27–65.

CHAPTER 21

Polycyclic Aromatic Hydrocarbons (PAHs)

MARINA A. TUDORAN,[1,2] ANA-MARIA PUTZ,[1,3] LAURA PITULICE,[1] and MIHAI V. PUTZ[1,2]

[1]*Laboratory of Structural and Computational Physical Chemistry for Nanosciences and QSAR, Biology-Chemistry Department, West University of Timisoara, Pestalozzi Street No. 44, Timisoara, RO–300115, Romania*

[2]*Laboratory of Renewable Energies-Photovoltaics, R&D National Institute for Electrochemistry and Condensed Matter, Dr. A. Paunescu Podeanu Str. No. 144, Timisoara, RO–300569, Romania*

[3]*Institute of Chemistry Timisoara of the Romanian Academy, 24 Mihai Viteazul Bld., Timisoara 300223, Romania*

21.1 DEFINITION

Polycyclic aromatic hydrocarbons (PAHs) represent a class of organic compounds with three or more fused benzene rings containing only carbon and hydrogen, produced by incomplete combustion or high-pressure processes, i.e., when complex organic substances are exposed to high temperatures or pressures.

21.2 HISTORICAL ORIGIN(S)

PAHs represent a class of organic compounds, having two or more bonded benzene rings in one of the three forms (Arey and Atkinson, 2003; Di-Toro et al., 2000): linear, angular or cluster arrangements (see Figure 21.1). They have an interesting characteristic: they bond with hydrogen atoms and forms aromatic rings similar to graphite (Putz et al., 2013). As a mechanism of formation, PAHs have different sources, such as products of incomplete combustion from man-made combustion sources, natural combustion sources or biological processes (Abdel-Shafy and Mansour, 2016). They are solid compounds, white, and pale yellow or colorless, with low vapor pressure and high boiling and melting points (Masih et al., 2012). They are low soluble in aqueous solutions and very soluble in organic solvents, highly lipophilic and present heat and corrosion resistance, light sensitivity and conductivity (Akyuz and Cabuk, 2010).

Even if PAHs are not chemically synthesized for industrial applications, they can be used as intermediaries in thermosetting plastics, photographic products, lubricating materials, agricultural, and pharmaceutical products (Kaminski et al., 2008), e.g., acenaphthene, pyrene, anthracene, and fluorine can be used in manufacture of pigments and dyes, acenaphthene, fluoranthene, and fluorene in pharmaceutical industry, phenanthrene, acenaphthene, and fluorine in pesticides (Abdel-Shafy and Mansour, 2016).

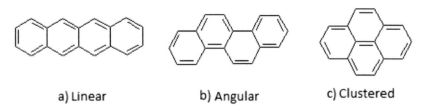

a) Linear   b) Angular   c) Clustered

FIGURE 21.1 a) Tetracene; b) Chrysene; c) Pyrene. Redrawn and adapted after Abdel-Shafy and Mansour (2016), Arey and Atkinson (2003), and Di-Toro et al., (2000).

There are several methods of removing PAHs from the environment, such as photochemical degradation or biodegradation (Abdel-Shafy et al., 2014; Perera et al., 2011), and from atmosphere (Delgado et al., 2010; Zhong and Zhu, 2013; Zhu et al., 2009) such as dry deposition or wet deposition. One of the most studied processes in literature is biodegradation, the PAHs needed to be in vapor phase or dissolved in order to be available for the bacteria (Cerniglia, 2003; Dandie et al., 2004; Fredslund et al., 2008), which exclude PAHs sorbet onto soil particles (Hatzinger and Martin, 1995; Kim et al., 2007; Rappert et al., 2006; Archana et al., 2008). One of the most important characteristics of PAHs bioavailability is their solubility, the aqueous solubility being related to their molecular weight (Thorsen et al., 2004; Chen and Aiken, 1999). In the photolysis process, the compound is destroyed in reactions which involve light absorption (Manahan, 1994), in case of PAHs, the absorbed light creates an unstable structural arrangement by the electrons excitation which allows several chemical and physical processes on the compound (Schwarzenbach et al., 1993). Chemical oxidation is a method less used for PAHs removal, can be natural or as part of treatment technologies and is correlated with the structure, molecular weight and physical state of the compound, the oxidizing agent and temperature (Abdel-Shafy and Mansour, 2013). Dry deposition occurs in the absence of precipitation, and the PAHs sorbet to atmospheric particles goes down to earth (Manahan, 1994; U.S.EPA, 2000; Chang et al., 2003; Bozlaker et al., 2008). On the other hand, wet deposition represents the process in which the vapor phase contaminants dissolute into precipitation (Dickhut and Gustafson, 1995; Wang et al., 2010), and also the removal from the atmosphere of the contaminants sorbed in particles.

21.3   NANO-SCIENTIFIC DEVELOPMENT(S)

The anthracene molecule consists in three linear benzene rings, with the C-H out-of-plane deformation mode exhibiting bands in 11–15 μm region, the presence of bands being related to the number of directly adjacent *H* atoms and occurs due to the coupling between the vibrating *H* atoms which are bonded to the neighboring *C* atoms from the ring (Saito et al., 2010). In order to obtain anthracene grains one can use the gas evaporation method, the so-called the smoke, which can also be used in producing nanometer-sized metal, semiconductor or oxide particles (Kimoto et al., 1963; Kaito, 1978). Using this technique, Saito, and his co-workers (2010) produce anthracene grains with size varying from 5 nm to 5 μm. From electron diffraction (ED) pattern one can observe Bragg reflection spots from the anthracene crystal, meaning that during evaporation, anthracene molecule does not broke. This molecule is a monocyclic crystal with angles of $\alpha = \gamma = 90°$ and $\beta = 103.57°$, the axes of $a = 0.9463$, $b = 0.6026$, and $c = 0.8550$ nm (Mason, 1964). When the He gas pressure is increased, the collision frequency increases with decreasing the mean free path leading to a slight increase in the size of the anthracene grains. By setting the temperature of the evaporation source at a slower rate of 2.5 K/s, on the amorphous carbon substrate were formed neuron-anthracene films along with anthracene grains (Saito et al., 2010). Regarding their stability, it is known from the literature that grains of PAHs with 25 carbon atoms formed by van der Waals forces can be easily broken by far-UV photon absorption (Léger and Puget, 1984; Tielens et al., 1999). PAH grains also have the property of photo-evaporation, resulting in free-flying PAHs in the region of photo-dissociation (Rapacioli et al., 2005; Sakon et al., 2007).

21.4   NANO-CHEMICAL APPLICATION(S)

Studying the adsorption of PAH molecules on carbon nanotubes become in center of scientists attention, as the interaction between carbon materials and PAH molecules leads to modifications in electronic properties of nano-tubes with application in the surface control of isolated tubes (Gotovac et al., 2007; Zhao et al., 2003). Starting from this considerate, Wesołowski and his co-workers (2011) use the molecular dynamics simulation (MD) to check proposed plausible adsorbed states of PAHs on isolated (13, 9) single-walled carbon nanotubes. The authors simulate the adsorption of several PAHs (e.g., naphthalene, anthracene, phenanthrene, tetracene, and pentacene) from toluene solution, and consider a single-walled carbon nanotube, fully flexible, modeled as infinite in axial dimension, and having the bonding field parameters taken from literature (Walther et al., 2001). In order to perform the MD simulations, they used OPLS-AA (All-Atom Optimized Potentials for Liquid Simulations) force field (Jorgensen et al., 1996) and GROMACS (GROningen MAchine for Chemical Simulations) package version 4.0.3 (van der Spoel et al., 2005) at the temperature of 200 and 298 K. For the molecules to which the distance between at least one atom of its molecule to the nearest tube atom was smaller than 0.5 nm, the authors (Wesołowski et al., 2011) determine the angular orientation of adsorbed PAH molecules and considered for that the molecule is parallel orientated when the angle equal with $0°$ and perpendicular orientated for the angle equal with $90°$. The orientation of PAHs molecules respecting the tube was determined for all adsorbed molecules at the assumed distance between all carbon atoms to the nearest carbon atoms which form a tube being smaller than 0.8, by projecting the coordinates the carbon atoms of the adsorbed molecules on a tube plane with a radius of 0.7495 nm (Wesołowski et al., 2011). The obtained results determined that the number of molecules with parallel orientation increase with increasing the number of the benzene ring in PAH molecule, and by decreasing the temperature the adsorbed molecules with flat orientation decrease (Wesołowski et al., 2011). Based on these results one can say that at a larger temperature and at high coverages, there is a larger number of molecules which will adopt vertical or slant orientation. Still, at the room temperature, the molecules can translate on the surface due to the fact that in this condition, the adsorption is mobile (Wesołowski et al., 2011).

Interaction of anthracene, known as carcinogenic and pollutant, with single wall carbon nanotubes (SWNTs), was also studied by Banaeian and

Mahdavian, (2015), in order to recognize, eliminate, and transform the PAH molecule in a less dangerous form. In their study, the geometrical structure of anthracene and the SWNT (8,8) with a length of 11.01 Å were optimized using GAMESS (General Atomic and Molecular Electronic Structure System) program package based on B3LYP/6–31G, and the electrical and thermo-dynamic properties of the simulated contact between these two structures were calculated using semi-empirical methods MNDO (Modified Neglect of Diatomic Overlap). First, the authors analyze the transfer of anthracene from the central axis of carbon nanotube, designed in 5 phases, simulated, and calculated and conclude that, at this stage, by rinsing the temperature one can remove and recycle the pollutant (Banaeian and Mahdavian, 2015). Another scenario is represented by the absorption of anthracene on the wall of carbon nanotube, where occur an electron exchange which leads to modifying the pollutant into less risky products (Banaeian and Mahdavian, 2015). In this case, the weaker thermal conductivity is founded along the vertical axes of carbon nanotubes and has maximum value along with their longitudinal axes. When anthracene is absorbed on the end of SWNT (8,8) there are several reactions which occur:

1. The anthracene is converted to toluene

$$C_{14}H_{10} - SWNT + 3H_2 \rightarrow 2C_7H_8 + SWNT \tag{1}$$

2. The toluene is converted to propane and 1,3 butadiene

$$C_7H_8 - SWNT + 2H_2 \rightarrow \left(C_3H_6 + C_4H_6\right) + SWNT \tag{2}$$

3. The propane is converted to CO_2

$$C_3H_6 - SWNT + 3O_2 \rightarrow 2CO_2 + SWNT \tag{3}$$

There were five steps in which the change and absorption of these reactions were computed and simulated: first two steps represent the approaching the absorbing of the pollutant by the carbon nanotubes, the third step represents the transition state, the fourth step consists in formation of a new product, while in the fifth step the pollutant is desorbed from the carbon nanotube surface (Banaeian and Mahdavian, 2015).

The results obtained by Banaeian and Mahdavian (2015) support the idea of using SWNT (8,8) to reduce and remove anthracene. From the thermo-dynamic properties, one can see that their interactions are exothermic and spontaneous and can be used in environmental purification.

21.5   MULTI-/TRANS-DISCIPLINARY CONNECTION(S)

PAHs are considered an important group of chemical carcinogens (Farmer et al., 2003; Baudouin et al., 2002; Barone et al., 1996), formed mostly from the combustion of the organic substances, but they can also be found in cigarette smoke or in diesel exhaust (Huetz et al., 2005). Carcinogenic compounds are also the metabolites of PAHs, such as benzo-ring diol epoxides. In general, diol epoxides represent a diastereomer pairs with the property that both epoxide oxygen and hydroxyl groups may be cis and trans, respectively, vice versa. Applied to PAH molecules, the epoxide group is situated beside the bay region, one of the commune example being the benzo[a]pyrene–7,8-diol–9,10-epoxide (BPDE) (Conney et al., 1994; Cooper et al., 1982; Shimada et al., 1999). The carcinogenic effect of the diol epoxide is determined by the fact that DNA is alkylating at the PAH angular bay region (Jerina et al., 1988), the major alkylation sites being the deoxyadenosine and the deoxyguanosine (Sayer et al., 1991). The nucleophilic attack at C_{10} of BPDE determines the adducts formation (the trans product) by opening the epoxide ring which gave stereochemical inversion. There are several types of cancer linked to the BPDE-DNA adducts due to "hot spot" creation in the p53 tumor suppressor gene, which further induces mutation on p53, such as the brain, stomach (Soussi et al., 2000) or lung cancer (Alexandrov et al., 2002). Still, there are studies which suggest that the covalent binding of BPDE to DNA can be prevented by the chemical reaction of ellagic acid (EA) with benyo[a]pyrene metabolites (Sayer et al., 1982; Wood et al., 1982). The reaction of EA with BPDE is presented in Figure 21.2.

EA may have an important role in cell cycle regulation in cancer cells due to its strong antioxidant activity (Festa et al., 2001; Narayanan et al., 1999) and can inhibit the benzo[a]pyrene CYP1A1-dependent activation with reducing the O6-methylguanine by methylating carcinogens (Barch and Fox, 1988; Teel, 1986). Moreover, EA appears to inhibit, besides PAH, aflatoxins, and also nitroso compounds. The inhibition of PAH-DNA adduction by EA appears to be enzyme independent (in part or entirely), as a consequence of the fact that the inhibition of BPDE-DNA adduction is made by the EA without the enzymatic influence (Smith and Gupta, 1999). But, even if the EA can prevent DNA alkylation by directly reacting with guanine, it has a greater affinity for DNA than for proteins (Whitley et al., 2003), meaning that, at a certain concentration the EA with becoming carcinogenic. However, the inhibition produced by EA to DNA-BPDE adduct formation is dose-dependent and has a narrow therapeutic-to-toxic ratio.

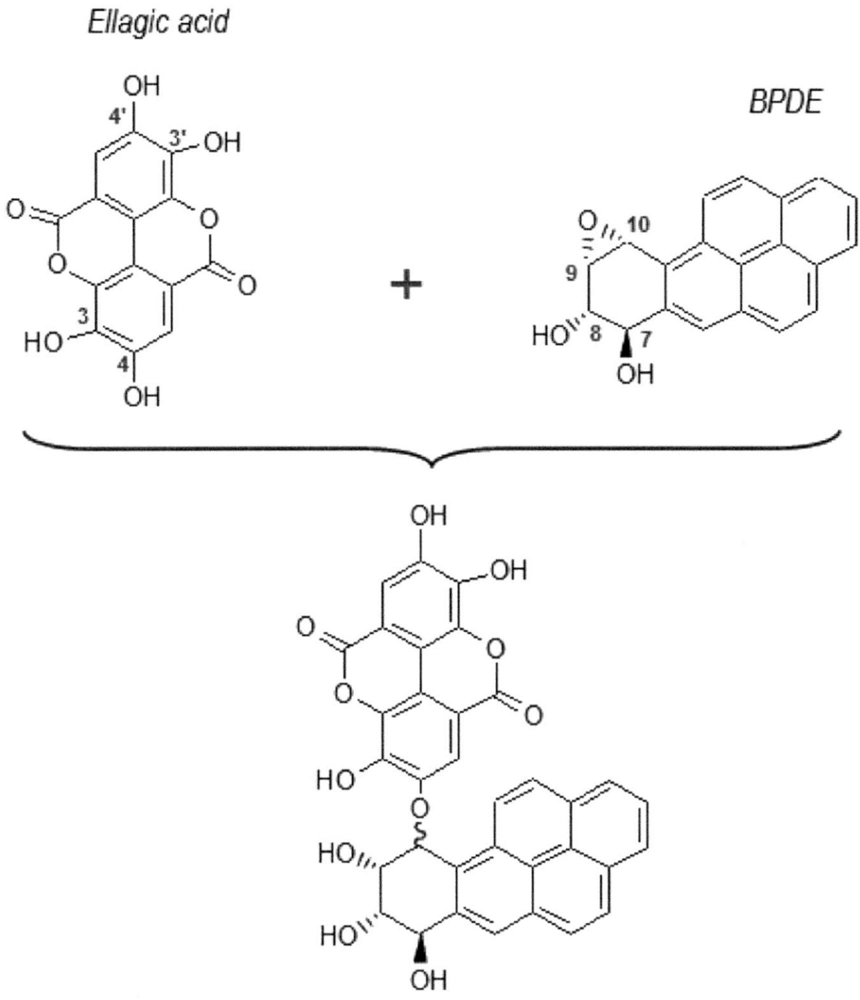

FIGURE 21.2   Ellagic acid in reaction with BPDE. Redrawn and adapted from Huetz et al. (2005).

Apart from their high toxicity and carcinogenic effects, PAHs are also known as a class of pollutants, their environmental persistence representing a major concern (Usman et al., 2016). A special interest was given to the soil contamination, several techniques being developed in order to remove from sediments or soil the persistent PAHs. An efficient technique is represented by chemical oxidation (Lee et al., 1998; Watts et al., 2002; Rivas, 2006; Ferrarese et al., 2008; Yap et al., 2011) using oxidants such as persulfate, permanganate, Fenton's reagent or ozone (ITRC, 2005), Fenton oxidation being the most efficient in PAHs degradation form the contaminated soil (Usman et al., 2016). The Fenton treatment is using a hydroxyl radical (•OH), highly reactive, generated from the hydrogen peroxide (H_2O_2) decomposition by iron, with the following reaction:

$$Fe^{II} + H_2O_2 \rightarrow Fe^{III}OH^{2+} + HO^{\bullet}$$

Although Fenton oxidation technique widely used, there are several limitations when it comes with its environmental application, one of the most important being the necessity of using traditional Fenton oxidation for removing PAHs from contaminating soils at acidic pH (Yap et al., 2011). In this case, one can use chelating agents in the so-called modified Fenton (MF) oxidation (Nam et al., 2001). Another major limitation is given by the PAHs characteristics, e.g., strong hydrophobicity (Riding et al., 2013), which determines a low availability for this type of molecules to be remediate (Flotron et al., 2005; Usman et al., 2012; Lemaire et al., 2013a, Choi et al., 2014). PAHs solubility in soils can be increased by using co-solvents like ethanol (Bonten et al., 1999), in order to increase their partitioning to the surfactant micelles hydrophobic cores and to reduce the surface tension between the aqueous phase and the compounds in soil.

21.6   OPEN ISSUES

All cellular life has a common feature of presenting boundaries with the property that they can self assemble as bilayers (Groen et al., 2012), consisting of amphiphilic molecules (phospholipids and PAHs mixed), their early provenance being linked with the extraterrestrial delivery of organic compounds or volcanism by synthesis through Fischer-Tropsch reactions (McCollom and Seewald, 2007; Rushdi and Simoneit, 2001: Simoneit, 2004). Studies in this area propose that monocarboxylic acids with 8 to 12 carbon atoms may be constituents of the primitive cell membranes based on the fact that, in aqueous dispersion and for pH similar with pK_a, pure fatty acids can self-assemble into vesicles by forming hydrogen bonds with the deprotonated and protonated head groups which are able to stabilize the bilayers structures (Monnard and Deamer, 2002, 2003). In a recent study (Cape et al., 2011), the authors determine that naphto[2,3a]pyrene and perylene can act as a primitive pigment system, being able to induce trans-membrane charge transport (Deamer, 1992). Another study made by Groen and his co-workers (2012) analyze the influence of the oxidized PAHs derivatives on the membrane. By incorporating several oxidized PAHs derivatives in fatty acid membranes, the authors studied if they can stabiles the membrane by decreasing the membrane permeability or the Critical Vesicle Concentration (CVC) to small solutes. The results obtained in a simulated prebiotic membrane sustain the idea that oxidized PAH derivatives have a stabilizing effect similar to cholesterol, as observed that 1-hydroxypyrene significantly reduces the dicationic acid permeability for small solute (Groen et al., 2012).

KEYWORDS

- **carcinogenesis**
- **ecotoxicity**
- **ellagic acid**
- **Fenton oxidation**
- **single wall carbon nanotubes**

REFERENCES AND FURTHER READING

Abdel-Shafy, H. I., & Mansour, M. S. M., (2013). Removal of selected pharmaceuticals from urine via Fenton reaction for agriculture reuse. *Journal of Sustainable Sanitation Practice, 17*, 20–29.

Abdel-Shafy, H. I., & Mansour, M. S. M., (2016). A review on polycyclic aromatic hydrocarbons: Source, environmental impact, effect on human health and remediation. *Egyptian Journal of Petroleum, 25*, 107–123.

Abdel-Shafy, H. I., Al-Sulaiman, A. M., & Mansour, M. S. M., (2014). Greywater treatment via hybrid integrated systems for unrestricted reuse in Egypt. *Journal of Water Process Engineering, 1*, 101–107.

Akyuz, M., & Cabuk, H., (2010). Gas-particle partitioning and seasonal variation of polycyclic aromatic hydrocarbons in the atmosphere of Zonguldak, Turkey. *Science of the Total Environment, 408*, 5550–5558.

Alexandrov, K., Cascorbi, I., Rojas, M., Bouvier, G., Kriek, E., & Bartsch, H., (2002). CYP1A1 and GSTM1 genotypes affect benzo[a]pyrene DNA adducts in smokers' lung: Comparison with aromatic/hydrophobic adduct formation. *Carcinogenesis, 23*, 1969–1977.

Arey, J., & Atkinson, R., (2003). Photochemical reactions of PAH in the atmosphere. In: Douben, P. E. T., (ed.), *PAHs: An Ecotoxicological Perspective* (pp. 47–63). John Wiley and Sons Ltd, New York.

Banaeian, Z., & Mahdavian, L., (2015). Thermodynamic study of polycyclic aromatic (anthracene) and SWNT nanofilters interaction. *Russian Journal of Applied Chemistry, 88*(12), 2056–2064.

Barch, D. H., & Fox, C. C., (1988). Selective inhibition of methylbenzylnitrosamine- induced formation of esophageal O6-methylguanine by dietary ellagic acid in rats. *Cancer Research, 48*, 7088–7092.

Barone, P. M. V. B., Camilo, A. Jr., & Galvao, D. S., (1996). Theoretical approach to identify carcinogenic activity of polycyclic aromatic hydrocarbons. *Physical Review Letters, 77*, 1186–1189.

Baudouin, C., Charveron, M., Tarroux, R., & Gall, Y., (2002). Environmental pollutants and skin cancer. *Cell Biology and Toxicology, 18*, 341–348.

Bonten, L. T. C., Grotenbuis, T. C., & Rulkens, W. H., (1999). Enhancement of PAH biodegradation in soil by physicochemical pretreatment. *Chemosphere, 38*, 3627–3636.

Bozlaker, A., Muezzinoglu, A., & Odabasi, M., (2008). Atmospheric concentrations, dry deposition, and air-soil exchange of polycyclic aromatic hydrocarbons (PAHs) in an industrial region in Turkey. *Journal of Hazardous Materials, 153*(3), 1093–1102.

Cerniglia, C. E., (2003). Recent advances in the biodegradation of polycyclic aromatic hydro-carbons by Mycobacterium species. In: Sasek V., (ed.), *The Utilization of Bioremediation to Reduce Soil Contamination: Problems and Solutions* (pp. 51–73). Kluwer Academic Publishers, The Netherlands.

Chang, K. F., Fang, G. C., Lu, C., & Bai, H., (2003). Estimating PAH dry deposition by measuring gas and particle phase concentrations in ambient air. *Aerosol and Air Quality Research, 3*(1), 41–51.

Chauhan, A., Fazlurrahman, O. J. G., & Jain, R. K., (2008). Bacterial metabolism of polycyclic aromatic hydrocarbons: Strategies for bioremediation. *Indian Journal of Microbiology, 48*(1), 95–113.

Chen, S. H., & Aiken, M. D., (1999). Salicylate stimulates the degradation of high-molecular-weight polycyclic aromatic hydrocarbons by Pseudomonas saccharophila P15. *Environmental Science & Technology, 33*(3), 435–439.

Choi, K., Bae, S., & Lee, W., (2014). Degradation of pyrene in cetylpyridinium chloride-aided soil washing wastewater by pyrite Fenton reaction. *Chemical Engineering Journal, 249*, 34–41.

Conney, A. H., Chang, R. L., Jerina, D. M., & Wei, S. J., (1994). Studies on the metabolism of benzo[a]pyrene and dose-dependent differences in the mutagenic profile of its ultimate carcinogenic metabolite. *Drug Metabolism Reviews, 26*, 125–163.

Cooper, C. S., Pal, K., Hewer, A., Grover, P. L., & Sims, P., (1982). The metabolism and activation of polycyclic aromatic hydrocarbons in epithelial cell aggregates and fibroblasts prepared from rat mammary tissue. *Carcinogenesis, 3*, 203–210.

Dandie, C. E., Thomas, S. M., Bentham, R. H., & McClure, N. C., (2004). Physiological characterization of Mycobacterium sp. strain 1B isolated from a bacterial culture able to degrade high-molecular-weight polycyclic aromatic hydrocarbons. *Journal of Applied Microbiology, 97*, 246–255.

Deamer, D. W., (1992). Polycyclic aromatic hydrocarbons: Primitive pigment systems in the prebiotic environment. *Advances in Space Research, 12*, 183–189.

Delgado, S. J. M., Aquilina, N., Baker, S., Harrad, S., Meddings, C., & Harrison, R. M., (2010). Determination of atmospheric particulate-phase polycyclic aromatic hydrocarbons from low volume air samples. *Analytical Methods, 2*, 231–242.

Di-Toro, D. M., McGrath, J. A., & Hansen, D. J., (2000). Technical basis for narcotic chemicals and polycyclic aromatic hydrocarbon criteria. I. Water and tissue. *Environmental Toxicology and Chemistry, 19*, 1951–1970.

Dickhut, R. M., & Gustafson, K. E., (1995). Atmospheric washout of polycyclic aromatic hydrocarbons in the southern Chesapeake Bay region. *Environmental Science & Technology, 29*, 1518–1525.

Farmer, P. B., Singh, R., Kaur, B., Sram, R. J., Binkova, B., Kalina, I., et al., (2003). Molecular epidemiology studies of carcinogenic environmental pollutants. Effects of polycyclic aromatic hydrocarbons (PAHs) in environmental pollution on exogenous and oxidative DNA damage. *Mutation Research/Fundamental and Molecular Mechanisms of Mutagenesis, 544*, 397–402.

Ferrarese, E., Andreottola, G., & Oprea, I. A., (2008). Remediation of PAH-contaminated sediments by chemical oxidation. *Journal of Hazardous Materials, 152*, 28–139.

Festa, F., Aglitti, T., Duranti, G., Ricordy, R., Perticone, P., & Cozzi, R., (2001). Strong antioxidant activity of ellagic acid in mammalian cells in vitro revealed by the comet assay. *Anticancer Research, 21*, 3903–3908.

Flotron, V., Delteil, C., Padellec, Y., & Camel, V., (2005). Removal of sorbed polycyclic aromatic hydrocarbons from soil, sludge, and sediment samples using the Fenton's reagent process. *Chemosphere, 59*, 1427–1437.

Fredslund, L., Sniegowski, K., Wick, L. Y., Jacobsen, C. S., De Mot, R., & Springael, D., (2008). Surface motility of polycyclic aromatic hydrocarbon (PAH)-degrading mycobacteria. *Research in Microbiology, 159*, 255–262.

Gotovac, S., Honda, H., Hattori, Y., Takahashi, K., Kanoh, H., & Kaneko, K., (2007). Effect of nanoscale curvature of single-walled carbon nanotubes on adsorption of polycyclic aromatic hydrocarbons. *Nano Letters, 7*, 583–587.

Groen, J., Deamer, D. W., Kros, A., & Ehrenfreund, P., (2012). Polycyclic aromatic hydrocarbons as plausible prebiotic membrane components. *Origins of Life and Evolution of Biospheres, 42*, 295–306.

Hatzinger, B. P., & Martin, A., (1995). Effect of aging of chemicals in soil on their biodegradability and extractability. *Environmental Science & Technology, 29*(2), 537–545.

Huetz, P., Mavaddat, N., & Mavri, J., (2005). Reaction between ellagic acid and an ultimate carcinogen. *Journal of Chemical Information and Modeling, 45*, 1564–1570.

Jerina, D. M., Cheh, A. M., Chadha, A., Yagi, J., & Sayer, J. M., (1988). In: Miners, J. O., Birkett, D. J., Drew, R., May, B. K., & McManus, M. E., (eds.), *Microsomes*, and *Drug Oxidations, Proceedings of the 7th International TAYLOR Symposium* (pp. 354–362). Taylor and Francis: London.

Jorgensen, W. L., Maxwell, D. S., & Tirado-Rives, J., (1996). Development and testing of the OPLS all-atom force field on conformational energetics and properties of organic liquids. *Journal of the American Chemical Society, 118*, 11225–11236.

Kaito, C., (1978). Coalescence growth of smoke particles prepared by a gas evaporation technique. *Japanese Journal of Applied Physics, 17*, 601–609.

Kaminski, N. E., Faubert, K. B. L., & Holsapple, M. P., (2008). In: Klaassen, C. D., (ed.), *Casarett*, and *Doull's Toxicology, The Basic Science of Poisons* (7th edn., Vol. 526). Mc-Graw Hill, Inc.

Kim, S. J., Kweon, O., Jones, R. C., Freeman, J. P., Edmondson, R. D., & Cerniglia, C. E., (2007). Complete and integrated pathway in Mycobacterium vanbaalenii PYR–1 based on systems biology. *Journal of Bacteriology, 189*, 464–472.

Kimoto, K., Kamiya, Y., Nonoyama, M., & Uyeda, R., (1963). An electron microscope study on fine metal particles prepared by evaporation in argon gas at low pressure. *Japanese Journal of Applied Physics, 2*, 702–713.

Lee, B. D., Hosomi, M., & Murakami, A., (1998). Fenton oxidation with ethanol to degrade anthracene into biodegradable 9, 10-anthraquinone: A pretreatment method for anthracene-contaminated soil. *Water Science and Technology, 38*, 91–97.

Leger, A., & Puget, J. L., (1984). Identification of the 'unidentified' IR emission features of interstellar dust? *Astronomy and Astrophysics, 137*, L5–L8.

Lemaire, J., Buès, M., Kabeche, T., Hanna, K., & Simonnot, M. O., (2013). Oxidant selection to treat an aged PAH-contaminated soil by in situ chemical oxidation. *Journal of Environmental Chemical Engineering, 1*, 1261–1268.

Manahan, S. E., (1994). *Environmental Chemistry*. CRC Press Inc.

Masih, J., Singhvi, R., Kumar, K., Jain, V. K., & Taneja, A., (2012). Seasonal variation and sources of polycyclic aromatic hydrocarbons (PAHs) in indoor and outdoor air in a semi-arid tract of northern India. *Aerosol and Air Quality Research, 12*, 515–525.

Mason, R., (1964). The crystallography of anthracene at 95°K and 290°K. *Acta Crystallographica, 17*, 547–555.

McCollom, T. M., & Seewald, J. S., (2007). Abiotic synthesis of organic compounds in deep-sea hydrothermal environments. *Chemical Reviews, 107*, 382–401.

Monnard, P. A., & Deamer, D. W., (2002). Membrane self-assembly processes: Steps toward the first cellular life. *Anatomical Record, 268*, 196–207.

Monnard, P. A., & Deamer, D. W., (2003). Preparation of vesicles from nonphospholipid amphiphiles. *Methods in Enzymology, 372*, 133–151.

Nam, K., Rodriguez, W., & Kukor, J. J., (2001). Enhanced degradation of polycyclic aromatic hydrocarbons by biodegradation combined with a modified Fenton reaction. *Chemosphere, 45*, 11–20.

Narayanan, B. A., Geoffroy, O., Willingham, M. C., Re, G. G., & Nixon, D. W., (1999). P53/P21(WAF1/CIP1) expression and its possible role in G1 arrest and apoptosis in ellagic acid treated cancer cells. *Cancer Letters, 136*, 215–221.

Perera, F. P., Wang, S., Vishnevetsky, J., Zhang, B., Cole, K. J., Tang, D., Rauh, V., & Phillips, D. H., (2011). Polycyclic aromatic hydrocarbons–aromatic DNA adducts in cord blood and behavior scores in New York City children. *Environmental Health Perspectives, 119*(8), 1176–1181.

Putz, M. V., Tudoran, M. A., & Putz, A. M., (2013). Structure properties and chemical-bio/ecological of PAH interactions: From synthesis to cosmic spectral lines, nanochemistry, and lipophilicity-driven reactivity. *Current Organic Chemistry, 17*, 2845–2871.

Rapacioli, M., Joblin, C., & Boissel, P., (2005). Spectroscopy of polycyclic aromatic hydrocarbons and very small grains in photodissociation regions. *Astronomy and Astrophysics, 429*, 193–204.

Rappert, S., Botsch, K. C., Nagorny, S., Francke, W., & Muller, R., (2006). Degradation of 2,3-diethyl–5-methylpyrazine by a newly discovered bacterium, Mycobacterium sp. strain DM–11. *Applied and Environmental Microbiology, 72*, 1437–1444.

Riding, M. J., Doick, K. J., Martin, F. L., Jones, K. C., & Semple, K. T., (2013). Chemical measures of bioavailability/bioaccessibility of PAHs in soil: Fundamentals to application. *Journal of Hazardous Materials, 261*, 687–700.

Rivas, F. J., (2006). Polycyclic aromatic hydrocarbons sorbed on soils: A short review of chemical oxidation based treatments. *Journal of Hazardous Materials, 138*, 234–251.

Rushdi, A. I., & Simoneit, B. R. T., (2001). Lipid formation by aqueous Fischer-Tropsch-type synthesis over a temperature range of 100 to 400°C. *Origins of Life and Evolution of Biospheres, 31*, 103–118.

Saito, M., Sakon, I., Kaito, C., & Kimura, Y., (2010). Formation of polycyclic aromatic hydrocarbon grains using anthracene and their stability under UV irradiation. *Earth Planets Space, 62,* 81–90.

Sakon, I., Onaka, T., Wada, T., Ohyama, Y., Matsuhara, H., Kaneda, H., et al., (2007). Properties of UIR bands in NGC 6946 based on mid-infrared imaging and spectroscopy with infrared camera on board AKARI. *Publications of the Astronomical Society of Japan, 59,* S483–S495.

Sayer, J. M., Chadha, A., Agarwal, S. K., Yeh, H. J. C., Yagi, H., & Jerina, D. M., (1991). Covalent nucleoside adducts of a benzo[a]pyrene 7,8-diol 9,10-epoxides: Structural reinvestigation and characterization of a novel adenosine adduct on the ribose moiety. *Journal of Organic Chemistry, 56,* 20–29.

Sayer, J. M., Yagi, H., Wood, A. W., Conney, A. H., & Jerina, D. M., (1982). Extremely facile reaction between the ultimate carcinogen benzo[a]-pyrene–7,8-diol 9,10-epoxide and ellagic acid. *Journal of the American Chemical Society, 104,* 5562–5564.

Schwarzenbach, R. P., Gschwend, P. M., & Imboden, D. M., (1993). *Environmental Organic Chemistry.* John Wiley and Sons, Inc, New York; p. 681.

Shimada, T., Gillam, E. M., Oda, Y., Tsumura, F., Sutter, T. R., Guengerich, F. P., & Inoue, K., (1999). Metabolism of benzo[a]pyrene to trans–7,8-dihydroxy–7,8-dihydrobenzo[a]pyrene by recombinant human cytochrome P450 1B1 and purified liver epoxide hydrolase. *Chem. Res. Toxicol., 12,* 623–629.

Simoneit, B. R. T., (2004). Prebiotic organic synthesis under hydrothermal conditions: An overview. *Advances in Space Research, 33,* 88–94.

Smith, W. A., & Gupta, R. C., (1999). Determining efficacy of cancer chemopreventive agents using a cell-free system concomitant with DNA adduction. *Mutation Research/ Fundamental and Molecular Mechanisms of Mutagenesis, 425,* 143–152.

Soussi, T., (2000). P53 antibodies in the sera of patients with various types of cancer: A review. *Cancer Research, 60,* 1777–1788.

Teel, R. W., (1986). Ellagic acid binding to DNA as a possible mechanism for its antimutagenic and anticarcinogenic action. *Cancer Letters, 30,* 329–336.

Thorsen, W. A., Gregory, C. W., & Damian, S., (2004). Bioavailability of PAHs: Effects of soot carbon and PAH source. *Environmental Science & Technology, 38*(7), 2029–2037.

Tielens, A. G. G. M., Hony, S., Van Kerckhoven, C., & Peeters, E., (1999). Interstellar and circumstellar PAHs, in the Universe, as seen by ISO. In: Cox, P., & Kessler, M. F., (eds.), *ESA SP–427* (p. 579).

U.S. EPA. (2000). *Deposition of Air Pollutants to the Great Waters: Third Report to Congress.* Office of air quality planning and standards. EPA–453/R–00–0005, http://www. epa.gov/ air/oaqps/gr8water/3rdrpt/index.html.

Van Der Spoel, D., Lindahl, E., Hess, B., Groenhof, G., Mark, A. E., & Berendsen, H. J. C., (2005). GROMACS, Fast, flexible, and free. *Journal of Computational Chemistry, 26,* 1701–1718.

Walther, J. H., Jaffe, R., Halicioglu, T., & Koumoutsakos, P., (2001). Carbon nanotubes in water: Structural characteristics and energetics. *Journal of Physical Chemistry B., 105,* 9980–9987.

Wang, Y., Li, P. H., Li, H. L., Liu, X. H., & Wang, W. X., (2010). PAHs distribution in precipitation at Mount Taishan: China. Identification of sources and meteorological influences. *Atmospheric Research, 95*(1), 1–7.

Watts, R. J., Stanton, P. C., Howsawkeng, J., & Teel, A. L., (2002). Mineralization of a sorbed polycyclic aromatic hydrocarbon in two soils using catalyzed hydrogen peroxide. *Water Research, 36,* 4283–4292.

Wesołowski, R. P., Furmaniak, S., Terzyk, A. P., & Gauden, P. A., (2011). Simulating the effect of carbon nanotube curvature on adsorption of polycyclic aromatic hydrocarbons. *Adsorption, 17*, 1–4.

Whitley, A. C., Stoner, G. D., Darby, M. V., & Walle, T., (2003). Intestinal epithelial cell accumulation of the cancer preventive polyphenol ellagic acids extensive binding to protein and DNA. *Biochemical Pharmacology, 66*, 907–915.

Wood, A. W., Huang, M. T., Chang, R. L., Newmark, H. L., Lehr, R. E., Yagi, H., Sayer, J. M., Jerina, D. M., & Conney, A. H., (1982). Inhibition of the mutagenicity of bay-region diol epoxides of polycyclic aromatic hydrocarbons by naturally occurring plant phenols: Exceptional activity of ellagic acid. *Proceedings of the National Academy of Sciences of the United States of America, 79*, 5513–5517.

Yap, C. L., Gan, S., & Ng, H. K., (2011). Fenton based remediation of polycyclic aromatic hydrocarbons-contaminated soils. *Chemosphere, 83*, 1414–1430.

Zhao, J., Lu, J. P., Han, J., & Yang, C. K., (2003). Noncovalent functionalization of carbon nanotubes by aromatic organic molecules. *Applied Physics Letters, 82*, 3746–3748.

Zhong, Y., & Zhu, L., (2013). Distribution, input pathway and soil-air exchange of polycyclic aromatic hydrocarbons in Banshan Industry Park, China. *Science of the Total Environment, 444*, 177–182.

Zhu, L., Lu, H., Chen, S., Amagai, T., Zhu, L., & Amagai, T., (2009). Pollution level, phase distribution and source analysis of polycyclic aromatic hydrocarbons in residential air in Hangzhou, China. *Journal of Hazardous Materials, 162*(2/3), 1165–1170.

CHAPTER 22

Pendentancy-Based Molecular Descriptors for QSAR/QSPR Studies

NAVEEN KHATRI,[1] HARISH DUREJA,[2] and A. K. MADAN[1]

[1]Faculty of Pharmaceutical Sciences, Pt. B.D. Sharma University of Health Sciences, Rohtak–124001, India, E-mail: madan_ak@yahoo.com

[2]Department of Pharmaceutical Sciences, M.D. University, Rohtak–124001, India

ABSTRACT

The drug discovery/development process in pharmaceutical industry is increasingly encountering high cost and competitiveness. In order to alleviate these problems, steps are being regularly undertaken to simultaneously reduce the cost and time required for the drug discovery/development process. [Quantitative] structure activity/property/toxicity relationship [(Q) SAR/QSPR/QSTR] based drug design has become one of the most standard and authoritative approaches used in drug design. The main problem in (Q) SAR pertains to the qualitative nature of chemical structures. Accordingly, one cannot have a direct correlation between qualitative chemical structures and quantitative biological activity. However, this problem can be easily overcome through quantification of chemical structures by making use of topology-based molecular descriptors (MDs). Pendentancy-based MDs constitute a very small fraction of topology-based MDs. In the present study, various pendentancy-based MDs have been defined, and their role in SAR/QSAR/QSPR/QSTR studies briefly reviewed.

22.1 INTRODUCTION

In the last decade, very few new drug applications (NDAs) were approved by FDA. This was despite large R&D investments and advancements in drug discovery technology. In addition, the new trend emerging out is a clear cut shift from primary care blockbuster mindset to specialty care and orphan drugs (Mullard, 2011). Many experts believe that this trend will continue and number of drug approvals per year for primary indications may not rise dramatically in near future (Mullard, 2012). The rising drug development cost has further aggravated the problem. According to a study conducted in 2014 by 'Tufts Center for the Study of Drug Development,' the average cost of launching a drug from concept stage to market has increased approximately from US $1020 million dollars to US $2550 million dollars (www.phrma.org, 2013; http://csdd.tufts.edu, 2014). Low productivity amalgamated with rising R&D costs has forced pharmaceutical industry to find out the underlying reasons and modify the process. The main reason of drug failure around 1990 was unacceptable pharmacokinetics of drugs in human beings (Graul et al., 2010). Then several preclinical screens were adopted by the industry to address absorption, permeability, metabolism, distribution, and excretion issues. The insufficient efficacy of the drug was the foremost reason for drug failures in Phase II and Phase III of clinical trials. Other reasons include failure due to strategic reasons (possibly due to competition or insufficient risk/benefit ratio), safety concerns, or insufficient margins. The majority of these failures were from therapeutic categories of neuroscience, cancer, cardiovascular, and metabolic (Kola and Landis, 2004; Graul et al., 2010; Graul and Cruces, 2011; Arrowsmith, 2011).

Therefore, the approaches which identify failure roots at an early stage of the drug development process will be of immense utility (Sussman and Kelly, 2003; Parenti and Rastelli, 2012) Thus, *in silico* approaches have become an important tool to reduce the cost and are being used as a versatile tool of priority experiments at every stage of the drug discovery/development process (Clark, 2006; Reddy et al., 2007). *In silico* drug design has found its applicability in almost every facet of the drug discovery process from its conventional application of lead discovery and optimization, towards target identification, target validation and preclinical study (ADME/Tox prediction).

In silico modeling of a drug initiates with quantification of chemical structure through the use of molecular descriptors (MDs) and leads to the development of a structure-activity/toxicity/property relationship. (Peltason and Bajorath, 2007; Baldi, 2010). (Q)SAR models can easily predict biological properties of newly designed compounds before their synthesis and allow removal of undesirable molecules at an early stage of their development (Sellasie, 2003). Various (Q)SAR approaches have been developed gradually since conventional Hansch and Fred Wilson's one or two-dimensional linear relationships. Based on the structural representation of the molecule or the way by which the descriptor values are derived the QSAR methods can be categorized into different classes from one dimension (1D) to the sixth dimension (6D) (Lille, 2007).

Topology-based MDs have been found to be immensely useful in the prediction of the diverse physical, chemical, biological, and pharmacokinetic properties for various compounds so as to facilitate designing a new drug molecule. A wide variety of these graph theoretical descriptors have been reported in literature for prediction of various properties of compounds. Many new topology based MDs have been specifically developed for [Q]SAR/QSPR/QSTR studies (Randić, 2003; Douglas et al., 2001; Mati and Victor, 1996; Joseph et al., 2005).

Most of the topology based MDs are related to either topological distance in the graph or to vertex adjacency (atom-atom connectivity). The distance matrix and adjacency matrix of molecular graphs constitute the roots of these topological indices (Sabljić and Trinajstić, 1981). Only a small number of MDs based on pendent matrix have been reported in literature. These include superpendentic index (Gupta et al., 1999), log of superpendentic index (Todeschini and Consonni, 2000), terminal wiener index (TWI) (Gutman et al., 2009), super-augmented pendentic indices (Dureja et al., 2009), super pendentic topochemical index, pendentic eccentricity index and pendentic eccentricity topochemical index (Goyal et al., 2010), etc. In the present study, various pendentancy-based indices have been defined, and their role in QSAR/QSPR studies briefly reviewed.

22.2 SUPERPENDENTIC INDEX ($\int^P$)

This index was reported by Gupta et al., (1999) and is defined as the square root of the sum of the products of non-zero elements in each row of the pendent matrix. The pendent matrix, D_P, of a graph G is a submatrix of distance matrix obtained by simply retaining the columns corresponding to pendent vertices.

$$\int^P (G) = \left(\sum_{i=1, j=1}^{n,m} \prod P_{(i.j)} \right)^{1/2} \tag{1}$$

where $P_{(i,j)}$ is length of the path that contains the least number of edges between vertex i and vertex j in graph G; m and n are maximum possible numbers of i and j, respectively.

This index was successfully applied to correlating 128 analogs of 4-substituted-2-guanidino thiazoles with anti-ulcer activity. The 82% accuracy of classification was found in the active range using moving average analysis (MAA) based model. This index has also been used in boiling point prediction (Goll and Jurs, 1999), antifungal activity (Doble and Kumar, 2006) and anti-microbial activity (Sabet et al., 2007) of the compounds.

22.3   LOG OF SUPERPENDENTIC INDEX (LOG $\int^P$)

Log of superpendentic index is logarithm of superpendentic index and has been reported by Todeschini and Consonni (2000). It is expressed as per following:

$$\log \int^P (G) = \log \left(\sum_{i=1,j=1}^{n,m} \prod P_{(i.j)} \right)^{1/2} \tag{2}$$

22.4   SUPERPENDENTIC TOPOCHEMICAL INDEX ($\int_c^P$)

Topochemical version of superpendentic index called superpendentic topochemical index was developed by Goyal et al., (2010) and is defined as the square root of the sum of the products of non-zero elements in each row of the chemical pendent matrix. The chemical pendent matrix, D_p^c, of a graph G is a sub-matrix of chemical distance matrix obtained by simply retaining the columns corresponding to pendent vertices.

$$\int_c^P (G) = \left(\sum_{i=1,j=1}^{n,m} \prod P_{c(i.j)} \right)^{1/2} \tag{3}$$

where $P_{c(i,j)}$ is length of the chemical path that contains the least number of edges between vertex i and vertex j in graph G; m and n are maximum possible numbers of i and j, respectively.

Goyal et al., (2010) also found correlation of this topochemical index with 5'-O-[(N-Acyl)sulfamoyl]adenosines and anti-tubercular activity. The overall accuracy of prediction of the order of 91.6% using MAA-based model. Active ranges in these models indicated exceptionally high potency in terms of anti-tubercular activity (Figure 22.1).

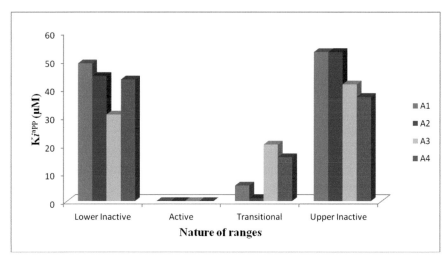

FIGURE 22.1 Average K*i*^app (μM) values of correctly predicted 5'-*O*-[(*N*-Acyl)sulfamoyl] adenosines in various ranges of the proposed MAA-based models (A1: Pendentic eccentricity topochemical index, A2: Super pendentic topochemical index, A4: Molecular connectivity topochemical index, A2: Wiener's topochemical index) (Goyal et al., 2010).

22.5 TERMINAL WIENER INDEX (TWI)

The TWI was reported by Gutman et al., (2009). It is defined as the sum of the distances between all pairs of the pendent vertices in the pendent matrix in a hydrogen suppressed molecular graph and is expressed as per following:

$$TWI = \left(\sum_{i=1, j=1}^{n} P_{(i,j)} \right) \qquad (4)$$

where $P_{(i,j)}$ is length of the path that contains the least number of edges between vertex *i* and vertex *j* in graph G.

22.6 SUPERAUGMENTED PENDENTIC INDICES 1–4

The superaugmented pendentic indices (denoted by $^{SA}\int^{P-1}$, $^{SA}\int^{P-2}$, $^{SA}\int^{P-3}$ and $^{SA}\int^{P-4}$) were reported by Dureja et al., (2009). These indices are defined as the summation of quotients, of the product of non-zero row elements in the pendent matrix and product of adjacent vertex degrees; and eccentricity raised to the power x of the concerned vertex, for all vertices in the hydrogen suppressed molecular graph. In general, these indices can be expressed as,

$$^{SA}\int^{P^x} = \sum_{i=1}^{n}\sum_{j=1}^{n}\left(\frac{P_{(ij)} * m_i}{(e_i)^x}\right) \tag{5}$$

The value of 'x' is 1, 2, 3 and 4 for $^{SA}\int^{P-1}$, $^{SA}\int^{P-2}$, $^{SA}\int^{P-3}$ and $^{SA}\int^{P-4}$ respectively, where p_i is the product of non-zero row elements in the pendent matrix, m_i is the product of degrees of all the vertices (v_j), adjacent to vertex i and easily obtained by multiplying all the non-zero row elements in augmentative adjacency matrix, d_i is the degree of the vertex v_i and e_i is the eccentricity of vertex v_i.

22.7　SUPERAUGMENTED PENDENTIC TOPOCHEMICAL INDEX-4

Topochemical version of superaugmented pendentic index-4 termed as superaugmented pendentic topochemical index-4 was developed by Khatri et al., (2015). It is defined as the summation of the quotients of the product of all the non-zero row elements in the chemical pendent matrix and product of chemical adjacent vertex degrees and the fourth power of the chemical eccentricity of the concerned vertex for all vertices in a hydrogen-suppressed chemical molecular graph. It can be expressed as:

$$^{SA}\int^{P-4}(G_{k,n}) = \sum_{i=1}^{n}\frac{p_i m_i}{e_i^4} \tag{6}$$

where p_{ic} is the chemical pendenticity and is obtained by multiplying all the non-zero row elements of the concerned vertex in the chemical pendent matrix, ΔPc, of a chemical graph $(G_k n)c$. ΔPc is a sub-matrix of the chemical distance matrix and is obtained by retaining the columns corresponding to pendent vertices. m_{ic} is the augmented chemical adjacency and is defined as the product of chemical degrees of all the vertices v_j adjacent to vertex v_i. e_{ic} is the chemical eccentricity of vertex v_i, and n is the number of vertices in graph G. Khatri et al., (2015) applied this index for the prediction of anti-HIV activity of purine nucleoside analogs. In MAA-based model, the overall accuracy of prediction achieved through this index was 96.9%. Average EC_{50} (µM) of correctly predicted compounds in active range was 0.141 and average value of selectivity index was 859.5 (Figure 22.2). Low EC_{50} and high selectivity index values indicate high potency and safety of compounds falling in the active range.

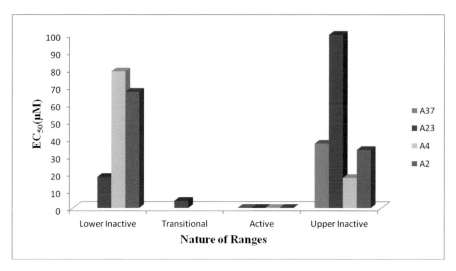

FIGURE 22.2 Average EC$_{50}$ (µM) values of anti-HIV activity of correctly predicted purine nucleoside analogs in various ranges of the MAA-based models (A37: Balaban-type index from *Z*-weighted distance matrix, A23: Superaugmented pendentic topochemical index-4, A4: Shape profile no. 20, A2: Spherocity index) (Khatri et al., 2015).

22.8 PENDENTIC ECCENTRICITY INDEX (ξ^P)

This index was developed by Goyal et al., (2010) and is defined as the summation of the quotients of the product of non-zero row elements in the pendent matrix and squared eccentricity of the concerned vertex, for all vertices in a hydrogen suppressed molecular graph. The eccentricity E_i of a vertex i in a graph G is the path length from vertex i to the vertex j that is farthest from i ($E_i = \max d(ij)$; j G). It is expressed as:

$$\xi^P = \sum_{i=1}^{n} \left\{ \prod_{i=1}^{m} P_{ij} \bigg/ E_i^2 \right\} \tag{7}$$

22.9 PENDENTIC ECCENTRICITY TOPOCHEMICAL INDEX (ξ_c^P)

Topochemical version of pendentic eccentricity index called pendentic eccentricity topochemical index was also developed by Goyal et al., (2010). It is defined as the summation of the quotients of the product of non-zero row elements in the chemical pendent matrix and squared chemical eccentricity of the concerned vertex, for all vertices in the hydrogen suppressed molecular graph. The chemical eccentricity E_{ic} of a vertex i in a graph G is the path length from vertex i to the vertex j that is farthest from i (E_{ic} = max d(ij); j G). It is expressed as:

$$\xi_c^P = \sum_{i=1}^{n} \left\{ \prod_{i=1}^{m} P_{icjc} \bigg/ E_{ic}^2 \right\} \tag{8}$$

This index was also used by Goyal et al., (2010) to establish correlation between a 5'-*O*-[(*N*-Acyl)sulfamoyl]adenosines and anti-tubercular activity. Using this index, they found overall accuracy of prediction of the order of 91.3% in MAA-based model. Active ranges in these models indicated exceptionally high potency in terms of anti-tubercular activity (Figure 22.1).

The representative applications of various pendentancy-based MDs have been enlisted in Table 22.1.

TABLE 22.1　Representative Applications of Pendent Based Indices

Name of index	Nature of activity/chemical series	Type of model	Reference
Superpendentic index	Antiulcer activity of 4-substituted-2-guanidino thiazoles	MAA-based model	Gupta et al., 1999
	Normal boiling points of hydrocarbons and heteroatom-containing organic compounds	Multiple linear regression and computational neural networks-based QSPR model	Goll and Jurs, 1999
	Fathead minnow acute aquatic toxicity of organic compounds	QSAR model	He and Jurs, 2005
	Antifungal activity of aromatic, salicylic derivatives, cinnamyl derivatives, etc.	QSAR models based on back propagation neural network and single/multiple regression	Doble and Kumar, 2006
	Anti- *T. pyriformis activity of* phenol derivatives	QSTR models	Pasha et al., 2007
	Kovats retention index of terpenoids	QSPR model and artificial neural networks model	Hemmateene-jad et al., 2007
	Antimicrobial activity of 3-hydroxypyridine-4-one and 3-hydroxypyran-4-one derivatives	QSAR models based on multiple linear regression	Sabet et al., 2007
	Angiotensin converting enzyme inhibitor (ACEI) activity and elastase substrate catalyzed kinetics (ESCK)	Genetic partial least squares (GPLS), support vector machine (SVM), and immune neural network (INN), based QSAR model	Liang et al., 2008

TABLE 22.1 *(Continued)*

Name of index	Nature of activity/chemical series	Type of model	Reference
	Antimalarial activity of artemisinin analogs	QSAR models	Srivastavaa et al., 2009
	Toxicity of aromatic nitro compounds	Multiple linear regression analysis based QSTR model	Pasha et al., 2009
	Antitubercular activity of 5'-O-[(N-Acyl)sulfamoyl] adenosines	Classification models based on decision tree, random forest and MAA	Goyal et al., 2010
	Antimicrobial activity of Mannich bases of glutarimides with sulfonamides and secondary amines	QSAR models	Manikpuria et al., 2010
	Anticancer activity of epipodophyllotoxin analogs	QSAR model	Naik et al., 2010
	Prostatic hyperplasia of aryl/ heteroaryl/aralkyl/aroyl piperazines	QSAR model	Sarswat et al., 2011
	Anticancer activity of Fe(III)-Salen-like complexes	QSAR models	Ghanbari et al., 2014
	Anti-HIV activity of HEPT compounds	QSAR models	Sarkar and Nandi, 2014
	Prediction of partition coefficient of 3-hydroxy pyridine-4-one derivatives	QSPR models	Shahlaei et al., 2014
	5-HT6 receptor antagonist activity of 3,4-dihydro–2H-benzo[1,4]oxazines	QSAR model	Choudhary et al., 2015
	Prediction of the Fate of Organic Compounds in the Environment	QSAR models	Mamy et al., 2015
Superpendentic topochemical index	Antitubercular activity of 5'-O-[(N-Acyl)sulfamoyl] adenosine derivatives	Classification models based on decision tree, random forest and MAA	Goyal et al., 2010
Terminal Wiener index	Anti-HIV Activity of Quinolone Carboxylic Acids	QSAR model	Senbagamalar and Babujee, 2013
Pendentic eccentricity index	Antitubercular activity of 5'-O-[(N-Acyl)sulfamoyl] adenosine derivatives	Classification models based on decision tree, random forest and MAA	Goyal et al., 2010

TABLE 22.1 *(Continued)*

Name of index	Nature of activity/chemical series	Type of model	Reference
Pendentic eccentricity topochemical index	Antitubercular activity of 5'-O-[(N-Acyl)sulfamoyl] adenosine derivatives	Classification models based on decision tree, random forest and MAA	Goyal et al., 2010
Superaug-mented pendentic topochemical index	Anti-HIV activity of purine nucleoside analogs	Classification models based on decision tree, random forest, support vector machine and MAA	Khatri et al., 2015

22.10 CONCLUSION

Ease of calculation, simplicity, exceptionally high discriminating power amalgamated with negligible degeneracy and high sensitivity towards branching as well as relative position(s) of substituent(s) has rendered pendentancy-based MDs as a potential tool in similarity/dissimilarity studies, characterization of structures, combinatorial library design, lead identification/optimization, pharmacokinetic relationship studies and quantitative structure-activity/property studies. Pendenticity based MDs have been successfully used for the development of quantitative structure-toxicity/activity/property relationship models utilizing different correlation and classification techniques. Successful use of pendenticity-based MDs for development of models with high degree of accuracy of prediction has been achieved for prediction of either toxicity or diverse biological activities like anti-ulcer, anti-microbial, anti-malarial, anti-fungal, and anti-tubercular activities.

KEYWORDS

- **decision tree**
- **distance matrix**
- **moving average analysis**
- **pendenticity**
- **random forest**
- **topological indices**

REFERENCES

Arrowsmith, J., (2011). Phase II failures: 2008–2010. *Nat Rev Drug Discov., 10*, 328.

Baldi, A., (2010). Computational approaches for drug design and discovery: An overview. *Syst. Rev. Pharm., 1*(1), 99–105.

Basak, C., (2007). *Mathematical Structure Descriptors: Development and Applications in Chemistry, Drug Discovery, Environmental Protection, and Bioinformatics.* Lecture notes for the 2nd Indo-US lecture series on discrete mathematical chemistry, Kerala. http://cmsintl.org/general announcements/Workshop On Graph Theory Proceedings.pdf (Accessed on 15/05/15).

Choudhary, M., Pilania, P., & Sharma, B. K., (2015). A QSAR study on 5-HT6 receptor antagonists: The 3,4-dihydro–2H-benzo[1,4]oxazines. *J. Chem. Pharm. Res., 7*(3), 2422–2433.

Clark, D. E., (2006). What has computer-aided molecular design ever done for drug discovery? *Expert Opin. Drug Discov., 1,* 103–110.

Doble, M., & Kumar, K. A., (2006). Experimental and modeling studies on antifungal compounds. *Central Eur. J. Chem., 4*, 428–439.

Douglas, H., Basak, C., & Shi, X., (2001). QSAR with few compounds and many features. *J. Chem. Inf. Comput. Sci., 41*, 663–670.

Dureja, H., Kinkar, C. D., & Madan, A. K., (2009). Superaugmented pendentic indices: Novel topological descriptors for QSAR/QSPR. *Sci. Pharm., 77,* 521–537.

Ghanbari, Z., Housaindokht, M. R., Izadyar, M., Bozorgmehr, M. R., Eshtiagh-Hosseini, H., Bahrami, A. R., Matin, M. M., & Khoshkholgh, M. J., (2014). Structure-activity relationship for Fe(III)-Salen-like complexes as potent anticancer agents. *The Sci. W. Journal,* p. 10.

Goll, E. S., & Jurs, P. C., (1999). Prediction of the normal boiling points of organic compounds from molecular structures with a computational neural network model. *J. Chem. Inf. Comput. Sci., 39*, 974–983.

Goyal, R. K., Dureja, H., Singh, G., & Madan, A. K., (2010). Models for antitubercular activity of 5'-O-[N-Acylsulfamoyl]adenosines. *Sci. Pharm., 78*, 791–820.

Graul, A., & Cruces, E., (2011). The year's new drugs & biologics, (2010). *Drugs Today (Barc.), 47*, 27–51.

Graul, A., Revel, L., Tell, M., Rosa, E., & Cruces, E., (2010). Overcoming the obstacles in the pharma/biotech industry: 2009 update. *Drug News Perspect, 23*, 48–63.

Gupta, S., Singh, M., & Madan, A. K., (1999). Superpendentic index: A novel topological descriptor for predicting biological activity. *J. Chem. Inf. Comp. Sci., 39*, 272–277.

Gutman, I., Furtula, B., & Petrović, M., (2009). Terminal Weiner index. *J. Math. Chem., 46*, 522–531.

He, L., & Jurs, P. C., (2005). Assessing the reliability of a QSAR model's predictions. *J. Mol. Graph Mod., 23*, 503–523.

Hemmateenejad, B., Javadnia, K., & Elyasi, M., (2007). Quantitative structure-retention relationship for the Kovats retention indices of a large set of terpenes: A combined data splitting-feature selection strategy. *Anal. Chim. Acta, 592*, 72–81.

http://csdd.tufts.edu/files/uploads/Tufts CSDD briefing on RD cost study- Nov 18, 2014.pdf (Accessed on 11/05/15).

http://www.phrma.org/sites/default/files/pdf/PhRMA%20Profile%202013.pdf (Accessed on 11/05/15).

Joseph, F., Philip, M., Lowell, H., & Lemont, B., (2005). QSAR modeling of carcinogenic risk using discriminant analysis and topological molecular descriptors. *Cur. Drug Dis. Tech., 2*, 55–57.

Khatri, N., Lather, V., & Madan, A. K., (2015). Diverse models for anti-HIV activity of purine nucleoside analogs. *Chem. Central J., 9*, 29.

Kola, L., & Landis, J., (2004). Can the pharmaceutical industry reduce attrition rates? *Nat. Rev. Drug Discov., 3*, 711–715.

Liang, L. Z., Rong, L. G., Mao, S., Ying, S. J., Bin, Y. S., Hu, M., et al., (2008). A novel vector of topological and structural information for amino acids and its QSAR applications for peptides and analogs, *Sci. China Ser. B-Chem., 51*, 946–957.

Lill, M. A., (2007). Multi-dimensional QSAR in drug discovery. *Drug Discov. Today, 12*, 1013–1017.

Mamy, L., Patureau, D., Barriuso, E., Bedos, C., Bessac, F., Louchart, X., Martin-laurent, F., Miege, C., & Benoit, P., (2015). Prediction of the fate of organic compounds in the environment from their molecular properties: A review. *Crit. Rev. Environ. Sci. Technol., 45*(12), 1277–1377.

Manikpuria, A. D., Joshib, S., & Khadikar, P. V., (2010). Synthetic, spectral, antimicrobial, and QSAR studies on novel Mannich bases of glutarimides. *J. Chil. Chem. Soc., 55*, 283–292.

Mati, K., & Victor, S., (1996). Quantum-chemical descriptors in QSAR/QSPR studies. *Chem. Rev., 96*, 1027–1043.

Mullard, A., (2011). 2010 FDA drug approvals. *Nat. Rev. Drug Discov., 10*, 82–85.

Mullard, A., (2012). 2011 in reflection. *Nat. Rev. Drug Discov., 11*, 6–8.

Naik, P. K., Dubey, A., & Kumar, R., (2010). Development of predictive quantitative structure-activity relationship models of epipodophyllotoxin derivatives. *J. Biomol. Screen., 15*, 1194–1203.

Parenti, M. D., & Rastelli, G., (2012). Advances and applications of binding affinity prediction methods in drug discovery. *Biotechnol. Adv., 30*(1), 244–250.

Pasha, F. A., Neaz, M. M., Cho, S. J., Ansari, M., Mishra, S. K., & Tiwari, S., (2009). *In silico* quantitative structure-toxicity relationship study of aromatic nitro compounds. *Chem. Biol. Drug Des., 73*, 537–544.

Pasha, F. A., Srivastava, H. K., Srivastava, A., & Singh, P. P., (2007). QSTR study of small organic molecules against tetrahymena pyriformis. *QSAR Comb. Sci., 26*, 69–84.

Peltason, L., & Bajorath, J., (2007). SAR index: Quantifying the nature of structure activity relationships. *J. Med. Chem., 50*, 5571–5578.

Randić, M., (2003). Chemical graph theory – Facts and fiction. *Ind. J. Chem., 42A*, 1207–1218.

Reddy, A. S., Pati, S. P., Kumar, P. P., Pradeep, H. N., & Sastry, G. N., (2007). Virtual screening in drug discovery – A computational perspective. *Curr. Protein Peptide Sci., 8*, 329–351.

Sabet, R., Fassihi, A., & Moeinifard, B., (2007). Preliminary MLR studies of antimicrobial activity of some 3-hydroxypyridine-4-one and 3-hydroxypyran-4-one derivatives. *Res. Pharm. Sci., 2*, 103–112.

Sabljić, A., & Trinajstić, N., (1981). Quantitative structure-activity relationship: The role of topological indices. *Acta Pharm. Jugosl., 31*, 198–214.

Sarkar, S., & Nandi, S., (2014). QSAR modeling of HEPT compounds: An attempt to anti-HIV drug design. *J. Comput. Met. Mol. Des., 4*(4), 15–25.

Sarswat, A., Kumar, R., Kumar, L., Lal, N., Sharma, S., Prabhakar, Y. S., et al., (2011). Aryl-piperazines for management of benign prostatic hyperplasia: Design, synthesis, quantitative structure-activity relationships, and pharmacokinetic studies. *J. Med. Chem., 54*, 302–311.

Sellasie, C. D., (2003). History of quantitative structure-activity relationships. In: Abraham, D. J., (ed.), *Burger's Medicinal Chemistry and Drug Discovery* (6th edn., pp. 1–48). John Wiley and Sons, New York.

Senbagamalar, J., & Babujee, J. B., (2013). Predicting anti-HIV activity of quinolone carboxylic acids -computation approach using topological indices. *EJBI, 9*(2), 9–13.

Shahlaei, M., Fassihi, A., Saghaie, L., & Zare, A., (2014). Prediction of partition coefficient of some 3-hydroxy pyridine-4-one derivatives using combined partial least square regression and genetic algorithm. *Res. Pharm. Sci., 9*(2), 143–153.

Srivastavaa, M., Singha, H., & Naik, P. K., (2009). Quantitative structure-activity relationship (QSAR) of artemisinin: The development of predictive *in vivo* antimalarial activity models. *J. Chemometrics, 23*, 618–635.

Sussman, N. L., & Kelly, J. H., (2003). Saving time and money in drug discovery- A pre-emptive approach. *Business Briefings: Future Drug Discov.*, 46–49.

Todeschini, R., & Consonni, V., (2000). *Handbook of Molecular Descriptors: Methods and Principal in Medicinal Chemistry* (Vol. 11). Wiley VCH, Weinheim.

CHAPTER 23

Photovoltaic System

MARIUS C. MIRICA,[1] MARINA A. TUDORAN,[1,2] and MIHAI V. PUTZ[1,2]

[1]*Laboratory of Renewable Energies-Photovoltaics, R&D National Institute for Electrochemistry and Condensed Matter, Dr. A. Paunescu Podeanu Str. No. 144, Timisoara, RO–300569, Romania*

[2]*Laboratory of Structural and Computational Physical Chemistry for Nanosciences and QSAR, Biology-Chemistry Department, West University of Timisoara, Pestalozzi Street No. 44, Timisoara, RO–300115, Romania*

23.1 DEFINITION

A photovoltaic (PV) system can be described as a system which converts directly the solar energy (the so-called PV effect), and consist in PV modules made of semiconductors as the main building blocks. The working process occurs in few steps: first, the semiconductor material releases the electrons from their atoms after exposing to light, the electrons are moving through the material carrying with them the electric current, the electricity being generated by the one-directional current flow. The generated energy can be either used directly or stored in batteries.

23.2 HISTORICAL ORIGIN(S)

The PV effect was demonstrated for the first time by 1839 in Edmond Becquerel (1839, 1841), as a result of an experiment in which he obtained electricity by using different type of light, including sunlight, to illuminate an electrode. The usual electrodes were from platinum, coated with AgCl or AgBr as coated materials for an improved effect, the best response being obtained for ultraviolet and blue light. He also obtained some results for silver electrodes. The photoconductivity of the element selenium was discovered by Willoughby Smith by several experiments in which he observes an increased conductivity of the selenium rods after exposed to light (Green, 1990; Benjamin, 1983; Shive, 1959; Wolf, 1976). In 1877, William Grylls Adams and Richard Evans Day (1877) conducted a series of experiments from which it was demonstrated the possibility of direct conversion of solar energy into electricity (Figure 23.1).

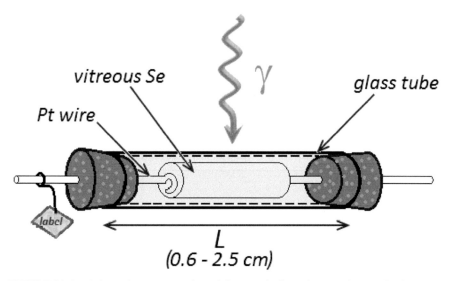

FIGURE 23.1   Schematic representation of the sample from the experiment of Adams and Day to investigate the photoelectric effect in Se. Redrawn and adapted after Adams and Day (1987) and http://pveducation.org/pvcdrom/manufacturing/first-photovoltaic-devices.

They also investigated the anomaly produced as a result of the photo-conductive effect in selenium using a device obtained by pushing heated platinum contacts inside small cylinders of vitreous selenium at the opposite ends (Figure 23.1) in order to determine if a current can be started in selenium using only the light action. By obtaining a positive result, they demonstrated for the first time the PV effect in all solid state systems. Also, the photogenerated current was explained by the crystallization process which occurred at the outer layer of the selenium bar and was induced by light (Green, 1990; Benjamin, 1983; Shive, 1959; Wolf, 1976) (Figure 23.2).

The first thin-film PV device was developed by Fritts (1883) using molten selenium compressed between plates consisting of different metals and a gold leaf. The thin selenium film obtained adhered only to one of the plate, and the gold leaf was pressed on the exposed selenium surface, this device having an area of 30 cm^2.

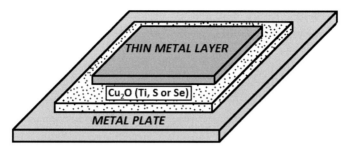

FIGURE 23.2 Schematic representation of an efficient photovoltaic device from the 1930s. Redrawn and adapted after http://pveducation.org/pvcdrom/manufacturing/first-photovoltaic-devices.

Another important discovery was made by Grondahl (1933), the rectifying action of the copper-cuprous oxide junction. In his work, he describes the PV cells and the copper-cuprous oxide rectifiers by observing the photoconductive effect which occurs in cuprous oxide layers that are grown on copper (Green, 1990; Benjamin, 1983; Shive, 1959; Wolf, 1976).

In 1931, Bergman (1931) realized improved selenium devices for commercial products with superior characteristics compared to the copper-based devices, and in 1939, Nix (Nix and Treptwo, 1939) propose a thallous-sulfide cell also with improved performance, these devices being among the most efficient copper-cuprous oxide devices of their times (Figure 23.2) (Green, 1990; Benjamin, 1983; Shive, 1959; Wolf, 1976).

The first solid state device developed in the late 1940s facilitate the developing of the first silicon solar cell with an efficiency of 6%, which was later assembled into modules on the PV devices (Chaar et al., 2011). The amorphous silicon (a-Si) was one of the first materials developed as an alternative to crystalline silicon, but its problems the light induces degradation were partially solved only recently. Nowadays, thin-film modules based on s-Si has efficiency of 6–8% range and due to their appearance are usually utilized for facade applications (Goetzberger and Hoffmann, 2005).

23.3 NANO-SCIENTIFIC DEVELOPMENT(S)

The solar energy represents a promising alternative to the conventional energy resources. The conversion of solar energy into electric power is realized using a PV) system based on PV cells arranged in modules. The conversion efficiency is related to the solar irradiance reaching the PV cell surface, the temperature and quality of the PV cell and the connections between PV modules and cell (Loschi et al., 2015; Seme et al., 2011; Prinsloo and Dobson, 2014).

The PV cell (Messenger and Ventre, 2004) consists in a pn junction or Schottky barrier device which produces electron-hole pairs when the incident atoms interact with the atom of the cell. Due to the electric field produced by the movement of the holes to the p-region and the electrons to the n-region of the cell. For an ideal cell, the I-V (current and voltage) characteristic equation has the following form:

$$I = I_l - I_o \left(e^{\frac{qV}{kT}} - 1 \right) \tag{1}$$

with I_l as cell current component due to photons, V as the cell voltage, $k = 1.38 \times 10^{-23}$ j/K, $q = 1.6 \times 10^{-19}$ coul and T as the cell temperature measured in Kelvin degrees.

The open circuit voltage (V_{OC}) can be determined from the equation (Messenger and Ventre, 2004):

$$V_{OC} - \frac{kT}{q} \ln \frac{I_l + I_o}{I_o} \cong \frac{kT}{q} \ln \frac{I_l}{I_o} \tag{2}$$

considering the normal case as $I_l \gg I_o$.

Starting from the fact that the cell power is obtained from the cell current multiplied by the cell voltage will give, the maximum cell power will have the formula (Messenger and Ventre, 2004):

$$P_{max} = I_m V_m = FF_{SC} V_{OC} \tag{3}$$

with I_m and V_m as the cell current and voltage respectively at maximum power and FF as the fill factor which measures the cell quality.

On illumination, only 20% of the irradiance is converted by the cell in electricity, and the rest is converted in heat producing the cell heating. The cell photocurrent depends on the intensity and the wavelength of the incident light, meaning that the constituent material has the property of efficiently convert the electricity at the appropriate spectrum. The cell diameter also varies in order to maximize the absorption, and some cells are coated with an anti-reflective coating in order to minimize the sunlight reflection from the cells (Messenger and Ventre, 2004).

The PV cells are connected in series in order to obtain adequate output voltage and form the PV module (Messenger and Ventre, 2004). The nominal operating cell temperature (NOCT) represents the cell temperature reached by cell at the open circuit at $G = 0.8 \, kW/m^2$, wind speed less than 1 m/s in an ambient temperature of 20°C at AM 1.5 irradiance conditions. When the irradiance and the ambient temperature are variable, the cell temperature can be estimated using the formula:

$$T_C = T_A + \left(\frac{NOCT - 20}{0.8} \right) G \tag{4}$$

The PV module electrical efficiency can be influenced by the module design and climatic parameters, e.g., an increase in PV temperature will determine a decrease in voltage, so the PV efficiency will decrease, a higher packing factor (i.e., the absorber area fraction which is occupied by the PV cell) will determine an electrical output increased per unit collector in the same time with increasing the module temperature, while increase in solar irradiance, being associated with a high number of electrons, will determine an increase in efficiency due to the face that there are more electron-hole pairs created and so more current is produced in the PV cell (Akarslan, 2012). In order to determine the cell temperature (T_c) on the PV cell efficiency (η_c) one can use the following linear regression (Skoplaki and Palyvos, 2009):

$$\eta_c = \eta_{ref} \left[1 - \beta_{ref} \left(T_c - T_{ref} \right) \right] \tag{5}$$

with T_c as the cell temperature, η_{reff} as the PV cell efficiency at the temperature T_{reff}; β_{reff} depends on the cell material and can be calculated with the formula (Akarslan, 2012):

$$\beta_{ref} = \frac{1}{T_0 - T_{ref}} \tag{6}$$

with T_0 as the maximum temperature which determines the PV cell efficiency to decrease to zero (Akarslan, 2012).

The PV array (Messenger and Ventre, 2004) consists of single modules connected in series producing a higher voltage, or in parallel producing higher currents. The parallel connection can be made by connecting in series each fuses with each series string of modules, in order to limit the excess current flowing from the rest of the strings in case one of them is failed. Another possibility of connection implies the production of a positive and negative voltage by the modules respecting the ground, so that a set three module will have a combined output which can feed a 3-phase inverter system.

The energy storage (Messenger and Ventre, 2004) can be made in batteries, by charging a large chemical capacitor, by producing hydrogen, by spinning a flywheel or by producing hydrogen. There are several type of batteries used on PV systems: *lead-acid storage batteries*, consisting in a lead oxide (PbO$_2$) anode and a lead cathode which are immersed in a solution of sulfuric acid; *nickel-cadmium storage batteries* which have cadmium oxide as cathode plates, nickel hydroxide as anode plates and potassium hydroxide as electrolyte; *nickel-zinc batteries*, a combination of Ni-Cd and Cu-Zn; *nickel-metal hydride,* having a metal hydride as cathode, the anode is the same as in Ni-Cd case and KOH as electrolyte (Messenger and Ventre, 2004).

Balance of System (BOS) is refereeing to the secondary components necessary in order to install the PV modules and arrays, containing power electronics which manage the output of PV arrays, wires which connect modules in series, mounting hardware and junction boxes for circuits merging. BOS requirements are related to environmental conditions, site-specific power needs and power storage (Shoro et al., 2013).

23.4 NANO-CHEMICAL APPLICATION(S)

The PV installations (Figure 23.3) can be classified in four major types: *off-grid* such as a standalone roof or ground-based systems, *off-grid distributed* such as industrial installations or power plants, *grid-centralized* such as large power plants and *grid distributed* such as roof or ground mounted small installations (Akarslan, 2012).

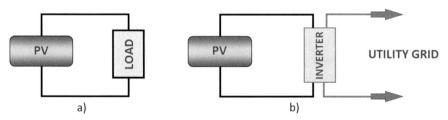

FIGURE 23.3 PV system: (a) directly connected to the load, (b) grid-connected system. Redrawn and adapted after Messenger and Ventre (2004).

The stand-alone PV powered systems are not connected to the electricity grid, having integrated a storage system in order to ensure electric power even at night or in times with low radiations. They can have a PV generator as an only power source, or auxiliary power source as additional generators with renewable energy or fossil fuel known as hybrid systems. Stand-alone systems are incorporated in watches and solar calculators, in traffic control systems or in the system for supplying buildings in remote areas and are of the two types: ac systems with an inverter or dc systems without or with a storage battery (Goetzberger and Hoffmann, 2005).

The solar watches have the advantage that works without a storage, but there is necessary an enough energy so that the display can be read. The products based on solar power usually contain a voltage DC/DC converter, small integrated PV modules and sometimes a charge controller and an energy storage unit (Goetzberger and Hoffmann, 2005).

On the other hand, the simplest Solar Home System can produce enough electrical energy to support the lighting, radio, TV, and a small refrigerator, consisting in a solar generator – the PV module, the charge controller, the lead battery and the DC appliance directly connected. A stand-alone system can be used to supply electricity for a single building when there is no connection to the public grid (Figure 23.4). Due to the fact that the radiation intensity change (day/night, summer/winter) the system needs to have a storage battery, in most cases, of the lead-acid type, and a charge controller which protects the battery of deep discharge and overcharging (Goetzberger and Hoffmann, 2005).

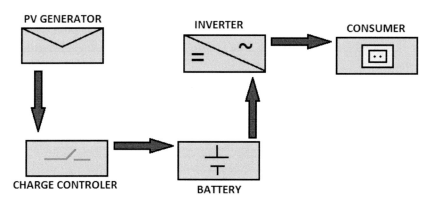

FIGURE 23.4   The working principle scheme of a PV system design to supply a building. Redrawn and adapted after Goetzberger and Hoffmann (2005).

The grid-connected PV systems are characterized by the fact that they are connected through an inverter to the public electricity due to the fact that the PV modules only produce sc power, and can be classified as central grid-connected systems and decentralized grid-connected systems. Both types present two main components: the PV module and the inverter. The central grid-connected PV systems are able to feed into the medium or high voltage directly by having an installed power up in the range of MW. The decentralized grid-connected PV systems are in general integrated into building facades or are installed on the building roof having a small power range. In this case, there is no need for an energy storage, the solar generator providing power in sunny days, and the excess energy is going into the public grid, while at night and overcast days the grid supply the house with energy (Goetzberger and Hoffmann, 2005).

23.5   MULTI-/TRANS- DISCIPLINARY CONNECTION(S)

The photovoltaic-thermal collectors (PV/T) were proposed for the first time in the 1970s, and have the advantage of generating both electricity and heat (Akarslan, 2012). The PV/T collectors are a hybrid solar collector in which the PV cells are incorporated in the absorber surface, and their working principle resembles the flat plate solar collectors, with the specification that the incident solar radiation in PV/T case is converted in electricity (Chow, 2010). If one considers the presence or absence of an auxiliary mechanism for solar radiation concentration, the collectors can be with concentrators, i.e., the amount of solar radiation is increased on the PV face using reflective devices, and without concentrators, known as the flat type (similar to PV panels) in which the collector area is directly reached by the solar radiation (Ramos et al., 2010).

A flat PV/T collector consists of a PV component made of doped crystalline silicone joined with a TH component behind it, and, by case the PV cell can be covered by glass (Ramos et al., 2010), including (Figure 23.5) the PV module, the heat extraction equipment (metallic plate made of pipes with role in heat absorption and cooling system for the PV module which is attached to it) and the thermal isolation (with role in minimizing the heat loss which may occur at the equipment walls).

There are several factors which need to be considered in determining the PV/T system performance, such as the configuration of the thermal flat plate module, the temperature, and the mass flow amount of the thermal fluid, the glass covers number and the control and supervisory system. In the case of PV/T collectors, the reduction in thermal resistance will determine increased coefficients of heat transfer which leads to increasing the thermal efficiency. By being easy to integrate into constructions and by occupying less space compared to thermal and PV modules used separate, the flat PV/T collectors are very promising for roof installation (Ramos et al., 2010).

Another combined system (Akarslan, 2012) was designed by Pande and his collaborators (2003), the Solar PV operated pump drip irrigation system, by taking into account several parameters, like the water requirements, the pump size, the pump pressure diurnal variation which occurs due to modification in pressure compensation and irradiance from the drippers. In his work, Badescu (2003) considered a complex time-dependent solar water pumping system with a DC motor, a battery, a PV array, and a centrifugal pump for which he analyzed the operating principles.

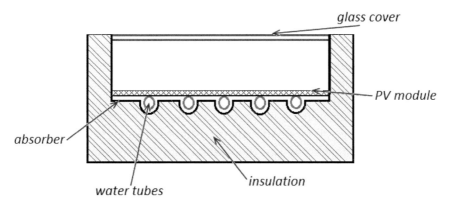

FIGURE 23.5 Schematic representation of a concentrating PV/T collector; Redrawn and adapted after Ramos et al., (2010) and Yang et al., (1997).

The PV technology can be used even in space, Seboldt, and his collaborators (2001) designing the so-called European Sail Tower, an Earth-orbiting Solar Power Satellite (SPS) of a small weight consisting in a tower-like orbital system. The PV cells generate a significant amount of electrical power which can be transmitted to Earth through the microwaves (Parida et al., 2011). There was also studied by Girish (2006) if it was possible to generate PV power by nighttime, especially in planetary bodies (e.g., the moon) with the aid of the reflected light energy flux from the planetary objects situated in their vicinity using the LILT (low-intensity-low-illumination) technology. There was also proven that the PV solar generators are optimal to provide the electrical power to satellites (Akarslan, 2012).

23.6 OPEN ISSUES

The efficiency of a PV system in time is influenced by the optimum performance of its components. In this context, there are several problems reported for the solar module in a PV system, even if this is the most reliable component (Goetzberger and Hoffmann, 2005):

- ***Degradation of cell/module interconnects:*** a change in geometry or structure of the joined ribbon-to-ribbon or cell-to-ribbon area determines an interconnect degradation for crystalline silicone modules; thermomechanical fatigue determine changes in solder-joint geometry with reducing the redundant solder joints number leading to decreased performance (Goetzberger and Hoffmann, 2005).
- ***Degradation of packaging materials:*** may appear during the normal service or when it is damaged the laminate package (Goetzberger and Hoffmann, 2005).
- ***Degradation of the semiconductor device:*** when present on the semiconductor material determine a loss in performance for field-aged modules; in crystalline cells, there is another form of degradation produced in the process of chemically assisted diffusion to the cell surface of the cell dopant (phosphorus); an increased concentration in phosphorus and the sodium movement from the soda lime glass substrate to the cell surface produce a low adhesional strength on the cell/encapsuled interface (Goetzberger and Hoffmann, 2005).
- ***Degradation of thin-film modules:*** the initial degradation of the amorphous silicon appears due to the Staebler-Wronski effect; the amorphous modules are influenced also by the seasons, degrading in winter and annealing in summer; still, the CIGS modules are assumed to be quite stable (Goetzberger and Hoffmann, 2005).
- ***Loss of adhesion:*** when the bonds between the material layers form the module laminate are breaking down, process known as delamination; when occurs at the front side leads to performance degradation due to the fact that the materials which transmit sunlight to the cells are optical decoupled; either side delamination leads to substituted-reverse-bias cell heating as a result or efficient heat dissipation interruption (Goetzberger and Hoffmann, 2005).
- ***Degradation caused by moisture intrusion:*** determine the apparition of corrosion and increased leakage of current; due to corrosion, the electrical connection between inverter and module can be interrupt, the semiconductor layers in thin film modules and the cell metallization in crystalline silicon modules are attacked producing a loss of electrical performance (Goetzberger and Hoffmann, 2005).
- ***Degradation due to dirt and dust on the module surface:*** determine a decrease in the yield energy; in order to avoid this type of degradation, the module surface needs to be cleaned periodically, i.e., when the inclination of module is lower than 5° (for an inclination more than 5° or 10° the leaves and the dust are washed by rain) (Goetzberger and Hoffmann, 2005).

ACKNOWLEDGMENT

This contribution is part of the research project "Fullerene-Based Double-Graphene Photovoltaics" (FG2PV), PED/123/2017, within the Romanian National program "Exploratory Research Projects" by UEFISCDI Agency of Romania.

KEYWORDS

- **electrical efficiency**
- **inverter**
- **open circuit voltage**
- **photovoltaic-thermal collector**
- **PV module**
- **thin-film solar cell**

REFERENCES AND FURTHER READING

Adams, W. G., & Day, R. E., (1877). The action of light on selenium. *Proceedings of the Royal Society* (Vol. A25, pp. 113–117). London.

Akarslan, F., (2012). Photovoltaic systems and applications. In: Şencan, A., (ed.), *Modeling, and Optimization of Renewable Energy Systems* (pp. 40–41). InTech, Croatia, Chapter 2.

Badescu, V., (2003). Time-dependent model of a complex PV water pumping system. *Renewable Energy, 28*, 543–560.

Becquerel, A. E., (1839). Recherches sur les effets de la radiation chimique de la lumiere solaire au moyen des courants electriques. *Comptes Rendus de L'Academie des Sciences* [Research on the effects of chemical radiation from the average light of electric currents. *Proceedings of the Academy of sciences*], 9, 145–149.

Becquerel, A. E., (1841). Memoire sur les effects d'electriques produits sous l'influence des rayons solaires. *Annalen der Physik und Chemie* [Memoir on the effects of the electrics produced under the influence of solar rays. *Annals of Physics and Chemistry*], 54, 35–42.

Benjamin, P., (1983). *Voltaic Cell, Its construction and Its Capacity,* 1st edn. (chapter XIV) Wiley, New York.

Bergmann, L., (1931). Uber eineneueselen-sperrschichtphotozelle. *Physikalische Zeitschrift* [About a new selenium barrier photocell. *Physical Journal*], 32, 286–288.

Chaar, L. E., Iamont, L. A., & Zein, N. E., (2011). Review of photovoltaic technologies. *Renewable and Sustainable Energy Reviews, 15*, 2165–2175.

Chow, T. T., (2010). A review on photovoltaic/thermal hybrid solar technology. *Applied Energy, 87*(2), 365–379.

Fritts, C. E., (1883). On a new form of selenium photocell. *American Journal of Science, 26*, 465–472.

Girish, T. E., (2006). Nighttime operation of photovoltaic systems in planetary bodies. *Solar Energy Materials & Solar Cells, 90*, 825–831.

Goetzberger, A., & Hoffmann, V. U., (2005). *Photovoltaic Solar Energy Generation* (pp. 155–158). Springer, Berlin, Chapter 10.

Green, M. A., (1990). *Photovoltaics: Coming of Age* (pp. 1–8). 21st IEEE photovoltaic specialists conference, Orlando, USA.

Grondahl, L. O., (1933). The copper-cuprous-oxide rectifier and photoelectric cell. *Review of Modern Physics, 5*, 141–168.

Loschi, H. J., Iano, Y., León, J., Moretti, A., Conte, F. D., & Braga, H., (2015). A review on photovoltaic systems: Mechanisms and methods for irradiation tracking and prediction. *Smart Grid and Renewable Energy, 6*, 187–208.

Messenger, R. A., & Ventre, J., (2004). *Photovoltaic Systems Engineering* (2nd edn.). CRC Press, Boca Raton, Florida.

Nix, F. C., & Treptwo, A. W., (1939). A thallous sulfide photo-EMF cell. *Journal of the Optical Society of America, 29*(11), 457–462.

Pande, P. C., Singh, A. K., Ansari, S., Vyas, S. K., & Dave, B. K., (2003). Design development and testing of a solar PV pump based drip system for orchards. *Renewable Energy, 28*, 385–396.

Parida, B., Iniyan, S., & Goic, R., (2011). A review of solar photovoltaic technologies. *Renewable and Sustainable Energy Reviews, 15*, 1625–1636.

Prinsloo, G. J., Dobson, R. T. (2014). *Solar Tracking: High Precision Solar Position Algorithms, Programs, Software, and Source-Code for Computing the Solar Vector, Solar Coordinates.* Sun angles in microprocessor, PLC, Arduino, PIC, and PC-based sun tracking devices or dynamic sun following hardware. (pp. 1–542). Stellenbosch: SolarBooks.

Ramos, F., Cardoso, A., & Alcaso, A., (2010). Hybrid photovoltaic-thermal collectors: A review. In: *Doctoral Conference on Computing, Electrical, and Industrial Systems (DoCEIS): Emerging Trends in Technological Innovation* (pp. 477–484).

Seboldt, W., Klimke, M., Leipold, M., & Hanowski, N., (2001). European sail tower SPS concept. *Acta Astronautica, 48*(5), 785–792.

Seme, S., Stumberger, G., & Vorsic, J., (2011). Maximum efficiency trajectories of a two-axis sun tracking system determined considering tracking system consumption. *IEEE Transactions on Power Electronics, 26*, 1280–1290.

Shive, J. N., (1959). *Semiconductor Devices*. Van Nostrand, New Jersey, Chapter 8.

Shoro, G. M., Hussain, D. M. A., & Sera, D., (2013). *Photovoltaic System in Progress: A Survey of Recent Development*s (Vol. 414, pp. 239–250). Communication technologies, information security, and sustainable development: Third international multi-topic conference, IMTIC, Jamshoro, Pakistan. Revised selected papers (2014) lecture notes in computer science. Springer.

Skoplaki, E., & Palyvos, J. A., (2009). On the temperature dependence of photovoltaic module electrical performance: A review of efficiency/power correlations. *Solar Energy, 83*, 614–624.

Wolf, M., (1976). *Historical Development of Solar Cells*. IEEE Press, New York.

Yang, M., Izumi, H., Sato, M., Matsunaga, S., Takamoto, T., Tsuzuki, K., Amono, T., & Yamaguchi, M., (1997). *A 3kW PV-Thermal System for Home Use*. Twenty-sixth IEEE photovoltaic specialists conference, 29 September - 03 October Anaheim, CA.

CHAPTER 24

Protein Surface

ADRIANA ISVORAN[1,2]

[1]*Department of Biology-Chemistry, West University of Timisoara, 16 Pestalozzi, Timisoara, Romania*

[2]*Advanced Environmental Research Laboratory, 4 Oituz, Timisoara, Romania*

24.1   DEFINITION

There is not a unique definition for the protein surface. We may refer to the van der Waals surface (vdWS), contact surface (CS), molecular surface (MS), or solvent accessible surface of a protein. All these terms are defined further.

24.2   HISTORICAL ORIGIN

The first approaches considering the protein's surface calculation ware made by Lee and Richards in 1970 years (Lee and Richards, 1971; Richards, 1977). They built a program for drawing the vdWS of a protein molecule based on the accessibility of atoms or groups of atoms to solvent molecules of a specified radius, the accessibility being proportional to surface area.

Another approach has been made by Connolly (1983), and it is the most known and popular one. Connolly defined the protein surface as all the points of the van der Walls surface that are touched by a solvent sphere with a given radius. Connolly also considered the points that cannot be reached by the solvent sphere that were substituted by points where the solvent touches simultaneously more than one atom of the structure, these points being considered the reentrant surface.

During the years, many computational tools have been developed to build and visualize the protein surface.

24.3   NANO-SCIENTIFIC DEVELOPMENT

There are a few possibilities to define and represent the protein surface: vdWS, CS, MS, and accessible solvent area also called the solvent accessible surface area (ASA).

vdWS is defined as the sum of exposed areas of the constituent atoms modeled as spheres with radii considered from their known van der Waals.

CS is defined as the surface containing all contact points of the molecule and a test particle represented by a sphere of radius R. When this sphere touches more than one atom we obtain the reentrant surface (Connoly, 1983).

MS is defined as the sum between the contact and reentrant surfaces (Connoly, 1983a). Solvent accessible surface is defined as the surface generated by the centers of the test particle of radius R rolling along the protein, and it is also called Lee-Richards MS (Lee and Richards, 1971; Richards, 1977). Usual, the radius of the probe sphere is considered 1.4 Å.

The different ways to define the surface area of a protein are illustrated in Figure 24.1.

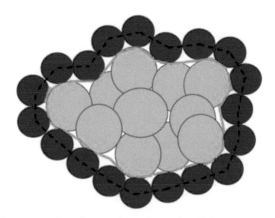

FIGURE 24.1   Representation of a protein surface using different ways to define it: the van der Waals surface (vdWSA – blue surface), the molecular surface (MS – surface delimited by the continuous green line) and the accessible solvent area (ASA – surface delimited by the dashed black line). Red sphere is the probe sphere.

CS, MS, and ASA values are dependent on parameters used for calculating them from the three-dimensional structure, especially on the probe sphere radius. A precise representation of the protein surface is important for studying and understanding diverse biological processes at the molecular level, such as protein folding and stability, protein-ligand, and protein-protein interactions. It also plays a central role in varied computational applications concerning protein crystal packing, enzyme catalysis, drug design, and molecular docking.

The shape and the physicochemical properties distribution on the protein MS manage the specific molecular interactions in protein-ligand complexes. Usually, the surface of a protein is irregular presenting a lot of cavities and channels having different shapes and sizes (Laskowski et al., 1996). Figure 24.2 reveals the surface of bovine serum albumin starting from its structural file taken from Protein Data Bank (Berman et al., 2000), code entry 4F5S (Bujacz, 2012), and visualized using Chimera computational tool (Petersen et al., 2004).

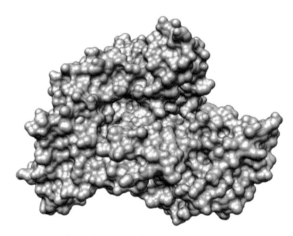

FIGURE 24.2 Representation of the continuous surface of bovine serum albumin (PDB code entry 4F5S).

The complexity of the protein surface may be described in terms of surface roughness, which is an important tool used to investigate and understand protein-ligand and protein-protein interactions (Petit and Bowie, 1999). In order to quantitatively describe the roughness of the protein surface, we may use the surface fractal dimension. The algorithm used to compute the surface fractal dimension of a protein was developed by Lewis and Rees (1985). According to this algorithm, the CS area depends on the radius of the probe molecule as given by

$$CS = N(R)R^2 \qquad (1)$$

where N is the number of molecules with radius R needed to cover the entire surface? This number decreases with increasing radius, the scaling law being

$$N(R) \sim R^{D_s} \tag{2}$$

and we obtain the scaling law for the CS of the protein

$$CS \sim R^{2-D_s} \tag{3}$$

The surface fractal dimension is proportional to the roughness of the protein surface, and it may be obtained from the slope of the line versus as

$$D_S = 2 - \frac{d\log(CS)}{d\log(R)} \tag{3}$$

with the surface fractal dimension (Lewis and Rees, 1985).

The protein surface has not a uniformly distributed roughness, and therefore it has not a constant fractal dimension over all scales. There are regions on the protein surface characterized by higher roughness and consequently having higher surface fractal dimensions, and these regions proved to be important for the selective recognition of the ligand. Other surface regions characterized by reduced roughness seem to be important for protein-protein interactions. These observations underline that protein surface shape presents local properties that are important for biological functions.

Not only roughness of some regions of protein surface is important for the protein biological activity, but also the chemical heterogeneity of these regions. From this point of view, the hydrophobicity and the electrostatic potential distribution on the protein surface are among the most important surface properties. For example, the biomolecular recognition is an event that is both geometrically-localized, and charge- and hydrophobicity-specific and its explanation requires taking into consideration all these properties.

Charges distribution on the protein surface controls many physiological processes involving electrostatically engaged ligand association and influences the catalytic rates and redox potentials (Honig and Nicholls, 1995). Mapping the electrostatic potential distribution of the protein surface helps to identify possible binding regions for charged ligands or to compare the electrostatic properties of distinct proteins or those of a native protein and its mutants (Miteva et al., 2005). Electrostatic potential distribution of the surface of the bovine serum albumin is presented in Figure 24.3a.

Previous studies have shown that hydrophobicity is a determinant property for the association of nonpolar ligands to the protein surface playing a very important role in molecular recognition (Eisenhaber and Argos, 1996). Mapping the distribution of hydrophobicity on the protein surface helps in identifying the highly hydrophobic cavities and prediction of binding of nonpolar groups with direct applications in drug design studies

(Giovambattista et al., 2008). Figure 24.3b presents the surface of bovine serum albumin colored by hydrophobicity.

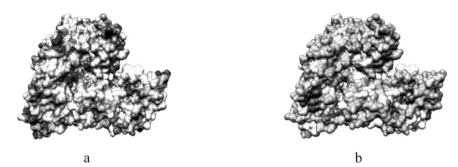

a   b

FIGURE 24.3 (a) Mapping the electrostatic potential and of distribution on the surface of bovine serum albumin: red regions are negatively charged, and blue regions are positively charged; (b) Mapping the hydrophobicity on the surface of bovine serum albumin: blue regions are hydrophilic, and orange regions are hydrophobic.

Protein surface analysis is a powerful tool that can be used in many applications: it allows a deeper understanding of the biochemical mechanisms of protein interaction with ligands, of enzyme catalysis or may indicate cases of possible convergent or divergent evolution.

24.4 NANO-CHEMICAL APPLICATION(S)

Conjugation of proteins to nanoparticles allows integration of biology and synthetic materials and has diverse applications in biotechnology, catalysis, and medicine. There are two approaches used to conjugate proteins to nanoparticles, both of them having strengths and limitations: the covalent linkage of the protein to the particle and non-covalent interactions between the particle and protein.

Covalent linkage of proteins to nanoparticles conducts to conjugates that are stable toward dissociation and offers control over protein reactivity and nanoparticles aggregation (Rana et al., 2010). Non-covalent conjugation of proteins to nanoparticles is a complementary strategy to covalent linkage enabling reversibility to the protein-nanoparticle complex and applications in areas of protein sensing and delivery (Rana et al., 2010). The properties of both covalent and non-covalent protein-nanoparticles conjugates strongly depend on the interfaces properties, the protein surface playing a crucial role.

24.5   MULTI-/TRANS-DISCIPLINARY CONNECTION(S)

The shape and chemical properties of the protein surface are essential for many phenomena such as: binding and diffusion of small molecules, absorption, heterogeneous catalysis, interactions with other proteins or nucleic acids and also interactions with solid materials. Protein interactions modulated by their surface properties are also essential for many biological and medical applications as they modulate cell adhesion, activate the biological cascades, and are of a central role to diagnostic assay and/or design of sensor devices assuring the biocompatibility. All these applications reflect the multi-transdisciplinary connections of the concept of the protein surface.

24.6   OPEN ISSUES

The great majority of biological processes involve protein-protein interactions and determination of participating residues expands our understanding of molecular mechanisms and contributes to the development of therapeutics. Experimental assessments to identifying interacting residues are costly and time-consuming and computational approaches could be applied.

In protein-protein interactions, it is important to predict protein interfaces, and structural features are significant discriminative attributes for this purpose. These features are associated with the protein spatial structure, solvent-ASA, and geometric shape of the protein surface (Esmaielbeiki et al., 2016). From this point of view, prediction of protein interface is strongly correlated to the computational building of the protein surface, and this representation denotes an actual challenge. Also, computational analysis and comparison of protein surfaces is another significant challenge as it allows a competent and precise functional characterization of proteins structures and biological functions.

KEYWORDS

- **contact surface**
- **molecular surface**
- **roughness**
- **solvent accessible surface**
- **van der Waals surface**

REFERENCES AND FURTHER READING

Berman, H. M., Westbrook, J., Feng, Z., Gilliand, G., Bhat, T. N., Weissig, H., Shindyalov, I. N., & Bourne, P. E., (2000). The protein data bank. *Nucl. Acids Res., 28*, 235–242. PMCID: PMC102472.

Connolly, M. L., (1983a). Analytical molecular surface calculation. *J. Appl. Cryst., 16*, 548–558. doi: 10.1107/S0021889883010985.

Connolly, M. L., (1983b). Solvent-accessible surfaces of proteins and nucleic acids. *Science, 221*(4612), 709–713. doi: 10.1126/science.6879170.

Connolly, M. L., (1986). Shape complementarity at the hemoglobin $\alpha_1\beta_1$ subunit interface. *Biopolymers, 25*(7), 1229–1247. doi: 10.1002/bip.360250705.

Connolly, M. L., (1993). The molecular surface package. *J. Mol. Graph., 11*(2), 139–141. *doi: 10.1016/0263–7855(93)87010–3.*

Eisenhaber, F., & Argos, P., (1996). Hydrophobic regions on protein surfaces: Definition based on hydration shell structure and a quick method for their computation. *Protein Eng., 9*(12), 1121–1133. doi: 10.1093/protein/9.12.1121.

Esmaielbeiki, R., Krawczyk, K., Knapp, B., Nebel, J. C., & Deane, C. M., (2016). Progress and challenges in predicting protein interfaces. *Briefings in Bioinformatics, 17*(1), 117–131. doi: 10.1093/bib/bbv027.

Giovambattista, N., Lopez, C. F., Rossky, P. J., & Debenedett, P. G., (2008). Hydrophobicity of protein surfaces: Separating geometry from chemistry. *PNAS, 105*(7), 2274–2279. doi: 10.1073/pnas.0708088105.

Honig, B., & Nicholls, A., (1996). Classical electrostatics in biology and chemistry. *Science, 268*(5214), 1144–1149. doi: 10.1126/science.7761829.

Laskowski, R. A., Luscombe, N. M., Swindells, M. B., & Thornton, J. M., (1996). Protein clefts in molecular recognition and function. *Protein Science, 5*, 2438–2452. PMID: 8976552.

Lee, B., & Richards, F. M., (1971). The interpretation of protein structures: Estimation of static accessibility. *J. Mol. Biol., 55*(3), 379–400. doi: 10.1016/0022–2836(71)90324-X.

Lewis, M., & Rees, D. C., (1985). Fractal surfaces of proteins. *Science, 230*, 1163–1165. doi: 10.1126/science.4071040.

Miteva, M. A., Tufféry, P., & Villoutreix, B. O., (2005). PCE: Web tools to compute protein continuum electrostatics. *Nucleic Acids Res., 33*, W372–W375. doi: 10.1093/nar/gki365.

Pettersen, E. F., Goddard, T. D., Huang, C. C., Couch, G. S., Greenblatt, D. M., Meng, E. C., & Ferrin, T. E., (2004). UCSF Chimera: a visualization system for exploratory research and analysis. *J. Comp. Chem., 25*, 1605–1612. doi: 10.1002/jcc.20084.

Pettit, F. K., & Bowie, J. U., (1999). Protein surface roughness and small molecular binding sites. *J. Mol. Biol., 285*, 1377–1382.

Rana, S., Yeh, Y. C., & Rotello, V. M., (2010). Engineering the nanoparticle-protein interface: Applications and possibilities. *Curr. Opin. Chem. Biol., 14*(6), 828–834. doi: 10.1016/j.cbpa.2010.10.001.

Richards, F. M., (1977). Areas, volumes, packing, and protein structure. *Annu. Rev. Biophys. Bioeng., 6*, 151–176. *doi: 10.1146/annurev.bb.06.060177.001055.*

CHAPTER 25

Radiation (Induced) Synthesis

LORENTZ JÄNTSCHI[1] and SORANA D. BOLBOACĂ[2]

[1]*Technical University of Cluj-Napoca, Romania*

[2]*Iuliu Haţieganu University of Medicine and Pharmacy Cluj-Napoca, Romania*

25.1 DEFINITION

Radiation (induced) synthesis (Radi-Synth) refers conducting of a synthesis in the presence of a radiation of a certain type (of massless or of massive particles; of electrically charged or neutral particles) and with a certain wavelength or energy. In a Radi-Synth process, products are favored against others by the use of the radiation. In this category, fits sound (induced) synthesis (or sono(-)chemistry), which usually initiates or enhances the chemical activity in solutions; ultrasound (assisted) synthesis (or ultrasonication), which have an extensive use in many industrial syntheses (including for solid phase products); microwave synthesis which usually increases the temperature in the process and gamma-radiation (induced) synthesis with use in preparation of certain solid-state nanocompounds.

25.2 HISTORICAL ORIGIN(S)

In conventional chemical synthesis or chemosynthesis, the transformation of the reactive into the products can be facilitated by the use of the catalysts. In the presence of a catalyst, less free energy is required to reach the transition state, but the total free energy from reactants to products does not change (IUPAC, 1997). Catalyzed reactions have lower activation energy (rate-limiting free energy of activation) than the corresponding uncatalyzed reaction, resulting in a higher reaction rate at the same temperature and for the same reactant concentrations. Kinetically, catalytic reactions are typical chemical reactions; i.e., the reaction rate depends on the frequency of contact of the reactants in the rate-determining step. The SI derived unit for measuring the catalytic activity of a catalyst is the katal, which are moles per second. The productivity of a catalyst can be described by the turn over number (or TON) and the catalytic activity by the turn over frequency (TOF), which is the TON per time unit.

Catalysts work by providing an (alternative) mechanism involving different transition state and lower activation energy (see Figure 25.1 adapted from Jäntschi, 2002).

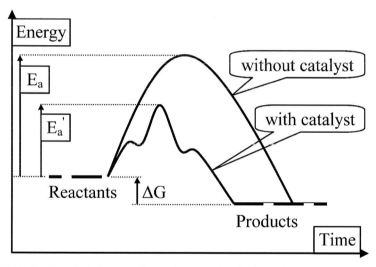

FIGURE 25.1   Generic influence of a catalyst to a chemical reaction.

Radiation interacts with the matter at different levels when is charged or not and when contain massive particles or not. Different types of radiation serve the purpose of the Radi-Synth (see Table 25.1).

TABLE 25.1   Different Types of Radiations Used in Radiation (Induced) Synthesis

Radiation	Mass	Charge	Examples of uses
α (=He^{2+})	Yes (4 a.u.)	2+	(Haertling et al., 2011)
n	Yes (1 a.u.)	0	(Viererbl et al., 2014)
p (=H^{1+})	Yes (1.a.u.)	1+	(Crowell et al., 2005)
β (=e^-)	No	1-	(Oulianov et al., 2007)
EM	No	No	(Khan et al., 2006)

The most extensive work was conducted on the use of the electromagnetic (EM) radiation (γ, X, UV, VIS, IR, μW, Radio). The following table connects the wavelength of the radiation with its typical effects on the matter (see Table 25.2 translated and adapted from Jäntschi, 2003).

TABLE 25.2 Effects on the Matter of Electromagnetic Radiations

Energy storage in matter	EM radiation		Wavelength (λ)	References
Molecular rotation	Radio		>1 m	Wardman, 1987
	Microwave (μW)		1 m	Dom et al., 2015
			1 mm	
	Far infrared		10^{-3} m	Delor et al., 2014
Molecular vibration			10^{-5} m	
	Near-infrared		10^{-6} m = 1 μm	
			700 nm	
Electronic excitation	Red	Visible	700–620 nm	Kumar and Francisco, 2015
(HOMO level)	Green		560–510 nm	Wang et al., 2015
	Violet		450–400 nm	Kitamura et al., 2014
	Ultraviolet (UV)		10^{-7} m	Huang et al., 2008
	UV of vacuum		10^{-8} m	Simakov, 2008
Excitation of the			10^{-9} m = 1 nm	
electronic core	X rays		10^{-10} m = 1 Å	Kameneva et al., 2015
			10^{-11} m	
Nucleus excitation			10^{-12} m = 1 pm	
	γ rays		10^{-12}–10^{-13} m	Hareesh et al., 2016
	Cosmic rays		<10^{-14} m	Bassez, 2015

25.3 NANO-SCIENTIFIC DEVELOPMENT(S)

It should be noted that it is a inverse proportionality between the wavelength of the radiation and its energy (E = h·c/λ, h = 6.626070040(81)×10^{-34} J·s, c = 299792458 m·s^{-1}, E is the energy and λ is the wavelength of the radiation), and therefore with the decreasing of the wavelength it is increased its energy and vice-versa.

In 1839, Becquerel (1839) discovered that electrical current flowed between a pair of electrodes if one electrode was illuminated with UV light, the phenomenon being later referred as the Becquerel effect (Prevenslik, 2003).

The interaction between high-energy radiation (γ-rays, X-rays, electrons, and ion beams) and matter produces a large number (~4×10^4 electrons per MeV of energy deposited) of non-thermal secondary low-energy electrons

(Kaplan and Miterev, 1987). Due to the inelastic collisions of these low-energy electrons with molecules and atoms, they become thermalized within approximately one picosecond (Mozumder and Hatano, 2004) and produce distinct energetic species such as free radicals that are the primary driving forces in a wide variety of radiation-induced reactions (Xu et al., 2014).

High-energy radiations (such as are γ-rays) shown a series of advantages for preparation of nanowires (Rana and Chauhan, 2014), colloidal metal nanoparticles (Henglein and Meisel, 1998), polymeric hydrogels (Burillo et al., 2012), graphene (Shahriary and Athawale, 2015), functionalized nanofibers (Han et al., 2015) and other nanostructures (Cui et al., 2014; Su et al., 2014).

The wet method for producing of metallic nanoclusters deposited on surfaces is based on the dissociation of the water molecules, succeeded by the dissociation of secondary alcohols (such as is isopropanol) or acids (such as is formic acid) also present in the prepared media (see Table 25.3, adapted from Belloni, 2006). Then the metal is deposited from its ions from solution alkalinized (to a pH = 11 with NH_4OH in Chettibi et al., 2006).

TABLE 25.3 Metallic Nanoclusters Growth Mechanism

Initiation	

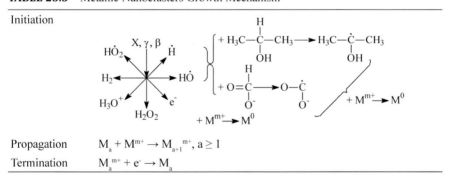

Propagation	$M_a + M^{m+} \rightarrow M_{a+1}{}^{m+}$, $a \geq 1$
Termination	$M_a{}^{m+} + e^- \rightarrow M_a$

Energetic excitation of a molecule may have followed routes to different intermediary products, as given below for electron-molecule interactions (Arumainayagam et al., 2010):

- Electron impact ionization ($e^- + AB \rightarrow 2e^- + AB^{+*}$), having a short (relatively to the molecule passing time) interaction time (10^{-16} s), the incident electrons being necessary to have above 10 eV (ionization potential of a typical molecule) with a maximum yield at about 100 eV (Deutsch et al., 2000). The ionized molecule may subsequently dissociate ($AB^{+*} \rightarrow A^+ + B^*$, or $AB^{+*} \rightarrow A^* + B^+$, or $AB^{+*} \rightarrow A^{+*} + B$ or $AB^{+*} \rightarrow A + B^{+*}$).

- Electron impact excitation ($e^- + AB \rightarrow e^- + AB^*$), when the incident electrons being necessary to have above 6 eV (excitation threshold for small organic molecules). A series of competing decay channels may be followed by the excited neutral molecule ($AB^* \rightarrow AB$ + energy (thermal or light); $AB^* \rightarrow A^* + B$; $AB^* \rightarrow A + B^*$; $AB^* \rightarrow A^+ + B^-$; $AB^* \rightarrow A^- + B^+$). Please note that the structures observed very infrequently at electron energies above the threshold for dipolar dissociation have usually been attributed to multiple electron-scattering prior to electron attachment (see Sambe et al., 1987).
- Electron attachment ($e^- + AB \rightarrow AB^-$), producing a transient negative ion and occurring at low energies of the incident electrons (below 15 eV). The electron takes a position in a LUMO level and usually has an antibonding character. The incoming electron's energy must lie in a restricted range defined by the Franck-Condon transition (Frank, 1926; Condon, 1926) to a discrete final state (AB^- or AB^{-*}) given that the molecular orbital associated with this state exists at a specific energy. Large cross-sections associated with electron attachment may be attributed to the resonant character of the process. Resonant scattering is characterized by interaction times longer than the typical molecular transit times (which are typically less than 10^{-15} s) while the lifetime of a temporary negative ion ranges from 10^{-15} *s* to 10^{-2} s. Depending on whether the resonance is above or below the corresponding ground state (0–5 eV) or core-excited state (5–15 eV) of the neutral molecule, resonances may be further classified as either open channel (shape) or closed channel (Feshbach) resonances, respectively (Feshbach, 1958).
- Autodetachment ($AB^{-*} \rightarrow e^- + AB$ or $AB^{-*} \rightarrow e- + AB^*$ as subsequent process of $e^- + AB \rightarrow AB^{-*}$). The autodetachment lifetimes vary from 10^{-14} *s* to 10^{-3} *s* (Illenberger, 1992), when as expected, complex molecules typically have long autodetachment lifetimes. Typically formed via resonances at near 0 eV, temporary negative ions for such molecules can often be detected by mass spectrometry because these ions do not undergo dissociation.
- Associative attachment and non-dissociative attachment ($AB^{-*} \rightarrow AB^-$ + energy), typical for clusters and condensed phase, when the energy is passed to neighboring molecules (as thermal energy). Associative attachment (resonance stabilization) involves the attachment of an electron to a molecule via capture into a transient negative ion state lying above the vacuum level while non-dissociative attachment, leads to the production of a long-lived negative ion that can be detected with

mass spectrometry. Such a mechanism were observed for molecules such as SF_6, C_6F_6, and C_{60} because their molecular complexity allows efficient intramolecular vibration redistribution (IVR) resulting in the formation of a metastable anion (Hotop et al., 2004).

- Radiative cooling ($AB^{-*} \rightarrow AB^- + \gamma$). This channel is open only to molecules with a positive electron affinity, and it involves relaxation via photon emission without being a significant competing one because radiative lifetimes are on the order of 10^{-8} s to 10^{-9} s (being much slower than the other ones).

It should be accounted that the branching ratio among the competing decay channels for the temporary negative ion is dependent on the state and phase (Illenberger, 2003). In addition, because of the large de Broglie wavelength (~12 Å for a 1 eV e⁻) of low-energy incident electrons, the interaction of such electrons with the condensed phase must be treated theoretically as a multiple scattering problems (Ingolfsson et al., 1996).

25.4   NANO-CHEMICAL APPLICATION(S)

Since the first report of microwave-assisted organic synthesis by the groups of Gedye et al., (1986) and Giguere et al., (1986), this technique becomes a tool for accelerating and controlling reactions, and increasing yields. Were introduced controlled, precise microwave reactors as an alternative to oil baths in an increasingly wide range of organic transformations (Mavandadi and Pilotti, 2006; Kappe and Dallinger, 2006).

The advantage of microwave technology in terms of heating can be applied to high-throughput techniques, such as solid-phase synthesis and polymer-assisted solution-phase synthesis. Synthesis of nano-sized structures may require the presence of a capping agent and/or of support. Table 25.4 contains recently communicated results of Radi-synth using microwaves for different classes of nano-sized structures.

Sonic waves find their application in direct exfoliation and dispersion of two-dimensional materials (graphene, h-BN, and MoS_2 were exfoliated in Kim et al., 2015), sonochemical synthesis catalyzed by nanocomposites (polyhydroquinolines are obtained in Zarnegar et al., 2015), obtaining of oxide nano-sized particles ($Cu_{0.88}Zn_{0.12}O$, Zn:CuO nano-sized particles were obtained in Eshedet al., 2014) and acceleration of reactions (in general) in solution (reviewed in Fadeev et al., 2010).

TABLE 25.4 Nano-Sized Structures Recently Developed With Microwaves

Group	Structure	Capping agent	Support	Reference
Metallic oxides	CeO_2	Citric acid	-	He et al., 2016
	CeO_2	Ethylene glycol	TiO_2	Lu et al., 2016
	CoO	-	-	Harish et al., 2016
	Fe_2O_3	-	-	Sun et al., 2016
	Fe_3O_4	-	polypyrrole	Yang et al., 2016
	MoO_3	-	-	Wang et al., 2016
	V_2O_5	Oxalic acid	-	Pan et al., 2016
	VO_x		SiO_2	Betiha et al., 2016
Double oxides	$BaFe_{12}O_{19}$	-	Ethyl cellulose	Nabiyouni and Bakhtiari, 2016
	$ZnAl_2O_4$	-	-	Quirino et al., 2016
Salts	$LiFePO_4$	Tetramethylene glycol	Graphene oxide	Lim et al., 2016
	$Mg_3(PO_4)_2$	-	-	Qi et al., 2016
	Nd^{3+}-KY_3F_{10}	-	-	Orlovskii et al., 2016
	CdS	Polyvinylpyrrolidone	-	Darwish et al., 2016
Metals	Au, Ag	curdlan biopolymer	-	El-Naggar et al., 2016
	Ag	l-Cysteine	-	Ma et al., 2016
Metallic mixes	Ru-Re	polyvinylpyrrolidone	Al_2O_3, SiO_2	Baranowska et al., 2016
	Ti-Ni-Sn	-	-	Lei et al., 2016
Others	I_xB_{10}-I_yB_{12}	-	-	Juhasz et al., 2016
	MWCNT	-	Ni	Bisht et al., 2016

The UV light (see Table 25.2) is often used for polymerization. In micro-fluidics, stop flow lithography (SFL), and optofluidic maskless lithography (OFML) can polymerize high-resolution 3D particles continuously by illuminating ultraviolet (UV) light on a static UV reactive fluid. Going further, Paulsen et al., have developed a new optofluidic fabrication method that relies on two sequential steps: (1) highly controllable inertial flow shaping in microfluidic channels (Amini et al., 2013) followed by (2) UV photopolymerization of the shaped fluid stream (Paulsen et al., 2015). The optofluidic device uses two sheath fluid streams of polyethylene glycol diacrylate (PEG-DA) from the side channels, sandwiching a photosensitive core fluid stream, PEG-DA with photoinitiator 2,2-dimethoxy–2-phenylacetophenone (DMPA).

High energy γ-rays (Lattach et al., 2013, 2014; Cui et al., 2014), as well as high energy electrons radiations (Coletta et al., 2015), are used to generate oxidizing species, hydroxyl radicals, through water radiolysis, facilitating thus that subsequently activated monomers to polymerize. It was enabled the preparation of nanostructured PEDOT (PEDOTox) and PPy conducting polymers (Cui et al., 2016).

25.5   MULTI-/TRANS-DISCIPLINARY CONNECTION(S)

Some studies have demonstrated that low-energy spin-polarized secondary electrons, produced by X-ray irradiation of a magnetized Permalloy substrate, can induce chiral selective chemistry, which may explain the creation of 'handedness' in biological molecules, one of the great mysteries of the origin of life (Rosenberg et al., 2008).

25.6   OPEN ISSUES

Ionizing radiation exemplifies one of the conundrums of modern science because at the same time can be lethal and life-preserving. This dilemma extends to its applications such as the nuclear power and radiation sterilization of food products.

25.7   MAIN TERMS

Radiolysis is the dissociation of molecules by nuclear radiation. It is the cleavage of one or several chemical bonds resulting from exposure to high-energy flux.

KEYWORDS

- **gamma radiation-induced synthesis**
- **microwave synthesis**
- **radiation chemistry**
- **sonication**
- **sonochemistry**
- **ultrasonication**

REFERENCES AND FURTHER READING

Amini, H., Sollier, E., Masaeli, M., Xie, Y., Ganapathysubramanian, B., Stone, H. A., & Di Carlo, D., (2013). Engineering fluid flow using sequenced microstructures. *Nat. Commun., 4*, 1826.

Arumainayagam, C. R., Lee, H. L., Nelson, R. B., Haines, D. R., & Gunawardane, R. P., (2010). Low-energy electron-induced reactions in condensed matter. *Surface Science Reports, 65*, 1–44.

Baranowska, K., Okal, J., & Tylus, W., (2016). Microwave-assisted polyol synthesis of bimetallic RuRe nanoparticles stabilized by PVP or oxide supports (γ-alumina and silica). *Applied Catalysis A: General, 511*, 117–130.

Bassez, M. P., (2015). Water, air, earth, and cosmic radiation. *Origins of Life and Evolution of Biospheres, 45*(1/2), 5–13.

Becquerel, E., (1839). Becquerel, E., (1839). Research on the effects of chemical radiation of solar light by means of electric currents (In French). *Compt R Acad Sci 9*, 145–149.

Belloni, J., (2006). Nucleation, growth, and properties of nanoclusters studied by radiation chemistry. *Application to Catalysis: Catalysis Today, 113*(3/4), 141–156.

Betiha, M. A., Rabie, A. M., Elfadly, A. M., & Yehia, F. Z., (2016). Microwave-assisted synthesis of a VO$_x$-modified disordered mesoporous silica for ethylbenzene dehydrogenation in the presence of CO$_2$. *Microporous and Mesoporous Materials, 222*, 44–54.

Bisht, A., Chockalingam, S., Panwar, O. S., Kesarwani, A. K., Singh, B. P., & Singh, V. N., (2016). Substrate bias induced synthesis of flowered-like bunched carbon nanotube directly on bulk nickel. *Materials Research Bulletin, 74*, 156–163.

Burillo, G., Castillo-Rojas, S., & Arrieta, H., (2012). Cu(II) immobilization in AAc/NIPAAm-based polymer systems synthesized using ionizing radiation. *Radiation Physics and Chemistry, 81*(3), 278–283.

Chettibi, S., Wojcieszak, R., Boudjennad, E. H., Belloni, J., Bettahar, M. M., & Keghouche, N., (2006). Ni-Ce intermetallic phases in CeO$_2$-supported nickel catalysts synthesized by γ-radiolysis. *Catalysis Today, 113*(3/4), 157–165.

Coletta, C., Cui, Z., Archirel, P., Pernot, P., Marignier, J. L., & Remita, S., (2015). Electron-induced growth mechanism of conducting polymers: A coupled experimental and computational investigation. *J. Phys. Chem. B., 119*, 5282–5298.

Condon, E., (1926). A theory of intensity distribution in-band systems. *Physical Review, 28*, 1182–1201.

Crowell, R. A., Shkrob, I. A., Oulianov, D. A., Korovyanko, O., Gosztola, D. J., Li, Y., & Rey-De-Castro, R., (2005). Motivation and development of ultrafast laser-based accelerator techniques for chemical physics research. *Nuclear Instruments and Methods in Physics Research, Section B: Beam Interactions with Materials and Atoms, 241*(1–4), 9–13.

Cui, Z., Coletta, C., Dazzi, A., Lefrancois, P., Gervais, M., Néron, S., & Remita, S., (2014). Radiolytic method as a novel approach for the synthesis of nanostructured conducting polypyrrole. *Langmuir, 30*(46), 14086–14094.

Cui, Z., Coletta, C., Rebois, R., Baiz, S., Gervais, M., Goubard, F., Aubert, P. H., Dazzi, A., & Remita, S., (2016). *Radiation Physics and Chemistry, 119*, 157–166.

Darwish, M., Mohammadi, A., & Assi, N., (2016). Microwave-assisted polyol synthesis and characterization of pvp-capped cds nanoparticles for the photocatalytic degradation of tartrazine. *Materials Research Bulletin, 74*, 387–396.

Delor, M., Scattergood, P. A., Sazanovich, I. V., Parker, A. W., Greetham, G. M., Meijer, A. J. H. M., Towrie, M., & Weinstein, J. A., (2014). Toward control of electron transfer in donor-acceptor molecules by bond-specific infrared excitation. *Science, 346*(6216), 1492–1495.

Deutsch, H., Becker, K., Matt, S., & Mark, T. D., (2000). *Int. J. Mass Spectrom., 197*(1), 37–69.

Dom, R., Chary, A. S., Subasri, R., Hebalkar, N. Y., & Borse, P. H., (2015). Solar hydrogen generation from spinel $ZnFe_2O_4$ photocatalyst: Effect of synthesis methods. *International Journal of Energy Research, 39*(10), 1378–1390.

El-Naggar, M. E., Shaheen, T. I., Fouda, M. M. G., & Hebeish, A. A., (2016). Eco-friendly microwave-assisted green and rapid synthesis of well-stabilized gold and core-shell silver-gold nanoparticles. *Carbohydrate Polymers, 136*, 1128–1136.

Eshed, M., Lellouche, J., Gedanken, A., & Banin, E., (2014). A Zn-doped CuO nanocomposite shows enhanced antibiofilm and antibacterial activities against Streptococcus mutans compared to nanosized CuO. *Advanced Functional Materials, 24*(10), 1382–1390.

Fadeev, G. N., Kuznetsov, N. N., Beloborodova, E. F., & Matakova, S. A., (2010). The influence of the acoustic resonance frequency on chemical reactions in solution. *Russian Journal of Physical Chemistry A., 84*(13), 2254–2258.

Feshbach, H., (1958). Unified theory of nuclear reactions. *Annals of Physics, 5*(4), 357–390.

Franck, J., (1926). Elementary processes of photochemical reactions. *Transactions of the Faraday Society, 21*, 536–542.

Gedye, R., Smith, F., Westaway, K., Ali, H., Baldisera, L., Laberge, L., & Rousell, J., (1986). The use of microwave ovens for rapid organic synthesis. *Tetrahedron Lett., 27*, 279–282.

Giguere, R. J., Bray, T. L., Duncan, S. M., & Majetich, G., (1986). Application of commercial microwave ovens to organic synthesis. *Tetrahedron Lett., 27*, 4945–4948.

Haertling, C., Usov, I., & Wang, Y., (2011). Outgassing from alpha particle irradiation of lithium hydride and lithium hydroxide. *Nuclear Instruments and Methods in Physics Research, Section B: Beam Interactions with Materials and Atoms, 269*(4), 444–451.

Han, J. M., Wu, N., Wang, B., Wang, C., Xu, M., Yang, X., Yang, H., & Zang, L., (2015). γ radiation-induced self-assembly of fluorescent molecules into nanofibers: A stimuli-responsive sensing. *Journal of Materials Chemistry C., 3*(17), 4345–4351.

Hareesh, K., Joshi, R. P., Dahiwale, S. S., Bhoraskar, V. N., & Dhole, S. D., (2016). Synthesis of Ag-reduced graphene oxide nanocomposite by gamma radiation assisted method and its photocatalytic activity. *Vacuum, 124*, 40–45.

Harish, S., Silambarasan, K., Kalaiyarasan, G., Narendra, K. A. V., & Joseph, J., (2016). Nanostructured porous cobalt oxide synthesis from $Co_3[Co(CN)_6]_2$ and its possible applications in a Lithium battery. *Materials Letters, 165*, 115–118.

He, D., Hao, H., Chen, D., Lu, J., Zhong, L., Chen, R., Liu, F., Wan, G., He, S., & Luo, Y., (2016). Rapid synthesis of nano-scale CeO_2 by microwave-assisted sol-gel method and its application for CH3SH catalytic decomposition. *Journal of Environmental Chemical Engineering, 4*(1), 311–318.

Henglein, A., & Meisel, D., (1998). Radiolytic control of the size of colloidal gold nanoparticles. *Langmuir, 14*(26), 7392–7396.

Hotop, H., Ruf, M. W., Allan, M., & Fabrikant, I. I., (2004). Resonance threshold phenomena in low-energy electron collisions with molecules and clusters. *Physica Scripta, T110*, 22–31.

Huang, L., Zhai, M. L., Long, D. W., Peng, J., Xu, L., Wu, G. Z., Li, J. Q., & Wei, G. S., (2008). UV-induced synthesis, characterization, and formation mechanism of silver nanoparticles in alkalic carboxymethylated chitosan solution. *J. Nanopart. Res., 10*, 1193–1202.

Illenberger, E., (1992). Electron-attachment reactions in molecular clusters. *Chem. Rev., 92*(7), 1589–1609.

Illenberger, E., (2003). Formation and evolution of negative ion resonances at surfaces. *Surf. Sci., 528*(1–3), 67–77.

Ingolfsson, O., Weik, F., & Illenberger, E., (1996). The reactivity of slow electrons with molecules at different degrees of aggregation: Gas phase, clusters, and condensed phase. *International Journal of Mass Spectrometry and Ion Processes, 155*(1), 1–68.

IUPAC, (1997). *"Catalyst" in Compendium of Chemical Terminology* (2nd edn.). Compiled by, A. D. McNaught and A. Wilkinson. Oxford: Blackwell Scientific Publications.

Jäntschi, L., (2002). *Chemical and Instrumental Analysis (in Romanian)*. Cluj-Napoca: UT Press.

Jäntschi, L., (2003). Physical chemistry. *Chemical and Instrumental Analysis (in Romanian)*, Cluj-Napoca: Academic Direct, p. 64.

Juhasz, M. A., Matheson, G. R., Chang, P. S., Rosenbaum, A., & Juers, D. H., (2016). Microwave-assisted iodination: Synthesis of heavily iodinated 10-vertex and 12-vertex boron clusters. *Synthesis and Reactivity in Inorganic, Metal-Organic, and Nano-Metal Chemistry 46*(4), 583–588.

Kameneva, S. V., Kobzarenko, A. V., & Feldman, V. I., (2015). Kinetics and mechanism of the radiation-chemical synthesis of krypton hydrides in solid krypton matrices. *Radiation Physics and Chemistry, 110*, 17–23.

Kaplan, I. G., & Miterev, A. M., (1987). Interaction of charged particles with molecular medium and track effects in radiation chemistry. *Advances in Chemical Physics, 68*, 255–386.

Kappe, C. O., & Dallinger, D., (2006). The impact of microwave synthesis on drug discovery. *Nat. Rev. Drug Discov., 5*, 51–63.

Khan, F., Ahmad, S. R., & Kronfli, E., (2006). γ-radiation-induced changes in the physical and chemical properties of lignocellulose. *Biomacromolecules, 7*(8), 2303–2309.

Kim, J., Kwon, S., Cho D. H., Kang, B., Kwon, H., Kim, Y., et al., (2015). Direct exfoliation and dispersion of two-dimensional materials in pure water via temperature control. *Nature Communications, 6*, 8294.

Kitamura, K., Ieda, N., Hishikawa, K., Suzuki, T., Miyata, N., Fukuhara, K., & Nakagawa, H., (2014). Visible-light-induced nitric oxide release from a novel nitrobenzene derivative cross-conjugated with a coumarin fluorophore. *Bioorganic and Medicinal Chemistry Letters, 24*(24), 5660–5662.

Kumar, M., & Francisco, J. S., (2015). Red-light-induced decomposition of an organic peroxy radical: A New source of the HO_2 radical. *Angewandte Chemie – International Edition*. doi: 10.1002/anie.201509311, *54*(52), 15711–15714.

Lattach, Y., Coletta, C., Ghosh, S., & Remita, S., (2014). Radiation-induced synthesis of nanostructured conjugated polymers in aqueous solution: Fundamental effect of oxidizing species. *Chemphyschem, 15*, 208–218.

Lattach, Y., Deniset-Besseau, A., Guigner, J. M., & Remita, S., (2013). Radiation chemistry as an alternative way for the synthesis of PEDOT conducting polymers under "soft" conditions. *Radiat. Phys. Chem., 82*, 44–53.

Lei, Y., Li, Y., Xu, L., Yang, J., Wan, R., & Long, H., (2016). Microwave synthesis and sintering of TiNiSn thermoelectric bulk. *Journal of Alloys and Compounds, 660*, 166–170.

Lim, J., Gim, J., Song, J., Nguyen, D. T., Kim, S., Jo, J., Mathew, V., & Kim, J., (2016). Direct formation of $LiFePO_4$/graphene composite via microwave-assisted polyol process. *Journal of Power Sources, 304*, 354–359.

Lu, X., Li, X., Qian, J., Miao, N., Yao, C., & Chen, Z., (2016). Synthesis and characterization of CeO2/TiO2 nanotube arrays and enhanced photocatalytic oxidative desulfurization performance. *Journal of Alloys and Compounds, 661*, 363–371.

Ma, Y., Pang, Y., Liu, F., Xu, H., & Shen, X., (2016). Microwave-assisted ultrafast synthesis of silver nanoparticles for detection of Hg^{2+}. *Spectrochimica Acta – Part A: Molecular and Biomolecular Spectroscopy, 153*, 206–211.

Mavandadi, F., & Pilotti, A., (2006). The impact of microwave-assisted organic synthesis in drug discovery. *Drug Discov. Today, 11*(3/4), 165–174.

Mozumder, A., & Hatano, Y., (2004). *Charged Particle and Photon Interactions with Matter.* New York: Marcel Dekker, p. 860.

Nabiyouni, G., & Bakhtiari, M., (2016). Microwave-assisted synthesis of BaFe12O19 nanoparticles and ethyl cellulose-based magnetic nanocomposite. *Synthesis and Reactivity in Inorganic, Metal-Organic, and Nano-Metal Chemistry, 46*, 163–167.

Orlovskii, Y. V., Vanetsev, A. S., Keevend, K., Kaldvee, K., Samsonova, E. V., Puust, L., et al., (2016). NIR fluorescence quenching by OH acceptors in the Nd^{3+} doped KY_3F_{10} nanoparticles synthesized by microwave-hydrothermal treatment. *Journal of Alloys and Compounds, 661*, 312–321.

Oulianov, D. A., Crowell, R. A., Gosztola, D. J., Shkrob, I. A., Korovyanko, O. J., & Rey-De-Castro, R. C., (2007). Ultrafast pulse radiolysis using a terawatt laser wakefield accelerator. *Journal of Applied Physics, 101*(5), 053102.

Pan, J., Li, M., Luo, Y., Wu, H., Zhong, L., Wang, Q., & Li, G., (2016). Microwave-assisted hydrothermal synthesis of V_2O_5 nanorods assemblies with an improved Li-ion batteries performance. *Materials Research Bulletin, 74*, 90–95.

Paulsen, K. S., Di Carlo, D., & Chung, A. J., (2015). Optofluidic fabrication for 3D-shaped particles. *Nat. Commun., 6*, 6976.

Prevenslik, T. V., (2003). The cavitation induced Becquerel effect and the hot spot theory of sonoluminescence. *Ultrasonics, 41*(4), 313–317.

Qi, C., Zhu, Y. J., Wu, C. T., Sun, T. W., Chen, F., & Wu, J., (2016). Magnesium phosphate pentahydrate nanosheets: Microwave-hydrothermal rapid synthesis using creative phosphate as an organic phosphorus source and application in protein adsorption. *Journal of Colloid and Interface Science, 462*, 297–306.

Quirino, M. R., Oliveira, M. J. C., Keyson, D., Lucena, G. L., Oliveira, J. B. L., & Gama, L., (2016). Synthesis of zinc aluminate with high surface area by microwave hydrothermal method applied in the transesterification of soybean oil (biodiesel). *Materials Research Bulletin, 74*, 124–128.

Rana, P., & Chauhan, R. P., (2014). Size and irradiation effects on the structural and electrical properties of copper nanowires. *Physica B: Condensed Matter, 451*, 26–33.

Rosenberg, R. A., Abu Haija, M., & Ryan, P. J., (2008). *Phys. Rev. Lett., 101*, 178301.

Sambe, H., Ramaker, D. E., Parenteau, L., & Sanche, L., (1987). Electron-stimulated desorption enhanced by coherent scattering. *Phys. Rev. Lett., 59*(4), 505–508.

Shahriary, L., & Athawale, A. A., (2015). Synthesis of graphene using gamma radiations. *Bulletin of Materials Science, 38*(3), 739–745.

Simakov, M. B., (2008). Asteroids and the origin of life – Two steps of chemical evolution on the surface of these objects. *Earth, Planets, and Space, 60*(1), 75–82.

Su, F., Miao, M., Niu, H., & Wei, Z., (2014). Gamma-irradiated carbon nanotube yarn as substrate for high-performance fiber supercapacitors. *ACS Applied Materials and Interfaces, 6*(4), 2553–2560.

Sun, T. W., Zhu, Y. J., Qi, C., Ding, G. J., Chen, F., & Wu, J., (2016). α-Fe$_2$O$_3$ nanosheet-assembled hierarchical hollow mesoporous microspheres: Microwave-assisted solvothermal synthesis and application in photocatalysis. *Journal of Colloid and Interface Science, 463*, 107–117.

Suslick, K. S., (2014). Mechanochemistry and sonochemistry: Concluding remarks. *Faraday Discussions, 170*, 411–422.

Viererbl, L., Klupák, V., Vinš, M., Lahodová, Z., Kolmistr, A., & Stehno, J., (2014). Radiation measurements after irradiation of silicon for neutron transmutation doping. *Radiation Physics and Chemistry, 95*, 389–391.

Wang, G., Yuan, D., Yuan, T., Dong, J., Feng, N., & Han, G., (2015). A visible light responsive azobenzene-functionalized polymer: Synthesis, self-assembly, and photoresponsive properties. *Journal of Polymer Science, Part A: Polymer Chemistry, 53*(23), 2768–2775.

Wang, L., Zhang, X., Ma, Y., Yang, M., & Qi, Y., (2016). Rapid microwave-assisted hydrothermal synthesis of one-dimensional MoO3 nanobelts. *Materials Letters, 164*, 623–626.

Wardman, P., (1987). The mechanism of radiosensitization by electron-affinity compounds. *Radiation Physics and Chemistry, 30*(5/6), 423–432.

Xu, H., Fang, H., Bai, J., Zhang, Y., & Wang, Z., (2014). Preparation and characterization of high-melt-strength polylactide with long-chain branched structure through γ-radiation-induced chemical reactions. *Industrial and Engineering Chemistry Research, 53*(3), 1150–1159.

Xu, H., Zeiger, B. W., & Suslick, K. S., (2013). Sonochemical synthesis of nanomaterials. *Chemical Society Reviews, 42*(7), 2555–2567.

Yang, R. B., Reddy, P. M., Chang, C. J., Chen, P. A., Chen, J. K., & Chang, C. C., (2016). Synthesis and characterization of Fe$_3$O$_4$/polypyrrole/carbon nanotube composites with tunable microwave absorption properties: Role of carbon nanotube and polypyrrole content. *Chemical Engineering Journal, 285*, 497–507.

Zarnegar, Z., Safari, J., & Kafroudi, Z. M., (2015). Co$_3$O$_4$-CNT nanocomposites: A powerful, reusable, and stable catalyst for sonochemical synthesis of polyhydroquinolines. *New Journal of Chemistry, 39*(2), 1445–1451.

CHAPTER 26

Receptor Binding

CORINA DUDA-SEIMAN,[1] DANIEL DUDA-SEIMAN,[2] and
MIHAI V. PUTZ[1,3]

[1]*Laboratory of Computational and Structural Physical Chemistry for
Nanosciences and QSAR, Biology-Chemistry Department,
Faculty of Chemistry, Biology, Geography at West University of
Timişoara, Pestalozzi Street No.16, Timişoara, RO–300115, Romania*

[2]*Department of Medical Ambulatory, and Medical Emergencies,
University of Medicine and Pharmacy "Victor Babes,"
Avenue C. D. Loga No. 49, RO–300020 Timisoara, Romania*

[3]*Laboratory of Renewable Energies-Photovoltaics, R&D National
Institute for Electrochemistry and Condensed Matter, Dr. A. Paunescu
Podeanu Str. No. 144, RO–300569 Timişoara, Romania,
Tel.: +40-256-592-638; Fax: +40-256-592-620,
E-mail: mv_putz@yahoo.com or mihai.putz@e-uvt.ro*

26.1 DEFINITION

A working definition of a receptor was published in 2003, considering that a receptor is represented by a "cellular macromolecule or an assembly of macromolecules, that is concerned directly and specifically in chemical signaling between and within cells." (Neubig et al., 2003) Ligands are compounds that generally bind specifically to a receptor. (http://groups.molbiosci.northwestern.edu/holmgren/Glossary/Definitions/Def-L/ligand.html) The ligand-receptor interaction is described as an interaction between a molecule, usually of extracellular origin and a structure of protein type localized at the level of the cellular wall or within it. (http://groups.molbiosci.northwestern.edu/holmgren/Glossary/Definitions/Def-L/ligand-receptor_interact.html).

26.2　HISTORICAL ORIGINS

One of the first observations that substances interact with different structures within human tissues was made by the physicist G.G. Stokes in the 19th century. He recognized that oxygen forms a complex with blood hemoglobin, being under certain condition whether added to, or removed from hemoglobin (Stokes, 1864; Klotz, 2004).

Paul Ehrlich has formulated the concepts of "chemical affinities" in terms of specific examples, such as toxin-antitoxin, or antigen-antiserum complex. His logical conclusion was that "a substance is not (biologically) active unless it is "fixed" (bound by a receptor)" (Ehrlich, 1913; Klotz, 2004).

26.3　NANO-CHEMICAL IMPLICATIONS

The chemistry of Ligand–Receptor interactions implies some restrictions: in the control mechanisms of binding of a ligand to its specific receptor, there intervene rather local biochemical processes than only pure physical processes (like diffusion) (Krohn, 2001).

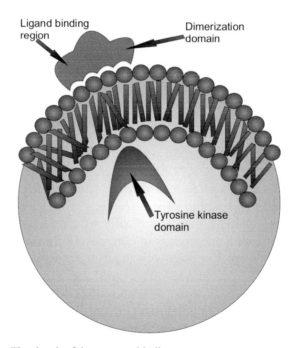

FIGURE 26.1　The sketch of the receptor binding.

G-protein-coupled receptors (GPCRs) play essential roles in intracellular signaling, therefore their implication in a wide range of diseases, comprising the main pathophysiological mechanisms: inflammation, from infective conditions to proliferative diseases (cancer). GPCRs record a high conformation variability, as well as a small exposed area of extracellular epitopes, providing difficulties in the approaches to isolate/prepare GPCR antigens and to design and obtain their specific antibodies (Jo, 2016). Activated GPCRs change their conformation in order to allow the recruitment and activation of proteins (modulating *G* proteins, scaffolding beta-arrestin) responsible for signal transmission from the plasma membrane to the in-cell. Monoclonal antibodies are highly selective and have an extended half-life (Delić et al., 2014). Mogamulizumab is a monoclonal antibody that targets the chemokine receptor CCR4, and it was recently approved for the treatment of adult *T* cell leukemia (Beck et al., 2012). Nanobodies (Nbs) is a novel class of antibodies-based therapeutics. Anti GPCR nanobodies imply chemokine receptors and modulate the GCPR function. Nanobodies have the potentiality to be used as targeting devices for toxic enzymes or engineered in order to circumscribe the effector functions to eradicate tumors (antibody-dependent cellular cytotoxicity). Mogamulizumab has been effectively modified by glycoengineering techniques increasing its capacity to exert specific antibody-dependant cellular cytotoxicity and therefore improving its effectiveness in adult *T* cell leukemia therapy (Delić et al., 2014).

In 1989, researchers have identified a novel type of antibodies, first in dromedaries and afterward in the species of *Camelidae* family. The particular aspects of these antibodies derive from the fact that they do not contain a light chain and also the lack of the first constant heavy domain. It was rapidly recognized the enlarged field of applicability of the isolated variable antigen-binding domains. Nanobodies are the smallest functional fragments of heavy chain antibodies (Vanlandschoot et al., 2011).

Nanobodies have a series of advantages over classic antibodies used in different therapy strategies: physical and chemical robustness; excellent stability even in extreme conditions (extreme temperatures, high pressure, low pH, the presence of proteases); nanobodies-based drugs can be administrated in several routes; excellent diffusion capacity and tissue penetration. Because of their small size, nanobodies have a rapid renal clearance. This disadvantage can be avoided, and thus its half-life can be prolonged, if nanobodies undergo several techniques, such as polyethylene glycol (PEG)-ylation, conjugation to the Fc domain of conventional antibody, coupling to abundant serum proteins (albumins). It has been demonstrated that albumins accumulate in areas of inflammation; thus, nanobodies coupled to serum albumins will be targeted to such areas (Steeland et al., 2016).

Nano-aggregates consist of a polymer-drug conjugated amphiphilic block copolymer. They are used for encapsulation of low water-soluble drugs by covalent conjugation and by physical encapsulation (technique that allows the incorporation of a large amount of drug). The active transport of nano-aggregates is achieved using vectors or ligands that bind specifically to receptors that are overexpressed in tumor cells (Sharma et al., 2013).

26.4   TRANS-DISCIPLINARY CONNECTIONS

Oncology is one of the most approached fields by researchers in terms of specific targeted therapy. So, it has been demonstrated that receptors (localized at the cell surface) are specific or are overexpressed by cancer cells (Retting, 1989; van den Eynde, 1998). Recent data concern nuclear medicine and optical molecular imaging because of their nanomolar sensitivity level and discrimination rate to identify and quantify cell-surface receptor concentrations in the tissue, as well as satellite dissemination, in order to target specific therapy (Tichauer et al., 2015).

26.5   OPEN ISSUES

The future of pharmaco-chemistry and medicine will be personalized therapy in order to target specific receptors expressed in specific pathological conditions. One of recent validated systems is the reconstituted high-density lipoprotein system as a drug delivery platform, obtaining bullet-shaped systems for lipophilic and hydrophilic drug delivery in several diseases (Simonsen, 2016).

Targeted therapies determine the increase of compliance to the treatment, a good quality of life, preserving, and improving therapy effectiveness and decreasing major side effects. Despite targeting affected tissues in the body, modern drugs express also a significant cytotoxic effect upon physiologically normal cells. New classes of drugs (proteins, peptides, DNA, siRNA) target specific cellular pathways. It is reasonable to design stimuli-responsive delivery systems based on smart biomaterials that respond to temperature or pH changes, protein, and ligand-binding, disease-specific degradation (Wanakule and Roy, 2012).

The immune system, its physiology and disturbances, is the key in understanding modern pathology: infective, inflammatory, and neoplastic diseases. In the era of modern therapy based on specific receptor-ligand interactions using smart therapy devices, it is crucial to correctly describe and understand nano-immuno-interactions (Fadeel, 2012).

KEYWORDS

- **ligand**
- **receptor**
- **receptor-ligand interaction**

REFERENCES AND FURTHER READING

Beck, A., & Reichert, J. M., (2012). Marketing approval of mogamulizumab: A triumph for glycoengineering. *MAbs, 4*, 419–425.

Delić, A. M., De Wit, R. H., Verkaar, F., & Smit, M. J., (2014). GPCR-targeting nanobodies: Research tools, diagnostics, and therapeutics. *Trends in Pharmacological Sciences, 35*(5), 247–255.

Ehrlich, P., (1913). Chemotherapeutics: Scientific principles. *Methods and results, Lancet, II*, 445–451.

Fadeel, B., (2012). Clear and present danger? Engineered nanoparticles and the immune system. *Swiss Med. Wkly., 142*, w13609.

Jo, M., & Jung, S. T., (2016). Engineering therapeutic antibodies targeting G-protein-coupled receptors. *Exp. Mol. Med., 48*, e207, doi: 10.1038/emm.2015.105.

Klotz, I. M., (2004). Ligand-receptor complexes: Origin and development of the concept. *The Journal of Biological Chemistry, 279*(1), 1–12.

Krohn, K. A., (2001). The physical chemistry of ligand-receptor binding identifies some limitations to the analysis of receptor images. *Nucl. Med. Biol., 28*(5), 477–483.

Ligand receptor interaction. http://groups.molbiosci.northwestern.edu/holmgren/Glossary/Definitions/Def-L/ligand-receptor_interact.html (accessed on 26 February 2019).

Ligand. http://groups.molbiosci.northwestern.edu/holmgren/Glossary/Definitions/Def-L/ligand. htm (accessed on 26 February 2019).

Neubig, R. R., Spedding, M., Kenakin, T., & Christopoulos, A., (2003). International Union of Pharmacology Committee on receptor nomenclature and drug classification. XXXVIII, update on terms and symbols in quantitative pharmacology. *Pharmacol. Rev., 55*, 597–606.

Retting, W. J., & Old, L. J., (1989). Immunogenetics of human cell surface differentiation. *Annu. Rev. Immunol., 7*, 481–511.

Sharma, V. K., Jain, A., & Soni, V., (2013). Nano-aggregates: Emerging delivery tools for tumor therapy. *Crit. Rev. Ther. Drug Carrier Syst., 30*(6), 535–563.

Simonsen, J. B., (2016). Evaluation of reconstituted high-density lipoprotein (rHDL) as a drug delivery platform - a detailed survey of rHDL particles ranging from biophysical properties to clinical implications. *Nanomedicine*, S1549–9634(16)30056–9. doi: 10.1016/j.nano.2016.05.009. [Epub ahead of print].

Steeland, S., Vandenbroucke, R. E., & Libert, C., (2016). Nanobodies as therapeutics: Big opportunities for small antibodies. *Drug Discovery Today.* http://dx.doi.org/10.1016/j.drudis.2016.04.003 (in press).

Stokes, G. G., (1864). On the reduction and oxidation of the coloring matter of the blood. *Proc. Roy. Soc., 13,* 355–364.

Tichauer, K. M., Wang, Y., Pogue, B. W., & Liu, J. T. C., (2015). Quantitative in vivo cellsurface receptor imaging in oncology: Kinetic modeling and paired-agent principles from nuclear medicine and optical imaging. *Phys. Med. Biol.,* 60(14), 239–269.

Van Den Eynde, B. J., & Scott, A. M., (1998). In: Roitt, D. P. J., & Roitt, I. M., (eds.), Encyclopedia of Immunology (pp. 2424–2431). (London: Academic).

Vanlandschoot, P., Stortelers, C., Beirnaert, E., Ibañez, L. I., Schepens, B., Depla, E., & Saelens, X., (2011). Nanobodies: New ammunition to battle viruses. *Antiviral Research,* 92, 389–407.

Wanakule, P., & Roy, K., (2012). Disease-responsive drug delivery: The next generation of smart delivery systems. *Current Drug Metabolism,* 13, 42–49.

CHAPTER 27

SMILES

MIHAI V. PUTZ and NICOLETA A. DUDAŞ

Laboratory of Computational and Structural Physical Chemistry for Nanosciences and QSAR, Biology-Chemistry Department, West University of Timişoara, Pestalozzi Str. No. 16, Timişoara 300115, Romania, E-mail: mihai.putz@e-uvt.ro, mv_putz@yahoo.com

27.1 DEFINITION

SMILES (Simplified Molecular Input Line Entry System) is a chemical notation system, based on principles of molecular graph theory, that allows rigorous structure specification by use of a very small and natural grammar. SMILES notation consists of a series of characters containing no spaces (ASCII codes of corresponding symbols), was designed for a modern chemical information processing and is well suited for the era of computers and internet.

27.2 HISTORICAL ORIGIN(S)

In the era of computers and the internet, with millions of chemical substances and chemical reactions, besides the conventional name of chemical compounds (in accordance with IUPAC) and notation conventional of equation of chemical reaction (using symbols and chemical formulas), it was absolutely necessary to introduce a new chemico-mathematic-informatic/computer scoring system to strengthen and develop the chemist-computer-internet connection, without losing any chemical information.

SMILES is a chemical notation system that continues and develops the Wiswesser Line Notation, the oldest method who represented the molecular structure as a linear string of symbols that could be efficiently read and stored by computer systems (Sliwoski et al., 2013; Weininger, 1988; Wiswesser, 1985). In the late 1980s, Weininger D. initiated SMILES-development by using the concept of a graph with nodes as atoms and edges as bonds to represent a molecule (Daylight Chemical Information Systems 2008a,b,c, 2013). It has quickly become widely accepted as a standard for exchange of molecular structures. (Weininger, 1988, 1990; Weininger et al., 1989; Daylight Chemical Information Systems, 2008a,b,c, 2013).

SMILES was developed as a proprietary specification by Daylight Chemical Information Systems, Daylight's SMILES Theory Manual is a standard for the SMILES language, and many independent SMILES software packages have been written (Daylight Chemical Information Systems, 2008a,b,c, 2013; Opensmiles).

With SMILES, molecules, and reactions can be specified using a string of characters that will represent atoms and bonds symbols. SMILES string of characters is humanly understandable, very compact, which contains the same information as is found in an extended connection table, and is easily managed by computer programs and stored in markup language fields (Daylight Chemical Information Systems, 2008a,b,c, 2013; Drefahl, 2011). In the SMILES language, there are two fundamental types of symbols: atoms and bonds; this symbols can specify a molecule's graph (its "nodes" and "edges") and assign "labels" to the components of the graph: what type of atom each node represents, and what type of bond each edge represents (Daylight Chemical Information Systems, 2008a,b,c, 2013; Sliwoski et al., 2013).

Nowadays, SMILES is most commonly used for storage and retrieval of compounds across multiple computer platforms, allow interpretation and generation of chemical notation, independent of the specific computer system in use (Weininger, 1988; 1990; Weininger et al., 1989, Daylight Chemical Information Systems, 2008a,b,c, 2013; Sliwoski et al., 2013).

27.3 NANO-SCIENTIFIC DEVELOPMENT(S)

Similar to conventional chemical notation, SMILES notation have rules for specifying SMILES for chemical structures, the specification of isomerism, substructures, and unique SMILES generation, basic SMILES grammar includes isotopic information, configuration about double bonds, chirality, etc. (Daylight Chemical Information Systems, 2008a,b,c, 2013). For the same chemical structure, there are usually many different but equally valid SMILES descriptions. (Weininger, 1988, 1990; Weininger et al., 1989; Daylight Chemical Information Systems, 2008a,b,c, 2013; Sliwoski et al., 2013). SMILES is unique for each structure, although dependent on the canonicalization algorithm used to generate it (Putz, 2013a; Dudaş, 2013a). A canonicalization algorithm exists to generate one special generic SMILES among all valid possibilities–the *unique SMILES*; *isomeric SMILES* is SMILES written with isotopic and chiral specifications, and a unique isomeric SMILES is known as an *absolute SMILES* (Daylight Chemical Information Systems, 2008a,b,c, 2013).

The five generic SMILES encoding rules for specifying SMILES for chemical structures (organic, inorganic, etc.) are corresponding to the specification of atoms, bonds, branches, ring closures, and disconnections (Daylight Chemical Information Systems, 2008a,b,c, 2013). These basic rules are listed in detail on http://www.daylight.com; we just mention some of the most important rules (see Figure 27.1) (Weininger, 1988; 1990; Weininger, et al., 1989, Daylight Chemical Information Systems, 2008a,b,c, 2013; Sliwoski et al., 2013; Brown, 2015):

- ***Atoms*** are represented by their atomic symbols enclosed in square brackets, [ ]. Atoms listed without square brackets, imply the presence of hydrogens, e.g.: [Mg] magnesium atom; [C] or [#6] is for carbon atom and *C* is for CH_4–methane molecule.
- ***Formal charges*** are specifically assigned as + or – followed by an optional digit inside the appropriate brackets.
- ***Hydrogen atoms*** usually are omitted (hydrogen-suppressed graphs), conventionally assumes that hydrogens make up the remainder of an atom's lowest normal valence (when hydrogen atoms are included is obtained hydrogen-complete graphs).
- ***Atoms in aromatic rings*** are specified by lower case letters.
- Single, double, triple, and aromatic *bonds* are represented by the symbols "-, =, #, and:," respectively. Adjacent atoms are assumed to be connected to each other by a single or aromatic bond (single and aromatic bonds may always be omitted).
- ***Branched*** systems are specified by enclosing them in parentheses.
- ***Cyclic structures*** are represented by breaking a ring at a single or aromatic bond and numbering the atoms on either side of the break with a number. The bonds are numbered in any order, designating ring opening (or ***ring closure***) bonds by a digit immediately following the atomic symbol at each ring closure.
- ***Disconnected*** compounds are separated by a period "." ionic bonds are considered disconnected structures with complementary formal charges.
- ***Isotopic*** specifications are indicated by preceding the atomic symbol with a number equal to the desired integral atomic mass. An atomic mass can only be specified inside brackets.

- **Stereochemistry around double bonds** is specified by the characters/ and\which are "directional bonds up and down" and can be thought of as kinds of single or aromatic bonds. These symbols indicate relative directionality between the connected atoms, and have meaning only when they occur on both atoms which are double bonded (see Figure 27.1).
- SMILES uses a very general type of *chirality* specification based on local chirality. SMILES can operate correctly on all types of *chiralities*: tetrahedral (TH), allene-like (AL), square-planar (SP), trigonal-bipyramidal (TB), and octahedral (OH). *Chirality at tetrahedral center*: the "chiral center" has four different connections and is TH–the two mirror images are known as "enantiomers." The symbol "@," the "visual mnemonic," indicates that the following neighbors are listed anticlockwise; "@@" indicates that the neighbors are listed clockwise. The chiral order of the ring closure bond is implied by the lexical order that the ring closure digit appears on the chiral atom (see Figure 27.1). SMILES handles the full range of chiral specification, including resolution of "reduced chirality" (where the number of enantiomers is reduced by symmetry) and "degenerate chirality" (where the center becomes non-chiral due to symmetrical substitution).
- *Reaction* SMILES Grammar is simple: **reactant '>'agent '>'product.** Any reaction must have exactly two > characters in it. Each of the ">"-separated components of a reaction must be a valid SMILES molecule (see Figure 27.1). SMILES does not indicate how the agent participates. Whether the agent is a solvent, catalyst, or performs another function within the reaction must be stored separately as data.

SMARTS (SMILES ARbitrary Target Specification) is an extension of SMILES that allows for variability within the represented molecular structures (Daylight Chemical Information Systems, 2008b). Using SMARTS, flexible, and efficient substructure-search specifications can be made in terms that are meaningful to chemists. (Daylight Chemical Information Systems, 2008b, 2013; Sliwoski et al., 2013). In SMARTS the labels for the graph's nodes and edges are extended to include "logical operators" and special atomic and bond symbols; these allow SMARTS atoms and bonds to be more general (Daylight Chemical Information Systems, 2008b, 2013). Examples of logical operators used by SMARTS are: "AND" (&–for high precedence;;–for low precedence), "OR" (,), and "NOT"(!). (Daylight Chemical Information Systems, 2008b, 2013; Sliwoski et al., 2013; Brown, 2015). Other symbols used by SMARTS are: atomic symbol [C,N] is an atom that can be aliphatic *C* or aliphatic N; [c,n] is an atom that can be aromatic *C* or aromatic N; the SMARTS bond symbol ~ (tilde) matches any bond; * (wildcard) matches to any atom; /? is directional bond "up or unspecified," \? is directional bond "down or unspecified"; [++] means an atom with a +2 charge. (Daylight Chemical Information Systems, 2008b, 2013; Sliwoski et al., 2013; Brown, 2015).

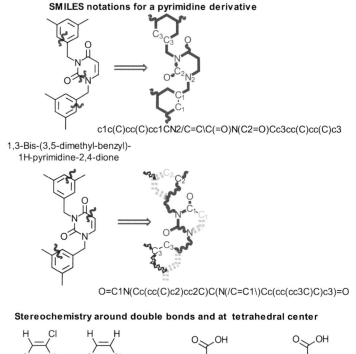

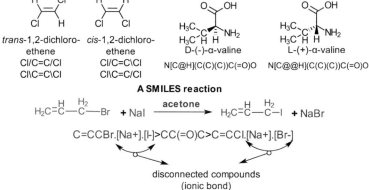

FIGURE 27.1 Application of SMILES encoding rules for specifying SMILES notation for some chemical structures and chemical reaction.

SMIRKS – A Reaction Transform Language, a subset of SMILES that encodes reaction transforms (is a hybrid of SMILES and SMARTS) (Daylight Chemical Information Systems, 2008c, 2013; Sliwoski et al., 2013; Brown, 2015). The language SMIRKS is defined for generic reactions, in order to meet the dual needs of them: expression of a reaction graph and expression of indirect effects (Daylight Chemical Information Systems, 2008c). It is a restricted version of reaction SMARTS involving changes in atom-bond patterns (Daylight Chemical Information Systems, 2008c). The rules for SMIRKS are (Daylight Chemical Information Systems, 2008c, 2013):

- The reactant and product sides of the transformation are required to have the same numbers and types of mapped atoms and the atom maps must be pairwise. However, non-mapped atoms may be added or deleted during a transformation.
- Stoichiometry is defined to be 1–1 for all atoms in the reactant and product for a transformation. Hence, if non-unit stoichiometry is desired, reactants or products must be repeated.
- Explicit hydrogens that are used on one side of a transformation must appear explicitly on the other side of the transformation and must be mapped.
- Bond expressions must be valid SMILES (no bond queries allowed).
- Atomic expressions may be any valid atomic SMARTS expression for nodes where the bonding (connectivity and bond order) doesn't change. Otherwise, the atomic expressions must be valid SMILES.

Curly SMILES (Curly-braces enhanced Smart Material Input Line Entry Specification) (Drefahl, 2011) was introduced as a chemical language for the specification of chemical materials and supramolecular structures. Curly-SMILES includes its own set of symbols, descriptors, and rules to denote entities and also modifies the well-established SMILES language (Drefahl, 2011). Molecular details and extra-molecular features (non-covalent interactions, attachment to a biomolecule, the surface of a substrate material or nanoparticle) are encoded, modifying the SMILES language into CurlySmiles (Drefahl, 2011). In CurlySMILES users can integrate shorthands such as aliases or compaction of repetitive structural units (Drefahl, 2011). A CurlySmiles notation is a string of dot-separated component notations, which can be a plain SMILES, an annotated SMILES, or a special format notation; SMILES code is strictly separated from other parts in the notation by using curly braces (Drefahl, 2011). A plain notation maintains the grammar and rules of the known SMILES language, and can be modified by introducing attributes (structural variations, details, and decorations) enclosed in curly braces (Drefahl, 2011). An annotation can be anchored to a particular atomic node or placed at the end of a SMILES component (Drefahl, 2011).

27.4 NANO-CHEMICAL APPLICATION(S)

Nowadays SMILES has become a part of languages for artificial intelligence or expert systems in chemistry and an entry system for chemical data. Usually, SMILES code is used as a key for database access and a mechanism for researchers to exchange chemical information (Daylight Chemical Information Systems, 2008a,b,c, 2013).

One example is QSARINS-Chem, a module in QSARINS (developed by Gramatica P. et al.), which is a software proposed for the development and validation of QSAR models; it includes several datasets of chemical structures and their corresponding endpoints (physicochemical properties and biological activities), where chemicals are accessible in different ways including SMILES (Gramatica et al., 2014).

The drug-minded protein interaction database (DrumPID) has been designed by Kunz et al., to provide fast, tailored information on drugs and their protein networks including indications, protein targets and side-targets (Kunz et al., 2016). DrumPID contains 1383 FDA-approved drugs, 4951 proteins, 4078 ortholog groups (clustered according to 993 unique COG/KOG–*Clusters of Orthologous Groups (COG)/EuKaryotic Orthologous Groups (KOG)*, 21120 orthologs from 67 different organisms) and over 1 million different protein interactions (various organisms); available stored data from>5000 FDA and non-FDA approved drugs (Kunz et al., 2016). The result page is very accessible to searches, and is divided into different sections: identifiers, biological properties, protein binding affinity and orthologous groups of targeted proteins, chemical properties (e.g., Lipinski's rules), pharmacological information and indications (Kunz et al., 2016). The *SMILES* of a drug can be searched or using *SMILES string* it can be screened for similar drug SMILES in this database (results include the substring matching and calculated similarity scores) (Kunz et al., 2016).

A lot of studies on drug-target interaction prediction use 2D-based compound similarity kernels such as SIMCOMP. Öztürk et al., (2016) based on the fact that the similarity between compounds can be computed using SMILES-based string similarity functions, they adapted and evaluated various SMILES-based similarity methods for drug-target interaction prediction and have investigated generating composite kernels by combining their best SMILES-based similarity functions with the SIMCOMP kernel. Also, the authors propose cosine similarity based SMILES kernels that make use of the Term Frequency (TF), and Term Frequency-Inverse Document Frequency (TF-IDF) weighting approaches. The more efficient SMILES-based similarity functions performed similarly to the more complex 2D-based SIMCOMP kernel in terms of AUC-ROC (Area Under the ROC Curve) scores (Öztürk et al., 2016). The authors concluded that using SMILES string as a molecular similarity kernel is easier, faster than the 2D-based SIMCOMP, more flexible (Öztürk et al., 2016).

SMILES fingerprint (SMIfp) was introduced by Schwartz et al., to perform ligand-based virtual screening; SMIfp is defined as a scalar fingerprint describing organic molecules by counting the occurrences of 34 different symbols in their SMILES strings, which creates a 34-dimensional chemical space and provides a new and relevant entry to explore the small molecule chemical space (Schwartz et al., 2013; Öztürk et al., 2016). In SMIfp space the aromatic and non-aromatic molecules are distinct because of the Lower/Upper case letter (Mannhold et al., 2015). Based on SMIfp similarity to a reference molecule it can be screen GBD (Global Burden of Disease) based to obtain a subset of a few million molecules which can be analyzed closer using more complex fingerprints or other scoring functions (e.g., docking) (Mannhold et al., 2015). With City-Block Distance–CBDSMIfp (using an online SMIfp-browser) some database (e.g., DrugBank, PubChem, etc.) can be searched, and ligand-based virtual screening with CBDSMIfp as similarity measure provides good AUC (area under the curve) values (Schwartz et al., 2013). Visualization of the SMIfp chemical space can be performed by principal component analysis and color-coded maps (Schwartz et al., 2013). These maps spread molecules according to their fraction of aromatic atoms, size, and polarity (Schwartz et al., 2013).

Truszkowski et al., sketched a complete cheminformatics roadmap that frames a mesoscopic Molecular Fragment Dynamics (MFD) simulation kernel to allow its efficient use and practical application, Molecular Fragment Cheminformatics is considered as an accelerator to propagate MFD and similar mesoscopic simulation techniques (Truszkowski et al., 2014). Mesoscopic simulation studies the structure, dynamics, and properties of large molecular ensembles with millions of atoms; its basic interacting units are molecular fragments, molecules or even larger molecular entities, the mesoscopic simulation kernel software use abstract matrix (array) representations for bead topology and connectivity (Truszkowski et al., 2014).

The basis of the cheminformatics approach is a SMILES-like line notation (*f*SMILES) with connected molecular fragments to represent a molecular structure (Truszkowski et al., 2014).

For SMILES notation have been found a lot of other uses, not only a key for database access, but also in computational chemistry, QSAR as an optimal descriptor or a hybrid optimal descriptor, for a description of some fractalic structure which can interact with a receptor or can encode the structure of the materials.

Putz and Dudaş used the SMILES structures not only as a graphical tool but also as an intermediate in the mechanistic picture of chemical ligand-biological receptor interaction yielding the recorded effect in the organism (Putz, and Dudaş, 2013a, 2013b; Putz et al., 2015). They used the 2D-to–1D representation of a genuine molecule to treat it as a sort of "transition state/meta-stable" structure favoring the binding with the receptor site of some uracil derivatives with anti-HIV activity (Putz and Dudaş, 2013a, 2013b; Putz et al., 2015). SMILES forms of a genuine molecules (which looks like a "fractalic chain" with some of its genuine chemical bonds broken) are considered by authors that are able to travel through cellular walls till docking with active sites of the target recep-tors (see Figure 27.2) (Putz and Dudaş, 2013a; 2013b; Putz et al., 2015). The SMILES molecular counterparts are used as the similarity criteria for selecting an appropriate molecular set for QSAR studies (SPECTRAL-SAR studies) (Putz and Dudaş, 2013a; 2013b; Putz et al., 2015). For every molecule of uracil derivative from the working series of compounds, the authors considered two aspects of their SMILES structure (Putz, and Dudaş, 2013a; 2013b; Putz et al., 2015):

> ➢ Maximum SMILES chains in LoSMoC (*Longest SMILES Molecular Chain*)–the resulting molecule displays a sort of 2D form of the original (genuine) molecule along the "fractalic" chain presumably responsible for best transport/transduction of ligand molecules through cellular (lipidic) walls, after which they may be released with a modified structure due to their further ionization resulting from interactions with cellular layers.

> ➢ A second modified structure of genuine molecule which favor the binding with a receptor in its pockets–the *Branching SMILES* (BraS), which provide ligand bond breakages such that many "bays" are formed, yet with consistent "arms" linking the short molecular "skel-eton" aiming to favor the binding with a receptor in its pockets.

After the QSAR analysis authors revealed the general mechanism of action for the Genuine-LoSMoC-BraS related configuration for uracils' anti-HIV ligands (Putz and Dudaş, 2013a; 2013b; Putz et al., 2015). Also, the authors made a double-variational procedure, combining the binding (affinity and total energy) with molecular conformation (by Genuine and LoSMoC and BraS SMILES forms) by available docking protocols for this uracil derivatives molecules with anti-HIV activities (Putz et al., 2015). This double-variational docking-SMILES analysis identified one molecule as "most potent"/"best anti-HIV molecule" as a result of intersection between minimum (in negative, so favoring the binding) affinity, global energy, and all tested configurations (Genuine, LoSMoC, and BraS), the common output for all inter-configuration variations (Putz et al., 2015).

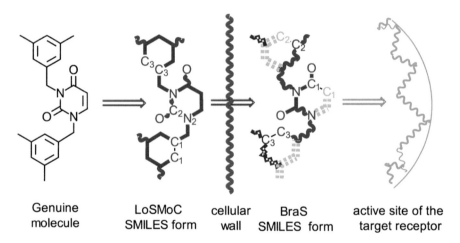

| Genuine molecule | LoSMoC SMILES form | cellular wall | BraS SMILES form | active site of the target receptor |

FIGURE 27.2   Schematic representation of the travel through cellular walls till docking with active sites of the target receptors by Genuine and LoSMoC and BraS SMILES forms.

Singh KP and Gupta S. encoded the structure of some materials in the SMILES form and calculated for nano-(Q)SAR modeling purposes a set of 32 molecular descriptors (topological, geometrical, and constitutional) based on this SMILES notations (Singh and Gupta, 2014; Oksel et al., 2015).

In several studies Toropov and Toporova et al., (Yilmaz et al., 2015; Toropov et al., 2005, 2010a,b,c, 2011a,b,c,d, 2012a,b, 2013, 2015; Toropova et al., 2010, 2011, 2012b, 2014, 2017; Nesměrák et al., 2013) used SMILES as an alternative to molecular graphs, they developed and calculated a number of SMILES-based descriptors and SMILES based descriptors in combination with topological descriptors using CORAL (CORrelations and Logic) software packages in order to develop/construct (Q)SAR/QSPR/Nano-QSAR (QFPR/QFAR) models for the prediction of the binding affinities of the fullerene, HEPT derivatives, amino-substituted nitrogen heterocyclic urea derivatives; for the development of models for carcinogenicity and anticancer activity, for the prediction of biodegradation of organic compounds, of water solubility, for octanol/water partition coefficient, for normal boiling points, and many others. These QSAR models are based on Monte Carlo optimization method where appropriate activity is treated as a random event (Toropova et al., 2014). Toropov and Toporova et al., conceived a "quasi-SMILES" approach which was applied for the development of model for membrane damage by ZnO and TiO_2 nanoparticles, mutagenicity of fullerene, mutagenic potential of multiwalled carbon nanotubes, and cytotoxicity of metal oxide nanoparticles to bacteria Escherichia coli (Toporova et al., 2017, 2014, 2012a; Toropov and Toropova, 2013, 2014, 2015). This approach can reflect all available eclectic data (e.g., size of particles, porosity, condition of synthesis, irradiation or absence of the irradiation, electromagnetic field, presence of pollutants, type of targets, e.g., cell, organ, organism, ecological system, etc., etc.) on nanomaterials augmented by features (impacts, conditions), and have clear influence on the performance of nanomaterials. (Toporova et al., 2017, Toropov et al., 2016; Toropov and Toropova, 2015).

The CORAL software free-packages is used by Toropov, A.A., Toropova, A.P. et al., in all listed above studies, and have three options for the selection of optimal descriptors: graph-based; SMILES-based; and hybrid descriptors (calculated using both graph and SMILES approaches) (Yilmaz et al., 2015; Toropov et al., 2005, 2010a,b,c, 2011a,b,c,d, 2012a,b, 2013, 2015; Toropova et al., 2010, 2011, 2012a,b, 2014, 2017; Nesměrák et al., 2013; Toropov and Toropova, 2013, 2014, 2015)

> The *optimal graph-based descriptor* based on so-called correlation weights (DCW) is calculated as the following (Eq. 1) (Yilmaz et al., 2015; Toropov et al., 2005, 2010a,b,c, 2011a,b,c,d, 2012a,b, 2013, 2015; Toropova et al., 2010, 2011, 2012a,b, 2014, 2017; Nesměrák et al., 2013; Toropov and Toropova, 2013, 2014, 2015):

$$^{Graph}DCW(T, N_{epoch})$$
$$= \sum CW(A_k) + \alpha \sum CW(^0EC_k) + \beta \sum CW(^1EC_k) + \gamma \sum CW(^2EC_k) + \delta \sum CW(^3EC_k) + \ldots \tag{1}$$

where T and N_{epoch} are parameters used in the Monte Carlo optimization method: T is threshold used for definition of rare molecular features and N_{epoch} is the number of the epochs of the Monte Carlo optimization; $\mathbf{A_k}$ is a chemical element (C, N, O, etc.) or atomic orbitals (1s1, 2p3, etc.); $^{0}\mathbf{EC_k}$, $^{1}\mathbf{EC_k}$,... is the hierarchy of the Morgan's extended connectivity.

➤ The *optimal SMILES-based descriptor* based on correlation weights is calculated (Eq. 2) (Yilmaz et al., 2015; Toropov et al., 2005, 2010a,b,c, 2011a,b,c,d, 2012a,b, 2013, 2015; Toropova et al., 2010, 2011, 2012a,b, 2014, 2017; Nesměrák et al., 2013; Toropov and Toropova, 2013, 2014, 2015):

$$^{SMILES}DCW(T, N_{epoch})$$
$$= \alpha \sum CW(S_k) + \beta \sum CW(SS_k) + \gamma \sum CW(SSS_k) + \delta * CW(ATOMPAIR) \quad (2)$$
$$+ x * CW(NOSP) + y * CW(HALO) + z * CW(BOND)$$

where:

- The coefficients α, β, γ, δ, x, y, z can be either 1 or 0: One indicates that the SMILES attribute is involved in the calculation of the DCW (Threshold) and zero indicates that the SMILES attribute is not involved. Combinations of values of different attributes provide the possibility of defining various versions of SMILES based optimal descriptors.
- The Sk, SSk, and SSSk are local SMILES attributes: Sk is the SMILES atoms: one symbol (C, O, P, etc.) or two symbols which cannot be examined separately (Cl, Br, etc.) and the SSk and SSSk are the SMILES attributes or combinations of two or three SMILES atoms (substrings of the SMILES string), respectively. If a SMILES is the sequence of SMILES atoms ABCDE, then one can show the Sk, SSk, and SSSk as the following: for the Sk the examples are: 'A,' 'B,' 'C,' 'D,' and 'E'; for the SSk the examples are: 'AB,' 'BC,' 'CD,' and 'DE'; for the SSSk the examples are: 'ABC,' 'BCD,' and 'CDE.' In some cases, two versions of the SMILES attribute (AB or BA) can be obtained. In this case, ASCII code for the definition of only one possibility (AB without BA) can be used. The same solution is applied for selection of only one of two combinations of three atoms (e.g., ABC without CBA).
- PAIR, NOSP, HALO, and BOND are global SMILES attributes which provide the possibility of carrying out an additional discrimination of substances into separated classes:
  - PAIR – reflects the presence of pairs of the listed features of molecule (e.g., the presence of pair from the list of F, Cl, Br, I, N, O, S, P, =, #, and @);

- NOSP – nitrogen, oxygen, sulfur, and phosphorus;
- HALO – fluorine, chlorine, and bromine;
- BOND – is related to the presence/absence of three categories of chemical bonds: double, triple, and stereospecific (=, #, @).

➢ The *hybrid SMILES-based descriptor* is calculated by CORAL software with SMILES-based and GRAPH-based descriptors (Eq. 3) (Yilmaz et al., 2015; Toropov et al., 2005, 2010a,b,c, 2011a,b,c,d, 2012a,b, 2013, 2015; Toropova et al., 2010, 2011, 2012a,b, 2014, 2017; Nesměrák et al., 2013; Toropov and Toropova, 2013, 2014, 2015):

$$^{Hybrid}DCW(T, N_{epoch}) = {}^{Graph}DCW(T, N_{epoch}) + {}^{SMILES}DCW(T, N_{epoch}) \tag{3}$$

With CW and DCW(T, N$_{epoch}$) calculated for compounds of training and test set, which give the preferable statistics for the calibration set, it can be calculated the model (Eq. 4) (Yilmaz et al., 2015; Toropov et al., 2005, 2010a,b,c, 2011a,b,c,d, 2012a,b, 2013, 2015; Toropova et al., 2010, 2011, 2012a,b, 2014, 2017; Nesměrák et al., 2013; Toropov and Toropova, 2013, 2014, 2015):

$$Endpoint = C_0 + C_1 * DCW(T, N_{epoch}) \tag{4}$$

Quasi-SMILES approach uses a similar method for calculation of SMILES-based descriptors, with the mention that in this approach is trying to reflect all available eclectic data.

KEYWORDS

- **daylight chemical information systems**
- **SMART**
- **SMILES (optimal descriptor) in QSAR/QSPR**
- **SMILES and database compounds**
- **SMILES notation/language**
- **SMIRK**
- **Weininger**

REFERENCES AND FURTHER READING

Brown, N., (2015). *In Silico Medicinal Chemistry: Computational Methods to Support Drug Design* (No. 8, pp. 26–30). Royal Society of Chemistry.

Daylight Chemical Information Systems, (2008a). *Daylight Theory: SMILES—A Simplified Chemical Language.* Available from: http://www.daylight.com/dayhtml/doc/theory/theory. smiles.html.

Daylight Chemical Information Systems, (2008b). *Daylight Theory: SMARTS—A Language for Describing Molecular Patterns.* Available from: http://www.daylight.com/dayhtml/doc/theory/theory.smarts.html.

Daylight Chemical Information Systems, (2008c). *Daylight Theory: SMIRKS—A Reaction Transform Language.* Available from: http://www.daylight.com/dayhtml/doc/theory/theory. smirks.html.

Daylight Chemical Information Systems, (2013). *Daylight Theory Manual.* Available from: http://www.daylight.com/dayhtml/doc/theory/.

Drefahl, A., (2011). Curly SMILES: A chemical language to customize and annotate encodings of molecular and nanodevice structures. *Journal of Cheminformatics*, *3*(1), 1.

Gramatica, P., Cassani, S., & Chirico, N., (2014). QSARINS-chem: Insubria datasets and new QSAR/QSPR models for environmental pollutants in QSARINS. *J. Comput. Chem.*, *35*(13), 1036–1044.

Kunz, M., Liang, C., Nilla, S., Cecil, A., & Dandekar, T., (2016). The drug-minded protein interaction database (DrumPID) for efficient target analysis and drug development. *Database*, baw041.

Mannhold, R., Kubinyi, H., & Folkers, G., (2015). In: Erlanson, D. A., & Jahnke, W., (eds.), *Fragment-Based Drug Discovery: Lessons and Outlook* (Vol. 67, pp. 63–65). John Wiley & Sons.

Nesměrák, K., Toropov, A. A., Toropova, A. P., Kohoutova, P., & Waisser, K., (2013). SMILES-based quantitative structure-property relationships for half-wave potential of N-benzylsalicyl thioamides. *Eur. J. Med. Chem.*, *67*, 111–114.

Oksel, C., Ma, C. Y., & Wang, X. Z., (2015). Current situation on the availability of nanostructure–biological activity data. *SAR and QSAR in Environmental Research*, *26*(2), 79–94.

Opensmiles–http://www.opensmiles.org/.

Öztürk, H., Ozkirimli, E., & Özgür, A., (2016). A comparative study of SMILES-based compound similarity functions for drug-target interaction prediction. *BMC Bioinformatics*, *17*(1), 128.

Putz, M. V., & Dudaş, N. A., (2013a). Variational principles for mechanistic quantitative structure–activity relationship (QSAR) studies: Application on uracil derivatives' anti-HIV action. *Struct. Chem.*, *24*, 1873–1893.

Putz, M. V., & Dudaş, N. A., (2013b). Determining chemical reactivity driving biological activity from SMILES transformations: The bonding mechanism of anti-HIV pyrimidines. *Molecules*, *18*, 9061–9116.

Putz, M. V., Dudaş, N. A., & Isvoran, A., (2015). Double variational binding-(SMILES) conformational analysis by docking mechanisms for anti-HIV pyrimidine ligands. *Int. J. Mol. Sci.*, *16*(8), 19553–19601. doi: 10.3390/ijms160819553.

Schwartz, J., Awale, M., & Reymond, J. L., (2013). SMIfp (SMILES fingerprint) chemical space for virtual screening and visualization of large databases of organic molecules. *Journal of Chemical Information and Modeling*, *53*(8), 1979–1989.

Singh, K. P., & Gupta, S., (2014). Nano-QSAR modeling for predicting biological activity of diverse nanomaterials. *RSC Advances*, *4*(26), 13215–13230.

Sliwoski, G., Kothiwale, S., Meiler, J., & Lowe, E. W. Jr., (2013). Computational methods in drug discovery. *Pharmacol. Rev., 66*(1), 334–395. doi: 10.1124/pr.112.007336.

Toropov, A. A., & Toropova, A. P., (2014). Optimal descriptor as a translator of eclectic data into endpoint prediction: Mutagenicity of fullerene as a mathematical function of conditions. *Chemosphere, 104*, 262–264.

Toropov, A. A., & Toropova, A. P., (2015). Quasi-QSAR for mutagenic potential of multi-walled carbon nanotubes. *Chemosphere*, *124*, 40–46.

Toropov, A. A., Toropova, A. P., & Benfenati, E., (2010b). QSPR modeling of normal boiling points and octanol/water partition coefficient for acyclic and cyclic hydrocarbons using SMILES based optimal descriptors. *Cent. Eur. J. Chem.*, *8*, 1047–1052.

Toropov, A. A., Toropova, A. P., Benfenati, E., Gini, G., Leszczynska, D., & Leszczynksy, J., (2013). CORAL: QSPR model of water solubility based on local and global SMILES attributes. *Chemosphere*, *90*, 877–880.

Toropov, A. A., Toropova, A. P., Benfenati, E., Gini, G., Leszczynska, D., & Leszczynski, J., (2011a). SMILES-based QSAR approaches for carcinogenicity and anticancer activity: Comparison of correlation weights for identical SMILES attributes. *Anticancer Agents Med. Chem. (Formerly Curr. Med. Chem. Anticancer Agents)*, *11*(10), 974–982.

Toropov, A. A., Toropova, A. P., Benfenati, E., Leszczynska, D., & Leszczynski, J., (2010a). SMILES-based optimal descriptors: QSAR analysis of fullerene-based HIV-1 PR inhibitors by means of balance of correlations. *J. Comput. Chem., 31*(2), 381–392.

Toropov, A. A., Toropova, A. P., Cappelli, C. I., & Benfenati, E., (2015). CORAL: Model for octanol/water partition coefficient. *Fluid Phase Equilib.*, *397*, 44–49.

Toropov, A. A., Toropova, A. P., Lombardo, A., Roncaglioni, A., Benfenati, E., & Gini, G., (2011b). CORAL: Building up the model for bioconcentration factor and defining it's applicability domain. *Eur. J. Med. Chem.*, *46*(4), 1400–1403.

Toropov, A. A., Toropova, A. P., Lombardo, A., Roncaglioni, A., De Brita, N., Stella, G., & Benfenati, E., (2012b). CORAL: The prediction of biodegradation of organic compounds with optimal SMILES-based descriptors. *Cent. Eur. J. Chem.*, *10*(4), 1042–1048.

Toropov, A. A., Toropova, A. P., Martyanov, S. E., Benfenati, E., Gini, G., Leszczynska, D., & Leszczynski, J., (2012a). CORAL: Predictions of rate constants of hydroxyl radical reaction using representation of the molecular structure obtained by combination of SMILES and graph approaches. *Chemometr. Intell. Lab Syst.*, *112*, 65–70.

Toropov, A. A., Toropova, A. P., Martyanov, S. E., Benfenati, E., Gini, G., Leszczynska, D., & Leszczynski, J., (2011c). Comparison of SMILES and molecular graphs as the representation of the molecular structure for QSAR analysis for mutagenic potential of polyaromatic amines. *Chemometr. Intell. Lab Syst.*, *109*(1), 94–100.

Toropov, A. A., Toropova, A. P., Mukhamedzhanoval, D. V., & Gutman, I., (2005). Simplified molecular input line entry system (SMILES) as an alternative for constructing quantitative structure-property relationships (QSPR). *Indian J. Chem., Sect A., 44*(8), 1545–1552.

Toropov, A. A., Toropova, A. P., Nesmerak, K., Veselinović, A. M., Veselinović, J. B., Leszczynska, D., & Leszczynski, J., (2016). Development of the latest tools for building up "nano-QSAR": quantitative features—property/activity relationships (QFPRs/QFARs). In: Leszczynski, J., & Shukla, M. K., (eds.), *Practical Aspects of Computational Chemistry IV* (pp. 353–396). Springer US.

Toropova, A. P., & Toropov, A. A., (2013). Optimal descriptor as a translator of eclectic information into the prediction of membrane damage by means of various TiO_2 nanoparticles. *Chemosphere*, *93*(10), 2650–2655.

Toropova, A. P., & Toropov, A. A., (2015). Mutagenicity: QSAR-quasi-QSAR-nano-QSAR. *Mini. Rev. Med. Chem.*, *15*(8), 608–621.

Toropova, A. P., Toropov, A. A., Benfenati, E., Gini, G., Leszczynska, D., & Leszczynski, J., (2011). CORAL: Quantitative structure–activity relationship models for estimating toxicity of organic compounds in rats. *J. Comput. Chemistry*, *32*(12), 2727–2733.

Toropova, A. P., Toropov, A. A., Benfenati, E., Gini, G., Leszczynska, D., & Leszczynski, J., (2012b). CORAL: Quantitative models for estimating bioconcentration factor of organic compounds. *Chemometr. Intell. Lab Syst.*, *118*, 70–73.

Toropova, A. P., Toropov, A. A., Benfenati, E., Puzyn, T., Leszczynska, D., & Leszczynski, J., (2014). Optimal descriptor as a translator of eclectic information into the prediction of membrane damage: The case of a group of ZnO and TiO_2 nanoparticles. *Ecotoxicol. Environ. Saf.*, *108*, 203–209.

Toropova, A. P., Toropov, A. A., Lombardo, A., Roncaglioni, A., Benfenati, E., & Gini, G., (2010). A new bioconcentration factor model based on SMILES and indices of presence of atoms. *Eur. J. Med. Chem.*, *45*(9), 4399–4402.

Toropova, A. P., Toropov, A. A., Rallo, R., Leszczynska, D., & Leszczynski, J., (2012a). Novel application of the CORAL software to model cytotoxicity of metal oxide nanoparticles to bacteria Escherichia coli. *Chemosphere*, *89*, 1098–1102.

Toropova, A. P., Toropov, A. A., Veselinović, A. M., Veselinović, J. B., Leszczynska, D., & Leszczynski, J., (2017). Quasi-SMILES as a novel tool for prediction of nanomaterials' endpoints. In: Speck-Planche, A., (eds.), *Multi-Scale Approaches in Drug Discovery From Empirical Knowledge to In Silico Experiments and Back* (pp. 191–221). Elsevier Ltd.

Toropova, A. P., Toropov, A. A., Veselinović, J. B., Miljković, F. N., & Veselinović, A. M., (2014). QSAR models for HEPT derivates as NNRTI inhibitors based on Monte Carlo method. *Eur. J. Med. Chem.*, *77*, 298–305. doi: 10.1016/j.ejmech.2014.03.013.

Truszkowski, A., Daniel, M., Kuhn, H., Neumann, S., Steinbeck, C., Zielesny, A., & Epple, M., (2014). A molecular fragment cheminformatics roadmap for mesoscopic simulation. *J. Cheminf.*, *6*(1), 45.

Weininger, D., (1988). SMILES, a chemical language, and information system. 1. Introduction to methodology and encoding rules. *J. Chem. Inf. Comput. Sci.*, *28*, 31–36.

Weininger, D., (1990). SMILES. 3. Depict. Graphical depiction of chemical structures. *J. Chem. Inf. Comput. Sci.*, *30*, 237–243.

Weininger, D., Weininger, A., & Weininger, J. L., (1989). SMILES. 2. Algorithm for generation of unique SMILES notation. *J. Chem. Inf. Comput. Sci.*, *29*, 97–101.

Wiswesser, W. J., (1985). Historic development of chemical notations. *J. Chem. Inf. Comput. Sci.*, *25*, 258–263.

Yilmaz, H., Sizochenko, N., Rasulev, B., Toropov, A. A., Guzel, Y., Kuzmin, V., Leszczynski, D., & Leszczynski, J., (2015). Amino-substituted nitrogen heterocycle ureas as kinase insert domain-containing receptor (KDR) inhibitors: Performance of structure-activity relationship approaches. *Journal of Food and Drug Analysis*, *23*(2), 168–175.

CHAPTER 28

Solar Cell

MARIUS C. MIRICA,[1] MARINA A. TUDORAN,[1,2] and MIHAI V. PUTZ[1,2]

[1]Laboratory of Renewable Energies-Photovoltaics, R&D National Institute for Electrochemistry and Condensed Matter, Dr. A. Paunescu Podeanu Str. No. 144, Timisoara, RO–300569, Romania

[2]Laboratory of Structural and Computational Physical Chemistry for Nanosciences and QSAR, Biology-Chemistry Department, West University of Timisoara, Pestalozzi Street No. 44, Timisoara, RO–300115, Romania

28.1 DEFINITION

A solar cell represents a photovoltaic cell which is used to convert the light from the sun in electricity through a direct mechanism.

28.2 HISTORICAL ORIGIN(S)

The dye-sensitized solar cells (DSC) were developed by Prof. Graetzel, by sensitizing the TiO_2 thin film with Ru(II) complex sensitizer he obtained an efficiency of about 7%. In order to manufacture DSC, there is not required advanced equipment, materials of high purity, high-temperature treatment or high vacuum, meaning that the cost is reduced compared to silicon solar cell (Lin et al., 2009).

Over the last years, scientists have made continuous development in increasing the efficiency of the silicon solar cells (Buchovska et al., 2017). They determine that the efficiency improvement is related to the wafer material quality and shown that by controlling the structural properties of the material, one can influence the cell performance (Müller and Friedrich, 2010). It was also proved that multi-crystalline silicon (mc-Si) material is more advantageous than the single crystalline material due to the low production cost and simplicity (Sarti and Einhaus, 2002; Müller et al., 2006).

By developing high-performance multi-crystalline (HPM) silicon wafers, there is a tendency of replacing the typical crystalline silicon materials from the products manufacturing, the first being superior to the second in conversion efficiency and power output (Buchovska et al., 2017; Tang et al., 2013; Zhu et al., 2014).

Conjugated polymer with thiophene rings care used in organic solar cells (Yilmaz et al., 2014) as electron donors in charge transport (Padinger et al., 2003; Li et al., 2007; Hoppe and Sariciftci, 2004; Seong et al., 2004). From this category poly(3-hcxylthiophene) (P3HT) is a common organic semiconductor and therefore can be also used in polymer solar cells. The efficiency of photovoltaic devices can be increased by mixing a fullerene molecule (C60) in the polymer film (bulk heterojunction (BH) blend) or by subliming the fullerene into the polymer in a heterojunction (bilayer) (Sariciftci et al., 1992, 1993; Sariciftci and Heeger, 1994; Yu and Heeger, 1995; Roman et al., 1997, 1998).

The small molecule organic solar cell (SMOSC) are one of the most promising among the organic type (Abe et al., 2013), having the power conversion efficiency (PCE) of about 8%, being used especially in low cost energy production (Forrest, 2004; Sun et al., 2012; Li et al., 2012; Zhou et al., 2013). They have superior properties, such as high tunability, are easy to synthesize and purify and present inherent monodispersity (Roncali, 2009; Walker et al., 2011; Beaujuge and Fréchet, 2011; Mishra and Bäuerle, 2012; Hatano et al., 2012; Yamamoto et al., 2013).

Due to the fact that polymer solar cell (Hu et al., 2016) have a high mechanic flexibility and can be manufactured at low temperature, they represent a good alternative to the classical silicon solar cells. Even if their PCE is about 11%, they can still be improved by using suitable device structure (Liu et al., 2013; He et al., 2012), developing suitable processing technique (Zhao et al., 2016; Padinger et al., 2003; Li et al., 2005), designing effective active materials (Mishra and Bauerle, 2012; Kan et al., 2014; Coughlin et al., 2014) or by using interface engineering (e.g., adjustment of the light absorption, tune the energy level alignment, improve the interfacial stability) (Kang et al., 2012, 2014; Tan et al., 2014; Ho et al., 2014; Yang et al., 2010).

Solution-processed polymer solar cells (Yuan et al., 2016) are most used in developing flexible electronic devices, the most studied being the BH devices with a blended film represented by a conjugated polymer and an active layer represented by fullerene derivative such as phenyl-C61-butyric acid methyl ester (PC61BM) (Amb et al., 2011; Price et al., 2011; Chu et al., 2011; He et al., 2011).

Among the photovoltaic systems, wafer-based solar cells (Piralaee and Asgari, 2016) represents a promising candidate in overcoming the energy crisis, but they have the disadvantage of being relatively expensive. In order to overcome the high price, one can replace the conventional wafer based cells with thin film solar cell (Catchpole and Polman, 2008; Akimov et al., 2009a). Another improvement could be made on the light trapping mechanism, a new method of increasing the optical absorption of the photoactive layer assume the usage of plasmonic nanostructures (Akimov et al., 2009b). Recent studies has been showing that the plasmonic solar cells performance can be increased if the metallic nanoparticles are deposited at the photoactive layer on top, the improvement being the results of the higher optical absorption exhibited by this film photoactive layer (Tsai et al., 2010; Pillai et al., 2007; Derkacs et al., 2006).

Quantum dot solar cells (QDSC) (Xu et al., 2017) are one of the most studied solar cells in the past decade, having several noticeable advantages, such as the light response can be easily tuned in a broad range and the size of the employed quantum dot can be adjusted by tuning the energy band gap. Studies in this area show that highly efficient photovoltaic devices can be constructed with lead sulfide quantum dots (PbS QDs) cells having an excitation Bohr radius of about 18 nm length (Wise, 2000; Luther et al., 2010). It was also determined that for QD solar cells passivized with molecular halides have an efficiency of 9.9% (Lan et al., 2016). In PbS QD solar cell, good electric contact between the buffer layer and the PbS QD film lead to efficient excitons splitting in the built-in field, this characteristic being of great importance in device manufacturing. A typical device with PbS QD solar cells presents an anode buffer represented by a p-type film (MoO_3) and a cathode buffer, which can be an n-type dens film (TiO_2 and CdS) (Maraghechi et al., 2013; Ning et al., 2014; Bhandari et al., 2013). Still, although these devices have high efficiency, their fabrication process is not appropriate for commercial applications. On the other hand, a low-cost fabrication process was obtained for an inverted structure consisting of a cathode buffer layer and a solution process anode (Xu et al., 2017).

28.3 NANO-SCIENTIFIC DEVELOPMENT(S)

The solar cells represent one of the main instruments obtaining solar energy, being both resourceful and clean and presenting a high availability. There are several classes of solar cells which have been recently developed, such as thin film solar cell, flexible solar cell or tandem solar cell (Lin et al., 2009). The advantage of using thin-film solar cells is given by the fact that the diffusion length is shortened form the photogenerated carriers and the recombination is decreased. This category includes: amorphous silicon solar cell, II-VI, and III-V group semiconductor solar cells, and organic solar cell, which, same as a dye-sensitized solar cell (DSC), do not use the p-n junction. The tandem solar cells consist of two or more solar cell units which are able to absorb light of different wavelength, leading to increased performance. Another class, flexible solar cells have the ability of being folded, making them easy to transport and are usually used in airplanes, architectures, helmets, solar automobiles, etc. As an absorber, these cells use flexible conductive substrates or conductive polymer semiconductor (Lin et al., 2009).

In order to characterize the solar cell, there are different parameters which can be used. For instance, the output electric power of a solar cell divided by its illumination power results in the efficiency of the solar cell, with the formula:

$$\eta = \frac{FF \times J_{SC} \times V_{OC}}{P_{in}} \tag{1}$$

with FF as the fill factor, J_{SC} as the short circuit current density, V_{OC} as the open circuit voltage and P_{in} as the illumination power obtained by the integration of the measured spectral irradiance by respecting the wavelength.

The short-circuit density can be defined as the integral of external quantum efficiency (EQE) multiplied with the illumination spectrum, respecting the wavelength, after the formula:

$$J_{SC} = q \int_{\lambda_1}^{\lambda_2} EQE(\lambda) * \Phi(\lambda) d\lambda \tag{2}$$

When the current density J_{ph} and the dark current density J_O are equal, the open circuit voltage can be determined from the diode equation as follow:

$$V_{oc} = \frac{nkT}{q} \ln\left(\frac{J_{ph}}{J_O} + 1\right) \tag{3}$$

Another parameter, the fill factor FF has the formula:

$$FF = \frac{V_{mmp} J_{mmp}}{V_{OC} J_{SC}} \tag{4}$$

with V_{mmp} as the voltage at maximum power point and J_{mmp} as the current density at maximum power point.

28.4 NANO-CHEMICAL APPLICATION(S)

Recently, one of the most popular sources of energy is represented by the solar energy, the solar cells being known for their flexibility, portability, and the availability for mass production in a continuous process. Among them, DSC are the most advantageous due to their low-cost production. By controlling the film thickness and the size of nanoparticles one can manufacture transparent solar cells, with a wide range of application, but with a lower conversion efficiency compared with the existing solar cells (Ahn et al., 2013; Kim et al., 2011; Papageorgiou et al., 1996). In their work, Kim, and his co-workers (2015) develop a new mechanism for improving the light conversion efficiency using a reflector in order to collect the light which otherwise is lost through the DSSC in the permeation process. In order to increase the DSSC efficiency, one can use a method which implies expanding the surface area of the semiconductor oxide, due to the fact that dye molecules absorbed as a single layer on the semiconductor are highest efficiency. Starting from this, the authors analyze a different method of increasing the scattering distance by reflecting the solar light in order to increase the number of electrons produces by the dye (Kim et al., 2015). The scattering distance of the solar light reflected on the dye layer from the DSSC was increased by using a micro-reflector, with a light angle (θ_l) set to the maximum. After that, the maximum value of the mirror angle (θ_m) was determined by measuring the light angle considering the aspect ratio of the width and height of the reflector micropattern (Figure 28.1).

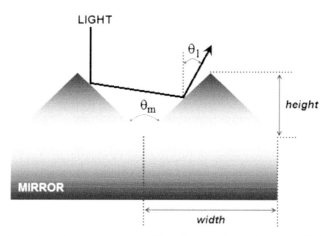

FIGURE 28.1 Schematic representation of the mirror angle and the light path. Redrawn and adapted from Kim et al., (2015).

For a value of 0.33 for the aspect ratio and 112.6° value of the mirror angle, the maximum value of the light angle was determined to be 67.5°, and, knowing that the dye layer thickness of the DSSC has the value of 40 μm, the authors determine that the largest amount of dye can be encountered by light when the scattering distance has the value of 104.5 μm (Kim et al., 2015). They also determined that the energy conversion efficiency exceeds with about 17% in the case of pyramid-shaped reflector compared with the case when a black plate was used (Kim et al., 2015).

The need of overcoming the efficiency limit is also applied to crystalline Si solar cell, and silicon nanostructures have been proposed as new materials, being different in the quantum confinement effect to the bulk crystalline Si (Lu et al., 1995; Vinciguerra et al., 2000; Zacharias et al., 2002; Priolo et al., 2014; Kim et al., 2006; Kurokawa et al., 2006). Due to the fact that the band gap energy of the silicon quantum dot (Si QD) is easy to be modify by changing the matrix material or by varying their size, Si QD are considered as promising material for the next generation solar cell (Zacharias et al., 2002; Kim et al., 2006; Park et al., 2009; Hao et al., 2009; Hong et al., 2011). Still, there is a limitation of the Si QD absorber layer given by the high series resistivity as a result of the insulating properties of the matrix material (Jiang and Green, 2006). In order to amend the carrier transport, one can dope method the Si QD layer, by producing a variation in the band structure. In their study, Kwak and his co-workers (2017) analyze the effect determined by the profile doping on the built-in electric field from a solar cell, using two different types of solar cells with Si QD heterojunction which present a p-type Si QD layer and were fabricated on substrates of the n-type crystalline Si, i.e., a Si QD solar cell having a single step doping profile of B, noted SS Si QDSC, and a Si QD solar cell having a double step doping profile of B, noted DS Si QDSC (Kwak et al., 2017). HR-TEM was used to investigate the formation of the two types of solar cells, results showing that SS Si QDSC (Quantum Dot Solar Cell) and the DS Si QDSC are no different in density and size, meaning that the concentration of boron does not affect the formation of Si QD in SiO$_2$ matrix. In other words, the double-step doping profile of B applied in the Si QD layer appears to improve the PCE, a further improvement being achieved if the band gap energy tuning is also applied (Kwak et al., 2017).

28.5   MULTI-/TRANS- DISCIPLINARY CONNECTION(S)

There are many types of solar cells with a flat panel, such as silicon and amorphous silicon solar cells or dye-sensitized solar cells (Chapin et al., 1954;

Green, 2000a, 2000b; Shah et al., 1999; O'Regan and Grätzel, 1991). Still, the lack of a solar tracker from these types of cells leads to a decrease in the energy production after a cosine function due to the continuously change during the day of the azimuth angles and the solar elevation (An et al., 1984). Another type of solar cell, dye-sensitized nanocrystalline solar cells (DSSC) (O'Regan and Grätzel, 1991; Grätzel, 2003) are now in the scientists attention as being more efficient and having a simple and a low-cost fabrication process. Starting from the DSSC operation principle, a new type of solar cell was fabricated by Liu and his co-workers (2006) with spiral photo-electrode, designed to capture the sunlight from 3-dimensional space (3D-cell) and to receive solar irradiance without the manifestation of the cosine losses. For the 3D-solar cell, the light excite the dye, and the photo-excitation of the sensitized determine the release of an electron in the TiO_2 conduction band which is then collected by the titanium wire (Liu et al., 2006). The iodide from the electrolyte reduces the oxidized dye to its original state, while the electrons passing through the load reduce the triiodide and regenerate the iodide at the counter-electrode. The titanium wires used as a substrate increase the efficiency of the 3D-cells, making them comparative with the solid-state dye-sensitized solar cells (Liu et al., 2006).

Another application of the solar cells is in astronomy, as the main energy in astronautics as part of the spacecraft (Lin, 1995; Wang, 2001; Green, 1992). In order to fulfill the working conditions in the aerospace, at 150 km from the Earth where the temperature have fluctuations, there is a big amount of energetic particles, and the Sun radiant intensity is 1.3 times higher than the Earth, active surface of the solar cells must be bond with anti-irradiating cover glass, so that the solar cell can be protected (Li, 2002; Crabb and Dollery, 1989; Nowland et al., 1991). If the physical characteristics of the solar cell are considered, the cover glass bonding to the solar cell can be made using one of the four kinds of technologies, such as Teflon bonding, cover glass adhesive bonding, static electric bonding and two sides array bonding (Norman, 2003; Mullaney et al., 1993). The cover-glass adhesive bonding starts with coating the solar cell surface, followed by the bonding of the solar cell with the anti-irradiating cover-glass. Next, the assembly is heated up and solidified, being finished by the testing process. But, even if this method has the easiest working principle operating in conventional method assumes overcoming some craft problems, such as for coating and bonding low precision, high intensity of labor, low efficiency, or, unequaled, and uncontrolled thickness of the adhesive layer (Fu et al., 2006). In their work Fu and his co-workers (2006) introduced the space solar cell bonding robot, consisting in a moving device with three-axis Cartesian coordinates, a bonding device, a computer control system, an orientation plate, and a coating device, so that the coating curve precision and the coating rate are increased. First, the adhesive is put in line on the solar cell and by adjusting the parameters of coating (velocity, height, and diameter of pinhead, space between tracks) is diffused evenly into a plane. After that, the cover glass is absorbed by the pneumatic sucker so that the solar cell and the cover glass are auto-bonded together. The control system follows a rectangular Cartesian coordinates in its moving and consists in the pneumatic system and the supervisory computer, so that the robot working position and the sensor messages can be feedback to the computer. Moreover, the computer can be also used in setting the working method and the robot parameters (Fu et al., 2006). By using the solar cell bonding robot, the authors determined an improvement in the efficiency and quality of the solar cell bonding compared to the conventional method. Apart from the high speed of bonding (2.4 times that of the conventional method), this way the health of operators is not at risk due to the poisonous adhesives (Fu et al., 2006).

A solar cell can be also used to measure the solar energy which is available below water surface at any depth (Muaddi, 2012), but the optimal one can be determined after a proper investigation of the solar cells efficiency and their specific response resulted from the spectral distribution which is available on a specific depth below water surface. In his work, Muaddi (2012) calculate the total solar cell photocurrent and the total solar energy from the digitized spectral response curve of the solar cells (Field, 1997), the solar spectrum[1] and the absorption coefficient of the pure water (Hale and Querry, 1973) using computational methods in order to predict the solar cell efficiency from different depth below water surface. In nature, in case of a normal incidence of solar spectrum radiation at the contact with the water surface, the incident spectrum takes two ways, a part is reflected, and the other part is transmitted through the water surface and follows the formula (Muaddi, 2012):

$$r = \frac{(n_2 - n_1)^2 + k_2^2}{(n_1 + n_2)^2 + k_2^2} \tag{5}$$

with n_1 and n_2 as the refractive indices of the first and the second medium, k as the coefficient of extinction for the second medium.

---

[1] Standard G173–03: Standard Tables for Reference Solar Spectral Irradiance at Air Mass 1.5: Direct Normal and Hemispherical for a 37 Degree Tilted Surface, American Society for Testing and Materials (ASTM).

In terms of the incident irradiance, I_0 and through the second medium, the transmitter part of the irradiance I_λ have the following integral form (Muaddi, 2012):

$$I_\lambda = \int_0^\infty I_{0(\lambda)}(1-r)d\lambda \tag{6}$$

Moving forward, at a depth x below water surface, one can determine the total spectral irradiance I_x following the equation:

$$I_x = \int_{\lambda_0}^\infty I_\lambda e^{-\mu_\lambda x} d\lambda \tag{7}$$

with μ_λ as the absorption coefficient of pure water and λ_0 as the spectrum short wavelength limit (Muaddi, 2012).

By exposing a solar cell to a solar radiation, there is a part of the incident spectrum which is absorbed by the cell, another part is reflected from the cell, and the rest of it is transmitted. Only a part of the absorbed spectrum, precisely, the one with energy higher than the cell's band gap energy E_g, can generate electron-hole pairs and follows the formula (Muaddi, 2012):

$$I = \int_0^{\lambda_g} I_{a(\lambda)} d\lambda \tag{8}$$

with λ_g representing the absorption edge equal with (hc/E_g) where h as the Plank's constant and c as the light speed; I_a representing the solar energy absorption fraction within the cell.

In order to obtain the solar spectral response for a cell at different depth in water one need to multiply the solar cell spectral response with the solar spectral distribution at a specific depth. The solar cell efficiency can be analyzed by determining the value of the ratio of the total photocurrent to the solar energy at a specific depth. Results of Muaddi study prove that for a solar cell, the efficiency increase with the depth in water until a certain depth after which starts to decrease (Muaddi, 2012).

28.6 OPEN ISSUES

The analysis of solar cells operation using quantum-size effects has been lately in the scientists' attention (Barnham and Duggan, 1989; Paxman et al., 1993; Ivanov et al., 2004). Recently, there are some theories which imply that in case of solar cells with structures of the type *p–i–n*, the direct-gap semiconductors are the only ones which can present an high electric field inside the *i* region when is considered the open circuit mode. On the other side, because a sufficiently lightly *i* region is necessary, the open circuit voltage decrease. In their paper, Sachenko, and Sokolovskii (2008) were studying comparatively the size-quantitized solar cells with AlGaAs, GaAs, and Si and usual solar cells from the photoconversion efficiency point of view, proving that size-quantitized solar cells are more advantageous than the usual solar cells only in a specific case, when the recombination centers density given by the building in the QWs do not present a major increase. They conduct the study in AM0 conditions and calculate the limiting photo-conversion limiting efficiency of AlGaAs, GaAs, and Si type solar cells having *p–i–n* structure and size quantitized insertions based on narrower gap materials. They also assume that for quantum wells (QWs) the carriers' life-times of nonequilibrium (noted with $_r$) exceed considerably their emission times (noted with τ_e). Starting from the fact that the emission mechanism of carrier for solar cells based on Si is thermal, meaning that the τ_e can be calculated using the expression (Raisky et al., 1997):

$$\tau_e = W \sqrt{2\pi m / kT} \, \exp\left(E_a / kT\right) \tag{9}$$

$$E_a \approx \left(E_b - E_g\right)/2 \tag{10}$$

$$E_g = E_{g0} + \pi^2 \hbar^2 / 2\mu W^2 \tag{11}$$

with *m* as the effective mass of the carrier in QW, *W* as the width of QW, E_a as the thermal activation energy which is equal with the narrow-gap insertion and the distance between the band edges, E_b as the barrier material's band gap, E_g as QW's effective band gap, E_{g0} as the QW's material band gap and μ as the reduced mass of hole and electron for QW (Sachenko and Sokolovskii, 2008).

For solar cells with GeSi, for $E_b \approx 0.15$ eV and $W \approx 10$ nm, the obtained value for τ_e is equal with 10^{-10} s. In case of direct gap solar cell, the minimal value of τ_e decrease in high electric field, its value being equal with 10^{-12} *s* due to the fact that in QW, the mechanism of carrier emission is combined (Nikolaev and Avrutin, 2003; Sachenko and Sokolovskii, 2008).

ACKNOWLEDGMENT

This contribution is part of the research project "Fullerene-Based Double-Graphene Photovoltaics" (FG2PV), PED/123/2017, within the Romanian National program "Exploratory Research Projects" by UEFISCDI Agency of Romania.

KEYWORDS

- **conductive polymer semiconductor**
- **dye sensitizer solar cell**
- **light conversion efficiency**
- **open circuit voltage**
- **silicon quantum dot solar cell**

REFERENCES AND FURTHER READING

Abe, Y., Yokoyama, T., & Matsuo, Y., (2013). Low-LUMO 56p-electron fullerene acceptors bearing electron-withdrawing cyano groups for small-molecule organic solar cells. *Organic Electronics, 14*, 3306–3311.

Ahn, S. H., Chun, D. M., & Chu, W. S., (2013). Perspective to green manufacturing and applications. *International Journal of Precision Engineering and Manufacturing, 14*(6), 873–874.

Akimov, Y. A., Koh, W. S., & Ostrikov, K., (2009b). Enhancement of optical absorption in thin-film solar cells through the excitation of higher-order nanoparticle Plasmon modes. *Optics Express, 17*(12), 10195–10205.

Akimov, Y. A., Ostrikov, K., & Li, E. P., (2009a). Surface plasmon enhancement of optical absorption in thin-film silicon solar cells. *Plasmonics, 4*(2), 107–113.

Amb, C. M., Chen, S., Graham, K. R., Subbiah, J., Small, C. E., So, F., & Reynolds, J. R., (2011). Dithienogermole as a fused electron donor in bulk heterojunction solar cells. *Journal of the American Chemical Society, 133*(26), 10062–10065.

An, Q. L., Cao, G. S., Li, G. X., et al., (1984). *Solar Cells Operating Principles and Technology* (in Chinese). Shanghai: Shanghai Science and Technology Press.

Barnham, K. W., & Duggan, G., (1989). A new approach to high-efficiency multi-bandgap solar cells. *Journal of Applied Physics, 67*(7), 3490–3493.

Beaujuge, P. M., & Fréchet, J. M. J., (2011). Molecular design and ordering effects in π-functional materials for transistor and solar cell applications. *Journal of the American Chemical Society, 133*(50), 20009–20029.

Bhandari, K. P., Roland, P. J., Mahabaduge, H., Haugen, N. O., Grice, C. R., Jeong, S., Dykstra, T., Gao, J., & Ellingson, R. J., (2013). Thin film solar cells based on the heterojunction of colloidal PbS quantum dots with CdS. *Solar Energy Materials and Solar Cells, 117*, 476–482.

Buchovska, I., Liaskovskiy, O., Vlasenko, T., Beringov, S., & Kiessling, F. M., (2017). Different nucleation approaches for production of high-performance multi-crystalline silicon ingots and solar cells. *Solar Energy Materials & Solar Cells, 159*, 128–135.

Catchpole, K. R., & Polman, A., (2008). Plasmonic solar cells. *Optics Express, 16*(26), 21793–21800.

Chapin, D. M., Fuller, C. S., & Pearson, G. L., (1954). New silicon p–n junction photocell for converting solar radiation into electrical power. *Journal of Applied Physics, 25*, 676–677.

Chu, T. Y., Lu, J., Beaupre, S., Zhang, Y., Pouliot, J. R. M., Wakim, S., Zhou, J., Leclerc, M., Li, Z., Ding, J., & Tao, Y., (2011). Bulk heterojunction solar cells using thieno[3,4-c] pyrrole–4,6-dione and dithieno[3,2-b: 2′,3′-d]silole copolymer with a power conversion efficiency of 7.3%. *Journal of the American Chemical Society, 133*(12), 4250–4253.

Coughlin, J. E., Henson, Z. B., Welch, G. C., & Bazan, G. C., (2014). Design and synthesis of molecular donors for solution-processed high-efficiency organic solar cells. *Accounts of Chemical Research, 47*, 257–270.

Crabb, R. L., & Dollery, A. A., (1989). Direct glassing of silicon solar cells. *European Space Power, 2*, 607−611.

Derkacs, D., Lim, S. H., Matheu, P., Mar, W., & Yu, E. T., (2006). Improved performance of amorphous silicon solar cells via scattering from surface plasmon polaritons in nearby metallic nanoparticles. *Applied Physics Letters, 89*(9), 093103.

Field, H., (1997). *Solar Cell Spectral Response Measurement Errors Related to Spectral Band Width and Chopped Light Waveform*. 26th IEEE Photovoltaic Specialists Conference, Anaheim, California.

Forrest, S. R., (2004). The path to ubiquitous and low-cost organic electronic appliances on plastic. *Nature, 428*, 911–918.

Fu, Z., Zhao, Y. Z., Liu, R. Q., & Dong, Z., (2006). A space solar cell bonding robot. *Frontiers of Mechanical Engineering in China, 3*, 360−363.

Grätzel, M., (2003). Dye-sensitized solar cells. *Journal of Photochemistry and Photobiology C: Photochemistry Reviews, 4*, 145–153.

Green, M. A., (1992). *Solar Cells: Operating Principles Technology and System Applications*. Englewood Cliffs, NJ, Sydney: Prentice Hall.

Green, M. A., (2000a). Photovoltaics: technology overview. *Energy Policy, 28*(14), 989–998.

Green, M. A., (2000b). The future of crystalline silicon solar cells. *Progress in Photovoltaics: Research and Applications, 8*(1), 127–139.

Guoxin, L., (2002). The progress of Shanghai spacecraft EPS technology in the 20th century. *Aerospace Shanghai, 3*, 42−48 (in Chinese).

Hale, G. M., & Querry, M. R. J., (1973). Optical constants of water in the 200-nm to 200-μm wavelength region. *Applied Optics, 12*(3), 555–563.

Hao, X. J., Cho, E. C., Flynn, C., Shen, Y. S., Park, S. C., Conibeer, G., & Green, M. A., (2009). Synthesis and characterization of boron-doped Si quantum dots for all-Si quantum dot tandem solar cells. *Solar Energy Materials and Solar Cells, 93*, 273–279.

Hatano, J., Obata, N., Yamaguchi, S., Yasuda, T., & Matsuo, Y., (2012). Soluble porphyrin donors for small molecule bulk heterojunction solar cells. *Journal of Materials Chemistry, 22*, 19258–19263.

He, Z., Zhong, C., Huang, X., Wong, W. Y., Wu, H., Chen, L., Su, S., & Cao, Y., (2011). Simultaneous enhancement of open-circuit voltage, short-circuit current density and fill factor in polymer solar cells. *Advanced Materials, 23*(40), 4636–4643.

He, Z., Zhong, C., Su, S., Xu, M., Wu, H., & Cao, Y., (2012). Enhanced power-conversion efficiency in polymer solar cells using an inverted device structure. *Nature Photonics, 6*, 591–595.

Ho, Y. C., Kao, S. H., Lee, H. C., Chang, S. K., Lee, C. C., & Lin, C. F., (2014). Investigation of the localized surface Plasmon effect from Au nanoparticles in ZnO nanorods for enhancing the performance of polymer solar cells. *Nanoscale, 7*, 776–783.

Hong, S. H., Kim, Y. S., Lee, W., Kim, Y. H., Song, J. Y., Jang, J. S., Park, J. H., Choi, S. H., & Kim, K. J., (2011). Active doping of *B* in silicon nanostructures and development of a Si quantum dot solar cell. *Nanotechnology, 22*, 425203.

Hoppe, H., & Sariciftci, N. S., (2004). Organic solar cells: An overview. *Journal of Materials Research, 19*(7), 1924–1945.

Hu, T., Jiang, P., Chen, L., Yuan, K., Yang, H., & Chen, Y., (2016). Amphiphilic fullerene derivative as effective interfacial layer for inverted polymer solar cells. *Organic Electronics, 37*, 35–41.

Ivanov, I. I., Skryschvsky, V. A., & Litvinenko, S. V., (2004). Numerical simulation of the photocurrent in the thin metal – silicon structures with quantum wells. *Ukrainian Journal of Physics, 49*, 917–920.

Jiang, C. W., & Green, M. A., (2006). Silicon quantum dot superlattices: Modeling of energy bands, densities of states, and mobilities for silicon tandem solar cell applications. *Journal of Applied Physics, 99*, 114902–1–114902–7.

Kan, B., Zhang, Q., Li, M., Wan, X., Ni, W., Long, G., Wang, Y., Yang, X., Feng, H., & Chen, Y., (2014). Solution-processed organic solar cells based on dialkylthiol-substituted benzodithiophene unit with efficiency near 10%. *Journal of the American Chemical Society, 136*, 15529–15532.

Kang, H., Hong, S., Lee, J., & Lee, K., (2012). Electrostatically self-assembled nonconjugated polyelectrolytes as an ideal interfacial layer for inverted polymer solar cells. *Advanced Materials, 24*, 3005–3009.

Kang, R., Oh, S. H., & Kim, D. Y., (2014). Influence of the ionic functionalities of polyfluorene derivatives as a cathode interfacial layer on inverted polymer solar cells. *ACS Applied Materials & Interfaces, 6*, 6227–6236.

Kim, M. S., Chun, D. M., Choi, J. O., Lee, J. C., Kim, K. S., Kim, Y. H., Lee, C. S., & Ahn, S. H., (2011). Room temperature deposition of TiO_2 using nanoparticle deposition system (NPDS): Application to Dye-Sensitized Solar Cell (DSSC). *International Journal of Precision Engineering and Manufacturing, 12*(4), 749–752.

Kim, T. W., Cho, C. H., Kim, B. H., & Park, S. J., (2006). Quantum confinement effect in crystalline silicon quantum dots in silicon nitride grown using SiH_4 and NH_3. *Applied Physics Letters, 88*, 123102.

Kim, Y. W., Kang, B. S., & Lee, D. W., (2015). Improving efficiency of dye-sensitized solar cell by micro-reflectors. *International Journal of Precision Engineering and Manufacturing, 16*(7), 1257–1261.

Kurokawa, Y., Miyajima, S., Yamada, A., & Konagai, M., (2006). Preparation of nanocrystalline silicon in amorphous silicon carbide matrix. *Japanese Journal of Applied Physics, 45*, L1064–L1066.

Kwak, G. Y., Lee, S. H., Jang, J. S., Hong, S., Kim, A., & Kim, K. J., (2017). Band engineering of a Si quantum dot solar cell by modification of B-doping profile. *Solar Energy Materials & Solar Cells, 159*, 80–85.

Lan, X., Voznyy, O., Kiani, A., Pelayo, G. A. F., Abbas, A. S., Kim, G., et al., (2016). Passivation using molecular halides increases quantum dot solar cell performance. *Advanced Materials, 28*, 299–304.

Li, G., Shrotriya, V., Huang, J., Yao, Y., Moriarty, T., Emery, K., & Yang, Y., (2005). High-efficiency solution processable polymer photovoltaic cells by self-organization of polymer blends. *Nature Materials, 4*, 864–868.

Li, G., Shrotriya, V., Yao, Y., Huang, J. S., & Yang, Y., (2007). Manipulating regioregular poly(3-hexylthiophene): [6,6]-phenyl-C_{61}-butyric acid methyl ester blends-route towards high-efficiency polymer solar cells. *Journal of Materials Chemistry, 17*, 3126–3140.

Li, Z., He, G., Wan, X., Liu, Y., Zhou, J., Long, G., Zuo, Y., Zhang, M., & Chen, Y., (2012). Solution processable rhodanine-based small molecule organic photovoltaic cells with a power conversion efficiency of 6.1%. *Advanced Energy Materials, 2*, 74–77.

Lin, H., Wang, W., Liu, Y., Li, X., & Li, J., (2009). New trends for solar cell development and recent progress of dye-sensitized solar cells. *Frontiers of Materials Science in China, 3*(4), 345–352.

Lin, L., (1995). Modern small satellites and its key technology. *Chinese Space Science and Technology, 15*(4), 37−43 (in Chinese).

Liu, Y., Chen, C. C., Hong, Z., Gao, J., Yang, Y., Zhou, H., Dou, L., Li, G., & Yang, Y., (2013). Solution-processed small-molecule solar cells: Breaking the 10% power conversion efficiency. *Scientific Reports, 3*, 3356.

Liu, Y., Shen, H., & Deng, Y., (2006). A novel solar cell fabricated with spiral photo-electrode for capturing sunlight 3-dimensionally. *Science in China Series E: Technological Sciences, 49*(6), 663–673.

Lu, Z. H., Lockwood, D. J., & Baribeau, J. M., (1995). Quantum confinement and light emission in SiO2/Si superlattices. *Nature, 378*, 258–260.

Luther, J. M., Gao, J., Lloyd, M. T., Semonin, O. E., Beard, M. C., & Nozik, A. J., (2010). Stability assessment on a 3% bilayer PbS/ZnO quantum dot heterojunction solar cell. *Advanced Materials, 22*, 3704–3707.

Maraghechi, P., Labelle, A. J., Kirmani, A. R., Lan, X., Adachi, M., Thon, S., Hoogland, S., Lee, A., Ning, Z., Fischer, A., Amassian, A., & Sargent, E., (2013). The donor-supply electrode enhances performance in colloidal quantum dot solar cells. *ACS Nano, 7*, 6111–6116.

Mishra, A., & Bäuerle, P., (2012). Small molecule organic semiconductors on the move: Promises for future solar energy technology. *Angewandte Chemie International Edition, 51*(9), 2020–2076.

Muaddi, J. A., (2012). An optimal solar cell for underwater measurement. *Applied Solar Energy, 48*(1), 20–23.

Mullaney, K., Dollery, A. A., Jones, G. M., & Bogus, K., (1993). An optimized Teflon bonding process for solar cell assemblies using Pilkington CMZ and CMG cover glasses. *Photovoltaic Specialists Conference, Conference Record of the Twenty Third IEEE*, 1392−1398.

Müller, A., Ghosh, M., Sonnenschein, R., & Woditsch, P., (2006). Silicon for photovoltaic applications. *Materials Science and Engineering: B., 134*, 257–262.

Müller, G., & Friedrich, J., (2010). Optimization and modeling of photovoltaic silicon crystallization processes, selected topics on crystal growth. *AIP Conference Proceedings, 1270*, 255–281.

Nikolaev, V. V., & Avrutin, A., (2003). Photocarrier escape time in quantum-well light-absorbing devices: Effects of electric field and well parameters. *IEEE Journal of Quantum Electronics, 39*, 1653–1660.

Ning, Z., Voznyy, O., Pan, J., Hoogland, S., Adinolfi, V., Xu, J., Li, M., et al., (2014). Air-stable n-type colloidal quantum dot solids. *Nature Materials, 13*, 822–828.

Norman, S., (2003). *Electrostatic Bonding* (Voo. 10, pp. 11−23). http:// www.twi. co.uk/ j32k/ protected/ band_3/ ksnrs002.html.

Nowland, M. J., Tobin, S. P., & Darkazalli, G., (1991). Direct cover glass bonding to GaAs and GaAs/Ge solar cells. *IEEE Photovoltaic Specialists Conference 22nd*, 1480−1484.

O'Regan, B., & Grätzel, M., (1991). A low-cost high-efficiency solar cell based on dye-sensitized colloidal TiO2 films. *Nature, 353*(24), 737–740.

Padinger, F., Rittberger, R. S., & Sariciftci, N. S., (2003). Effects of postproduction treatment on plastic solar cells. *Advanced Functional Materials, 13*, 85–88.

Papageorgiou, N., Athanassov, Y., Armand, M., Bonho, P., Pettersson, H., Azam, A., & Grätzel, M., (1996). The performance and stability of ambient temperature molten salts for solar cell applications. *Journal of the Electrochemical Society, 143*(10), 3099–3108.

Park, S., Cho, E., Song, D., Conibeer, G., & Green, M. A., (2009). n-type silicon quantum dots and p-type crystalline silicon heteroface solar cells. *Solar Energy Materials and Solar Cells*, 93684–93690.

Paxman, M., Nelson, J., Braun, B., Connolly, J., Barnham, K. W. J., Foxon, C. T., & Roberts, J. S., (1993). Modeling the spectral response of the quantum well solar cell. *Journal of Applied Physics, 74*, 614–621.

Pillai, S., Catchpole, K. R., Trupke, T., & Green, M. A., (2007). Surface Plasmon enhanced silicon solar cells. *Journal of Applied Physics, 101*(9), 093105.

Piralaee, M., & Asgari, A., (2016). Modeling of optimum light absorption in random plasmonic solar cell using effective medium theory. *Optical Materials, 62*, 399–402.

Price, S. C., Stuart, A. C., Yang, L., Zhou, H., & You, W., (2011). Fluorine substituted conjugated polymer of medium band gap yields 7% efficiency in polymer-fullerene solar cells. *Journal of the American Chemical Society, 133*(12), 4625–4631.

Priolo, F., Gregorkiewicz, T., Galli, M., & Krauss, T. F., (2014). Silicon nanostructures for photonics and photovoltaics. *Nature Nanotechnology, 9*, 19–32.

Raisky, O. J., Wang, W. B., Alfano, R. R., Reynolds, Jr. C. L., & Swaminathan, V., (1997). Investigation of photoluminescence and photocurrent in InGaAsP/InP strained multiple quantum well heterostructures. *Journal of Applied Physics, 81*(1), 394–399.

Roman, L. S., Andersson, M. R., Yohannes, T., & Inganäs, O., (1997). Photodiode performance and nanostructure of polythiophene/C_{60} blends. *Advanced Materials, 9*(15), 1164–1168.

Roman, L. S., Mammo, W., Pettersson, L. A. A., Andersson, M. R., & Inganäs, O., (1998). High quantum efficiency polythiophene. *Advanced Materials, 10*(10), 774–777.

Roncali, J., (2009). Molecular bulk heterojunctions: An emerging approach to organic solar cells. *Accounts of Chemical Research, 42*(11), 1719–1730.

Sachenko, A. V., & Sokolovskii, I. O., (2008). Comparative analysis of limiting photoconversion efficiency of usual solar cells and solar cells with quantum wells. *Semiconductors, 42*(10), 1219–1227.

Sariciftci, N. S., Braun, D., Zhang, C., Srdanov, V., Heeger, A. J., & Wudl, F., (1993). Semiconducting polymer-buckminsterfullerene heterojunctions – diodes, photodiodes, and photovoltaic cells. *Applied Physics Letters, 62*, 585–587.

Sariciftci, N. S., Smilowitz, L., Heeger, A. J., & Wudl, F., (1992). Photoinduced electron transfer from a conducting polymer to buckminsterfullerene. *Science, 258*(5087), 1474–1476.

Sariciftci, N., & Heeger, A. J., *US Patent No: 5331183*. Conjugated polymer - acceptor heterojunctions; diodes, photodiodes, and photovoltaic cells. 19 July.

Sarti, D., & Einhaus, R., (2002). Silicon feedstock for the multi-crystalline photovoltaic industry. *Solar Energy Materials and Solar Cells, 72*, 27–40.

Seong, J. Y., Chung, K. S., Kwak, S. K., Kim, Y. H., Moon, D. G., Han, J. I., & Kim, W. K., (2004). Performance of low-operating voltage P3HT-based thin-film transistors with anodized Ta2O5 insulator. *Journal of the Korean Physical Society, 45*, S914–S916.

Shah, A., Torres, P., Tscharner, R., Wyrsch, N., & Keppner, H., (1999). Photovoltaic technology: The case for thin-film solar cells. *Science, 285*(30), 692–698.

Sun, Y., Welch, G. C., Leong, W. L., Takacs, C. J., Bazan, G. C., & Heeger, A. J., (2012). Solution-processed small-molecule solar cells with 6.7% efficiency. *Nature Materials, 11,* 44–48.

Tan, Z. A., Li, L., Li, C., Yan, L., Wang, F., Xu, J., Yu, L., Song, B., Hou, J., & Li, Y., (2014). Trapping light with a nanostructured CeOx/Al back electrode for high-performance polymer solar cells. *Advanced Materials Interfaces, 1,* 1400197.

Tang, X. H., Francis, L. A., Gong, L. F., Wang, F. Z., Raskin, J. P., Flandre, D., Zhang, S., You, D., Wu, L., & Dai, B., (2013). Characterization of high-efficiency multi-crystalline silicon in industrial production. *Solar Energy Materials and Solar Cells, 117,* 225–230.

Tsai, F. J., Wang, J. Y., Huang, J. J., Kiang, Y. W., & Yang, C. C., (2010). Absorption enhancement of an amorphous Si solar cell through surface plasmon-induced scattering with metal nanoparticles. *Optics Express, 18*(2), A207–A220.

Vinciguerra, V., Franzò, G., Priolo, F., Iacona, F., & Spinella, C., (2000). Quantum confinement and recombination dynamics in silicon nanocrystals embedded in Si/SiO2 superlattices. *Journal of Applied Physics, 87,* 8165.

Walker, B., Kim, C., & Nguyen, T. Q., (2011). Small molecule solution-processed bulk heterojunction solar cells. *Chemistry of Materials, 23,* 470–482.

Wang, R., (2001). A study on proton irradiation effects of homemade solar cell for space use. *Journal of Beijing Normal University (Natural Science), 37*(4), 507–510 (in Chinese).

Wise, F. W., (2000). Lead salt quantum dots: The limit of strong quantum confinement. *Accounts of Chemical Research, 33,* 773–780.

Xu, W., Tan, F., Liu, Q., Liu, X., Jiang, Q., Wei, L., Zhang, W., Wang, Z., Qu, S., & Wang, Z., (2017). Efficient PbS QD solar cell with an inverted structure. *Solar Energy Materials & Solar Cells, 159,* 503–509.

Yamamoto, T., Hatano, J., Nakagawa, T., Yamaguchi, S., & Matsuo, Y., (2013). Small molecule solution-processed bulk heterojunction solar cells with inverted structure using porphyrin donor. *Applied Physics Letters, 102,* 013305/1-4.

Yang, T., Cai, W., Qin, D., Wang, E., Lan, L., Gong, X., Peng, J., & Cao, Y., (2010). Solution-processed zinc oxide thin film as a buffer layer for polymer solar cells with an inverted device structure. *The Journal of Physical Chemistry C., 114,* 6849–6853.

Yilmaz, C. N., Safak-Boroglu, M., Bilgin-Eran, B., & Günes, S., (2014). Bilayer polymer/fullerene solar cells with a liquid crystal. *Thin Solid Films, 560,* 1–76.

Yu, G., & Heeger, A. J., (1995). Charge separation and photovoltaic conversion in polymer composites with internal donor/acceptor heterojunctions. *Journal of Applied Physics, 78*(7), 4510–4515.

Yuan, J., Lu, K., Ford, M., Bazan, G. C., & Ma, W., (2016). Dual structure modifications to realize efficient polymer solar cells with low fullerene content. *Organic Electronics, 32,* 187–194.

Zacharias, M., Heitmann, J., Scholz, R., Kahler, U., Schmidt, M., & Bläsing, J., (2002). Size-controlled highly luminescent silicon nanocrystals: A SiO/SiO$_2$ superlattice approach. *Applied Physics Letters, 80,* 661.

Zhao, J. L. Y., Yang, G., Jiang, K., Lin, H., Ade, H., Ma, W., & Yan, H., (2016). Efficient organic solar cells processed from hydrocarbon solvents. *Nature Energy, 1,* 15027.

Zhou, J. Y., Zuo, Y., Wan, X. J., Long, G. K., Zhang, Q., Ni, W., et al., (2013). Solution-processed and high-performance organic solar cells using small molecules with a benzodithiophene unit. *Journal of the American Chemical Society, 135*(23), 8484–8487.

Zhu, D., Ming, L., Huang, M., Zhang, Z., & Huang, X., (2014). Seed-assisted growth of high-quality multi-crystalline silicon in directional solidification. *Journal of Crystal Growth, 386,* 52–56.

CHAPTER 29

Spectral-SAR

MIHAI V. PUTZ,[1,2] ANA-MARIA PUTZ,[1,3] CORINA DUDA-SEIMAN,[1] and DANIEL DUDA-SEIMAN[4]

[1]*Laboratory of Computational and Structural Physical Chemistry for Nanosciences and QSAR, Biology-Chemistry Department, Faculty of Chemistry, Biology, Geography at West University of Timișoara, Pestalozzi Street No. 16, Timișoara, RO–300115, Romania*

[2]*Laboratory of Renewable Energies-Photovoltaics, R&D National Institute for Electrochemistry and Condensed Matter, Dr. A. Paunescu Podeanu Str. No. 144, RO–300569 Timișoara, Romania, Tel.: +40-256-592-638; Fax: +40-256-592-620, E-mail: mv_putz@yahoo.com or mihai.putz@e-uvt.ro*

[3]*Institute of Chemistry Timisoara of the Romanian Academy, 24 Mihai Viteazul Bld., Timisoara 300223, Romania*

[4]*Department of Medical Ambulatory, and Medical Emergencies, University of Medicine and Pharmacy "Victor Babes," Avenue C. D. Loga No. 49, RO–300020 Timisoara, Romania*

29.1 DEFINITION

Spectral-SAR (S-SAR) represents a new approach to study quantitative structure-activity relationships, in an exclusively algebraic manner, replacing the old multi-regressional model. This method highlights structural descriptors as vectors in a generic data space using the Gram-Schmidt algorithm, therefore, S-SAR will be represented in a total orthogonal space. The coordinates between data and orthogonal space will be processed, and the S-SAR equation is obtained under a simple form determinant-dependant for each chemical-biological considered interaction. Applicability of methods such as S-SAR (analytical equations) lies in the similarity of obtained results when reference is made to standard multivariate statistical correlation results.

As such, in this context of S-SAR's it is useful and necessary to introduce spectral norm as a valid substitute for the correlation factor, thus having the advantage of different designs of SAR methods by accepting the principle of minimum spectral ways (Putz and Lacrama, 2007; Putz et al., 2009).

29.2 HISTORICAL ORIGIN(S)

Although the main goal of QSAR studies is to find structural parameters that correlate best way possible with the activity/property of targeted interactions, there is a variety of mathematical and computational methods for achieving this. In other words, QSAR seeks to identify the most appropriate way of quantifying the causes so that they can be reflected in measurements with maximum accuracy and minimum errors. Phenomenological, these methods can be named conceptually as 'classic' (Simon et al., 1984; Putz and Lacrama, 2007; Putz et al., 2008, 2009, 2011; Ivanciuc et al., 2000; Putz, 2006). All these methods are represented in Figure 29.1.

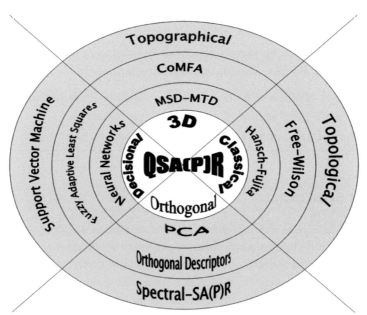

FIGURE 29.1 Generic representation of quantitative structure-activity relationships/-QSA(P)R-classical methods, 3D, decision-making methods and orthogonal multivariate analysis of chemical-biological interactions. Legend: MSD-MTD = minimal steric differences-minimal topological differences; CoMFA = Compared molecular field analysis; PCA = principal component analysis (Putz and Lacrama, 2007).

29.3   NANO-CHEMICAL IMPLICATIONS

The essential principle of relations structure-activity resides in the indepen-dence of the structural parameters considered in Table 29.1.

TABLE 29.1   Summary of Main S-SAR Descriptors

Activity	Variables predictor structural				
y_1	x_{11}	...	x_{1k}	...	x_{1M}
y_2	x_{21}	...	x_{2k}	...	x_{2M}
⋮	⋮	⋮	⋮	⋮	⋮
y_N	x_{N1}	...	x_{Nk}	...	x_{NM}

This feature can be used to quantify the SAR via an orthogonal space. To accomplish a SAR study according to the above-mentioned principle, the columns containing structural data Table 29.1 will be transformed into an orthogonal abstract space, where all necessary predictor variables are independent (Figure 29.2). To solve the S-SAR problem, results will be correlated with initial data using a transformation of coordinates. The analytical procedure is developed in three simple steps.

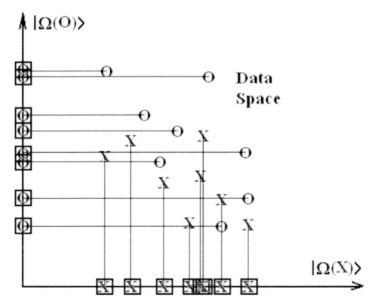

FIGURE 29.2   Generic representation of data in space containing the vectorial set $\{|X\rangle, |O\rangle\}$ in the $\{|\Omega(X)\rangle, |\Omega(O)\rangle\}$ orthogonal space (Putz and Lacrama, 2007).

Initially, Table 29.1 is reconsidered as Table 29.2, which, in order to be complete, the units column $|X_0\rangle = |1\ 1\ \dots\ 1\rangle$ has been added to consider the coefficients of the free term (b_0) in the system (1).

TABLE 29.2 Spectral (Vectorial) Version of SAR Descriptors in Table 29.1

Activity	Predictor structural variables					
$|Y\rangle$	$|X_0\rangle$	$|X_1\rangle$	$\dots$	$|X_k\rangle$	$\dots$	$|X_M\rangle$
y_1	1	x_{11}	$\dots$	x_{1k}	$\dots$	x_{1M}
y_2	1	x_{21}	$\dots$	x_{2k}	$\dots$	x_{2M}
$\vdots$	$\vdots$	$\vdots$	$\vdots$	$\vdots$	$\vdots$	$\vdots$
y_N	1	x_{N1}	$\dots$	x_{Nk}	$\dots$	x_{NM}

$$Y = \begin{pmatrix} y_1 \\ y_2 \\ \vdots \\ y_N \end{pmatrix}, E = \begin{pmatrix} e_1 \\ e_2 \\ \vdots \\ e_N \end{pmatrix}, B = \begin{pmatrix} b_0 \\ b_1 \\ b_2 \\ \vdots \\ b_M \end{pmatrix}, X = \begin{pmatrix} 1 & x_{11} & x_{12} & \cdots & x_{1M} \\ 1 & x_{21} & x_{22} & \cdots & x_{2M} \\ \vdots & \vdots & \vdots & \vdots & \vdots \\ 1 & x_{N1} & x_{N2} & \cdots & x_{NM} \end{pmatrix} \tag{1}$$

Moreover, given that the columns are now regarded as vectors in the data space, one searches the "spectral" decomposition of the $|Y\rangle$ activity vector using the $\{|X_0\rangle, |X_1\rangle, \dots, |X_k\rangle, \dots, |X_M\rangle\}$ considered structural vectors:

$$|Y\rangle = b_0|X_0\rangle + b_1|X_1\rangle + \dots + b_k|X_k\rangle + \dots + b_M|X_M\rangle + |e\rangle \tag{2}$$

Equation (2) is actually a correspondent of spectral decomposition of multi-linear equation, hence the name of this approach.

The next step is building a vector algorithm so that the $|e\rangle$ residual vector can be sent to zero in equation (2) to meet the conditions (3) of minimizing errors.

$$X^T (Y - XB) = 0 \tag{3}$$

To achieve minimal errors in equation (2), one will move to transformation of basic data $\{|X_0\rangle, |X_1\rangle, \dots, |X_k\rangle, \dots, |X_M\rangle\}$ into orthogonal data, like $\{|\Omega_0\rangle, |\Omega_1\rangle, \dots, |\Omega_k\rangle, \dots, |\Omega_M\rangle\}$.

With these details, the consecrated Gram-Schmidt procedure is approached, well known in quantum chemistry (when it is searched for an orthogonal basis for a set of orthogonal bases in spectral decomposition of the function of atomic and molecular wave) (Daudel et al., 1983). Before applying it effectively, the generalized scalar product must be introduced by the basic rule:

$$\langle \Psi_l | \Psi_k \rangle = \sum_{i=1}^{N} \psi_{il} \psi_{ik} = \langle \Psi_k | \Psi_l \rangle \tag{4}$$

obtaining a real number of two N-dimensional arbitrary vectors,

$$|\Psi_l\rangle = |\psi_{1l} \quad \psi_{2l} \quad ... \quad \psi_{Nl}\rangle \text{ and } |\Psi_k\rangle = |\psi_{1k} \quad \psi_{2k} \quad ... \quad \psi_{Nk}\rangle$$

Orthogonal conditions require that the (5) scalar product type (3.9) to be zero; the $\{|\Omega_0\rangle, |\Omega_1\rangle, ..., |\Omega_k\rangle, ..., |\Omega_M\rangle\}$ orthogonal base can be built starting from the $\{|X_0\rangle, |X_1\rangle, ..., |X_k\rangle, ..., |X_M\rangle\}$ set.

i)   One chooses

$$|\Omega_0\rangle = |X_0\rangle \tag{5}$$

ii) Then, choosing $|X_1\rangle$ as the next vector to be transformed, we can state that

$$|\Omega_1\rangle = |X_1\rangle - r_0^1 |\Omega_0\rangle, \, r_0^1 = \frac{\langle X_1 | \Omega_0 \rangle}{\langle \Omega_0 | \Omega_0 \rangle} \tag{6}$$

$\langle \Omega_0 | \Omega_1 \rangle = 0$, making sure that $|\Omega_0\rangle$ and $|\Omega_1\rangle$ are orthogonal.

iii) Next, steps (i) and (ii) are repeated, until the vectors $|\Omega_0\rangle$, $|\Omega_1\rangle$,..., $|\Omega_{k-1}\rangle$ are orthogonally built. We can, for instance, transform $|X_k\rangle$ then in:

$$|\Omega_k\rangle = |X_k\rangle - \sum_{i=0}^{k-1} r_i^k |\Omega_i\rangle, \, r_i^k = \frac{\langle X_k | \Omega_i \rangle}{\langle \Omega_i | \Omega_i \rangle} \tag{7}$$

so that the $|\Omega_k\rangle$ vector is orthogonally on the previous vectors.

Step (iii) is repeated and extended until the last orthogonal predictor vector $|\Omega_M\rangle$ is obtained.

Using the Graham-Schmidt algorithm, the starting predictor vectorial base $|X_0\rangle, |X_1\rangle, ..., |X_k\rangle |X_M\rangle$ with the orthogonal one $\{|\Omega_0\rangle, |\Omega_1\rangle, ..., |\Omega_k\rangle, ..., |\Omega_M\rangle\}$, by extracting suitably from the original vectors the unwanted orthogonal contribution. Note that the above procedure applies to any arbitrary order of original vectors to be orthogonalized. In the built orthogonal space, the $|Y\rangle$ activity vector reaches the real spectral decomposition form:

$$|Y\rangle = \omega_0 |\Omega_0\rangle + \omega_1 |\Omega_1\rangle + ... + \omega_k |\Omega_k\rangle + ... + \omega_M |\Omega_M\rangle \tag{8}$$

The residual vector in Eq. (2) disappeared in Eq. (8) as it has no structural significance in the abstract orthogonal base. Alternatively, it can be stated that in the abstract orthogonal space, the $|e\rangle$ residual vector can be identified with the vector with all $|0,0,...0\rangle$ components, which is always perpendicular on all other vectors of the orthogonal base. Thus, the Graham-Schmidt algorithm, via its specific recursive rule, absorbs or processes the minimizing conditions of the errors in Eq. (2) to simplify the identification with the origin of the orthogonal data space.

At this point there is no residual vector in Eq. (8). We can consider that the SAR problem is practically resolved when the new coefficients (ω_0, ω_1,..., ω_k,..., ω_M) in equation (8) are determined. These new coefficients can be immediately deduced considering the orthogonal features of the spectral decomposition (8) and considering that:

$$\langle \Omega_k | \Omega_l \rangle = 0 \ , \ k \neq l \tag{9}$$

is a condition given by type of vectors from the built orthogonal basis. So, each coefficient is obtained like the scalar product between their specific predictor vectors and the activity vector (10):

$$\omega_k = \frac{\langle \Omega_k | Y \rangle}{\langle \Omega_k | \Omega_k \rangle}, k = \overline{0, M} \tag{10}$$

With the given coefficients by Eq. (10), it is obtained the spectral expansion of the activity vector in an orthogonal base of the type (8). This does not mean that those coefficients have been found that directly bind the activity with predictor vectors, as Eq. (2) requires. This goal is achieved in the final stage of the present SAR algorithm. It consists in the reversal from the orthogonal database to the initial one using the coordinates transforming system:

$$\begin{cases} |Y\rangle = \omega_0 |\Omega_0\rangle + \omega_1 |\Omega_1\rangle +...+ \omega_k |\Omega_k\rangle +...+ \omega_M |\Omega_M\rangle \\ |X_0\rangle = 1 \cdot |\Omega_0\rangle + 0 \cdot |\Omega_1\rangle +...+ 0 \cdot |\Omega_k\rangle +...+ 0 \cdot |\Omega_M\rangle \\ |X_1\rangle = r_0^1 |\Omega_0\rangle + 1 \cdot |\Omega_1\rangle +...+ 0 \cdot |\Omega_k\rangle +...+ 0 \cdot |\Omega_M\rangle \\ \cdots\cdots\cdots\cdots\cdots\cdots\cdots\cdots\cdots\cdots\cdots\cdots\cdots\cdots\cdots\cdots \\ |X_k\rangle = r_0^k |\Omega_0\rangle + r_1^k |\Omega_1\rangle +...+ 1 \cdot |\Omega_k\rangle +...+ 0 \cdot |\Omega_M\rangle \\ \cdots\cdots\cdots\cdots\cdots\cdots\cdots\cdots\cdots\cdots\cdots\cdots\cdots\cdots\cdots\cdots \\ |X_M\rangle = r_0^M |\Omega_0\rangle + r_1^M |\Omega_1\rangle +...+ r_k^M |\Omega_k\rangle +...+ 1 \cdot |\Omega_M\rangle \end{cases} \tag{11}$$

The first equation in Eq. (11) reproduces the entire (2) spectral decomposition. The other equations in Eq. (11) are conventional rewritings of the (5)–(7) Graham-Schmidt algorithm.

Finally, the Eq. (11) system is algebraically correct only and just only the extended associated determinant disappears (Putz and Lacrama, 2007),

$$
\begin{vmatrix}
|Y\rangle & \omega_0 & \omega_1 & \cdots & \omega_k & \cdots & \omega_M \\
|X_0\rangle & 1 & 0 & \cdots & 0 & \cdots & 0 \\
|X_1\rangle & r_0^1 & 1 & \cdots & 0 & \cdots & 0 \\
\vdots & \vdots & \vdots & \vdots & & \vdots & \\
|X_k\rangle & r_0^k & r_1^k & \cdots & 1 & \cdots & 0 \\
\vdots & \vdots & \vdots & \vdots & & \vdots & \\
|X_M\rangle & r_0^M & r_1^M & \cdots & r_k^M & \cdots & 1
\end{vmatrix} = 0
\tag{12}
$$

This being the condition derived from the theorem which states that *a system that has a column as a linear combination of all other has an associated determinant that is zero.* (Danko et al., 1983)

We believe that the involvement of multilinear regression standard on structural selected descriptors and on their combinations is iterative (Soskic et al., 1997) as it unjustifiably complicates the minimization of residual errors in previous orthogonal approaches (Klein et al., 1997), resulting in a larger number of used intercorrelations. Moreover, a purely heuristic method used to find an orthogonal set of descriptors, even with a minimal correlation factor, loses the real sense of the orthogonalized set of descriptors, compared with the original in relation with the original problem that must be solved. On the contrary, using the orthogonal study in the proposed manner, the S-SAR method brings a new, innovative model to solve completely the SAR problem, related to the measured activities (or observed properties) with structural descriptors in an algebraic, direct, and transparent fashion than it does the "standard" multilinear regression method.

In the proposed S-SAR analysis, the order of parameters that are orthogonalized is eliminated, this problem existing in all previous methods that use orthogonal descriptors. (Klein et al., 1997) This stems from the fact that all structural descriptors are spectral developed simultaneously on orthogonal basis, according to Eq. (11), avoiding mutual iterative correlations (of orthogonal descriptors) in heuristic orthogonalization methods – when considered orthogonalization order becomes essential.

Extending the observation to the first column, the final solution of the basic SAR problem in Table 29.1 is given by the determinant (12), also with respect to the principle of included and independent error minimization. It is noteworthy that the present method is much simpler computationally than the classic version, although the proposed approach is not a new concept by considering spectral (orthogonal) data space expansion of input (both activities and descriptors) using the (11) system. The explanation resides in the fact that the discussed method has nothing in common with the calculation of matrix coefficients $B = \left(X^T X \right)^{-1} X^T Y$ on the basis of minimizing statistical errors, the latter being a completely implicit and non-transparent procedure.

The present method allows direct writing of the S-SAR solution (equation) through the development $(M + 2)$-dimensional determinant of type (12) – whose components are activity, structural vectors between Gram-Schmidt and spectral decomposition coefficients, r_k^l and ω_k, respectively.

Both standard classical QSAR studies and spectral ones, although considering different mathematical principles, have similar results, since it applies the following theorem [125]: if a X matrix of $Y = XB + E$ form, with the size $N \times (M + 1)$, $N > M + 1$, has linear independent columns, in other words – orthogonal as in spectral approach, there is a unique matrix Q of size $N \times (M + 1)$ with orthogonal columns and R triangular matrix $(M + 1) \times (M + 1)$ dimensional with the elements of the main diagonal equal with 1, as it can be identified in the first small determinant in Eq. (12), so that the X matrix that can be factored as follows:

$$X = QR. \tag{13}$$

When combining equation (13) with the optimal equation (3), applying the rules of algebraic operations with matrices, there is obtained the vector B of the form estimates with following form:

$$B = \left(Q^T Q \right)^{-1} Q^T Y \tag{14}$$

In a matrix arrangement close to normal, based on the $B = \left(X^T X \right)^{-1} X^T Y$ equation. When comparing matrices $X^T X$ and $Q^T Q$ in previous two equations, it is clear that the last case provides clearly a diagonal form certainly easy to handle (to find its reverse) when searching for B coefficients of SAR vectors.

Based on these considerations, the presented S-SAR technique is a valid approach to solve problems in chemistry, QSAR, pharmacology, and related molecular fields.

S-SAR method provides a real alternative to the standard QSAR at every level of modeling, independently of the number of descriptors, compounds or of the order of orthogonalization of considered parameters. This feature highlights the fundamental, analytical, and computational advantages of the method S-SAR, with fewer steps and total analyticity of the provided structure-activity equation, by a simple explicit determinant.

In principle, the implementation procedure is circumscribed to the regression model that prescribes the expansion of activity, with the generic form:

$$A = B_0 + B_1 \left(\begin{array}{c} electronic \\ parameter \end{array} \right) + B_2 \left(\begin{array}{c} hydrophobic \\ parameter \end{array} \right) + B_3 \left(\begin{array}{c} steric \\ parameter \end{array} \right) + ... \quad (15)$$

This series is explicit when correlation coefficients are determined by analytical correlations of molecular parameters involved. When the series (15) is truncated after the first four coefficients, classic Hansch QSAR is obtained (Chiriac et al., 2002). In addition to the already known traditional regression method for determining the coefficients of correlation of expression (15) (Miller et al., 2005), in the following the Spectral-SAR method is applied (Putz, 2006) at different trophic levels, from the enzymatic to the mono- and multi-cellular one.

Consequently, we can also use the spectral norm of measured or predicted activity (Putz, 2009; Putz et al., 2009),

$$\left\| Y \right\rangle^{MEASURED/}_{PREDICTED} \right\| = \sqrt{\sum_{i=1}^{N} (y_i^2)^{MEASURED/}_{PREDICTED}} \quad (16)$$

As a general tool with which many models can be compared, no matter the size and degree of multilinearity, given that all are reduced to a single number.

In this way, it is achieved the desideratum of QSAR approach of chemical and biological interactions for the subsequent supply of a conceptual basis to compare different models and endpoints. Moreover, beyond accurately reproducement of QSAR's statistics standard, the current S-SAR allows an alternate calculation way of the correlation factor using the concept of spectral norm. Thus, the so-called algebraic correlation factor is defined as the ratio of the spectral norm of the predicted activity versus the measured one (Putz, 2009; Putz et al., 2009):

$$r^{ALGEBRAIC}_{S-SAR} = \frac{\left\| Y \right\rangle^{PREDICTED} \right\|}{\left\| Y \right\rangle^{MEASURED} \right\|} \quad (17)$$

Essentially, the above-described S-SAR procedure is circumscribed to the consecrated Hansch quantitative structure-activity relationship, which prescribes the generic form of activity expansion (15).

Besides the already traditional regression methods for determining corre-lation coefficients of the expression (15) (Miller et al., 2005), the spectral orthogonal S-SAR method has many practical advantages, being able to provide the key to treating the so-called spectral analysis of the activity itself by introducing the spectral norm and its principle of minimum way, thus providing the image of the molecular mechanism of the considered system (Putz, 2009; Putz et al., 2009).

KEYWORDS

- **minimum spectral ways**
- **spectral norm**
- **S-SAR**

REFERENCES AND FURTHER READING

Chiriac, A., Mracec, M., Oprea, T. I., Kurunczi, L., & Simon, Z., (2002). *Quantum Biochemistry and Specific Interactions*. The QSAR and quantum chemistry group of Timişoara, Romania, Ed. Mirton, Timisoara.

Danko, P. E., Popov, A. G., & Kozhevnikova, T. Y. A., (1983). *Higher Mathematics in Problems and Exercises* (Vol. I.), Mir Publishers: Moscow.

Daudel, R., Leroy, G., Peeters, D., & Sana, M., (1983). *Quantum Chemistry* (Vol. 5). John Wiley & Sons, New York.

Ivanciuc, O., Taraviras, S. L., & Cabrol-Bass, D., (2000). Quasi-orthogonal basis sets of molecular graph descriptors as chemical diversity measure. *J. Chem. Inf. Comput. Sci., 40*, 126–134.

Klein, D. J., Randić, M., Babić, D., Lučić, B., Nikolić, S., & Trinajstić, N., (1997). Hierarchical orthogonalization of descriptors. *Int. J. Quant. Chem., 63*, 215–222.

Miller, J. N., & Miller, J. C., (2005). *Statistics and Chemometrics for Analytical Chemistry* (5th edn.). Pearson Education Limited.

Putz, M. V., (2006). A spectral approach of the molecular structure-biological activity relationship Part I. The general algorithm, Annals of West University of Timişoara. *Series of Chemistry, 15*(3), 159–166.

Putz, M. V., & Lacrămă, A. M., (2007). Introducing spectral structure-activity relationship (S-SAR) analysis. Application to ecotoxicology. *Int. J. Mol. Sci., 8*(5), 363–391.

Putz, M. V., Duda-Seiman, C., Duda-Seiman, D. M., & Putz, A. M., (2011). Turning SPECTRAL-SAR into 3D-QSAR analysis. Application on H^+K^+-ATPase inhibitory activity, Chapter 33. In: Putz, M. V., (ed.), *Advances in Chemical Modeling* (pp. 435–451). Nova Science Publishers, Inc.

Putz, M. V., Duda-Seiman, C., Duda-Seiman, D., & Putz, A. M., (2008). Turning SPECTRAL-SAR into 3D-QSAR analysis. Application on H+K+-ATPase inhibitory activity. *International Journal of Chemical Modeling, 1*, 45–61.

Putz, M. V., Putz, A. M., Lazea, M., & Chiriac, A., (2009). Spectral vs. statistic approach of structure-activity relationship. Application on ecotoxicity of aliphatic amines. *Journal of Theoretical and Computational Chemistry, 8*(6), 1235–1251.

Simon, Z., Chiriac, A., Holban, S., Ciubotariu, D., & Mihalas, I., (1984). *Minimum Steric Difference*. The MTD method for QSAR studies. Research Studies Press, Letchworth (UK); distributed by John Wiley & Sons, Inc., 58B Station Road, Letchworth, Hertsfordshire, United Kingdom SG6 3BE. 173 + ix pp..

Soskić, M., Plavsić, D., & Trinajstić, N., (1997). Inhibition of the Hill reaction by 2-methylthio-4,6-bis (monoalkylamino)-1,3,5-triazines A QSAR study. *J. Mol. Struct. (Theochem), 394*, 57–65.

CHAPTER 30

Stem Cells

ANA-MATEA MIKECIN[1] and GRDISA MIRA[2]

[1]*Rudjer Boskovic Institute, Division of Molecular Medicine, Bijenicka 54, 10000 Zagreb, Croatia, E-mail: Ana-Matea.Mikecin@irb.hr*

[2]*Rudjer Boskovic Institute, 10000 Zagreb, Croatia, E-mail: grdisa@irb.hr*

30.1 DEFINITION

Stem cells (SC) are unique cells present in multicellular organisms that, unlike other cells, have the potential to differentiate into specialized cells as well as to divide via mitosis to produce more SC. Embryonic stem cells (ESC) are present in an embryo which forms from a fertilized egg. ESCs differentiate in three germ layers–ectoderm, endoderm, and mesoderm–which give rise to all the tissues and organs present in an organism. Adult SC (somatic SC) are undifferentiated cells found in differentiated tissue or organ. The maintenance and repair of many adult tissues are ensured by adult SC, which resides at the top of the cellular hierarchy of these tissues. While ESCs are defined by their origin, the origin of adult SC is still under debate.

A progenitor cell is a cell that has a potential to differentiate into a specific type of cell, however, compared to a SC it is more differentiated and does not have a potential to become any cell of the living organism. Unlike SC, progenitor cells cannot replicate indefinitely, but only a limited number of times.

Adult (differentiated) cells can be genetically reprogrammed to acquire SC-like properties by forced expression of genes and factors important for maintaining the defining properties of embryonic SC. Those cells are called *induced pluripotent stem cells* (iPSCs) and meet the defining criteria for pluripotent SC. However, it is still not known if sESCs and iPSCs differ in a clinically significant way (Yamanaka, 2012).

Cancer stem cells (CSC) are cells found in certain types of cancer that possess characteristics of normal SCs. Through the process of self-renewal and differentiation CSCs can give rise to all the cell types present in a certain cancer sample and are therefore considered tumorigenic. CSCs are relatively quiescent and hard to target via commonly used anticancer therapies. They continue to reside inside the body after the therapeutic treatment and pose a source of cancer cells that causes relapse as well as metastasis (Beck and Blanpain, 2013). So far, CSC has been confirmed to be present in leukemia, melanoma, multiple myeloma as well as brain, breast, colon, ovary, pancreas, and prostate cancer (Kreso and Dick, 2014).

30.2 HISTORICAL ORIGIN(S)

The term "SC" was coined in the late 19th century by German biologist Ernst Haeckel. He used the phrase to describe a fertilized egg that becomes an organism. He used the same expression for a single-celled organism that was an ancestor cell to all living things in history.

In 1958 Martin Evans of Cardiff University, UK, then at the University of Cambridge, was first to identify ESC in mice and cultivate them *in vitro*. He also used ESC to create specific gene modifications in mice. In 2007, he shared the Nobel Prize in Physiology or Medicine for work in the development of the knockout mouse and contribution to the efforts to develop new treatments for illnesses in humans.

SC transplantation was pioneered from the 1950s through the 1970s by a team at the Fred Hutchinson Cancer Research Center led by Donnall E. Thomas. They used bone-marrow-derived SCs to repopulate the bone marrow and produce new blood cells after irradiation therapy in leukemia patients. This work was later recognized with a Nobel Prize in Physiology or Medicine.

The next great breakthrough in SC field happened in 1998 when James Thomson of the University of Wisconsin in Madison and John Gearhart of Johns Hopkins University in Baltimore isolated for the first time human ESC and cultured them in the lab.

The new millennium brought an explosion of SC-related research, and the advancements in it continue to grow exponentially bringing promise to patients with neurodegeneration, brain, and spinal cord injury, heart disease, diseases of hematopoietic cells including leukemia, diabetes, deafness as well as blindness and vision impairment. Also, SCs can be used to treat infertility, orthopedic patients as well as those with HIV/AIDS.

In 2012 Nobel Prize was awarded to John B. Gurdon and Shinya Yamanaka for their discoveries that the specialization of cells is reversible and how intact mature cells in mice could be reprogrammed to become immature SCs, respectively. The interesting fact is that the research on reversible properties

of cell specialization was carried out in 1962 while the research on reprogramming of mature cells to become immature ones in the early 2000s.

30.3   NANO-SCIENTIFIC DEVELOPMENT(S)

SC self-renewal and differentiation are influenced by various signals which
are tightly regulated by the stem-cell niche. Stem-cell niche is a microenvi-
ronment where the SC is found, and that is responsible for maintaining SCs
in pluripotent form with the potential to self-renew. In general, SC fate is
regulated by SC's niche (Morrison and Spradling, 2008). Maintenance of
the niche is achieved through environmental cues as well as certain signaling
pathways that include Wnt (Nusse, 2008), JAK-STAT (Beebe et al., 2010),
Notch (Androutsellis-Theotokis et al., 2006) and Hedgehog (Conia et al.,
2013) signaling pathways. These signaling pathways are closely linked to
developmental processes and regulation of SC self-renewal. They are also
frequently found deregulated in cancer.

Signaling molecules, such as growth factors and Wnt proteins influence
the process of differentiation and are communicated to SCs through niche.
An equally important role plays internal signals and is governed by SC's
genes. These signaling pathways are integral to the generation of specific
differentiated cells; however, the signals that determine the cell type and
destination of a certain SC are not entirely elucidated.

On the other hand, there are signals that can result in the reprogramming
of a differentiated cell which can then be transformed into an iPSC which
is a stem-like cell. iPSCs can be generated through forced expression of
transcription factors such as OCT4, SOX2, and Nanog (Boiani and Schöler,
2005) which regulate the expression of selected induction genes and are
used to create pluripotent cells.

30.4   NANO-CHEMICAL APPLICATION(S)

SCs are used in deciphering the complex events that occur during the development of a multicellular organism where the primary goal is to reveal how
an undifferentiated SC becomes a differentiated one that forms tissues and
organs. Cancer and birth defects are considered to arise, at least in part, due
to impaired molecular pathways present in SCs. It is therefore believed that
a more thorough understanding of these processes may yield information
about how such diseases arise and might suggest new strategies for therapy.

SCs are also used in stem-cell related therapy among which bone marrow transplant is the most widely used one. Moreover, SCs hold great potential for the further development of regenerative medicine.

30.5 MULTI-/TRANS-DISCIPLINARY CONNECTION(S)

Development in molecular biology and medicine as well as technology in generally enabled advancements that were in SC field in the past 50 years. The same advancements, on the other hand, provide a potential for development of regenerative medicine as well as therapy for many different diseases which include immunological disorders, cancer, and HIV/AIDS using knowledge obtained on SCs.

30.6 OPEN ISSUES

The controversy around SC research is centered on the moral implications of destroying human embryos used to obtain embryonic SCs. The research opened many questions that deal with morals and religion besides research. Some of them are: Does life begin at fertilization, in the womb, or at birth? Is a human embryo equivalent to a human child? Does a human embryo have any rights? Might the destruction of a single embryo be justified if it provides a cure for a countless number of patients? Since ES cells can grow indefinitely in a dish and can, in theory, still grow into a human being, is the embryo really destroyed?

However, with the advancements in the related research area, alternatives to human ESC became readily available, and the debate over SC research is becoming increasingly irrelevant. However, ethical questions regarding hES cells may not entirely go away because some human embryos will still be needed for certain types of research.

KEYWORDS

- adult stem cells
- cancer stem cell
- embryonic stem cells
- progenitor cell
- stem cells

REFERENCES AND FURTHER READING

Androutsellis-Theotokis, A., Leker, R. R., Soldner, F., Hoeppner, D. J., Ravin, R., Poser, S. W., Rueger, M. A., Bae, S. K., Kittappa, R., & McKay, R. D. G., (2006). Notch signaling regulates stem cell numbers in vitro and in vivo. *Nature, 442*, 823–826.

Beck, B., & Blanpain, C., (2013). Unraveling cancer stem cell potential. *Nature Reviews Cancer, 13*, 727–738.

Beebe, K., Lee, W. C., & Micchelli, C. A., (2010). JAK/STAT signaling coordinates stem cell proliferation and multilineage differentiation in the Drosophila intestinal stem cell lineage. *Developmental Biology, 338*, 28–37.

Boiani, M., & Schöler, H. R., (2005). Regulatory networks in embryo-derived pluripotent stem cells. *Nature Reviews Molecular Biology, 6*, 872–881.

Conia, S., Infanteb, P., & Gulinob, A., (2013). Control of stem cells and cancer stem cells by Hedgehog signaling: Pharmacologic clues from pathway dissection. *Biochemical Pharmacology, 85*, 623–628.

Kreso, A., & Dick, J. E., (2014). Evolution of the cancer stem cell model. *Cell Stem Cell, 14*, 275–291.

Morrison, S. J., & Spradling, A. C., (2008). Stem cells and niches: Mechanisms that promote stem cell maintenance throughout life. *Cell, 132*, 598–611.

Nusse, R., (2008). WNT signaling and stem cell control. *Cell Research, 18*, 523–527.

Yamanaka, S., (2012). Induced pluripotent stem cells: Past, present, and future. *Cell Stem Cell, 10*, 678–684.

CHAPTER 31

Steric Taft Parameters

BOGDAN BUMBĂCILĂ[1] and MIHAI V. PUTZ[1,2]

[1]*Laboratory of Computational and Structural Physical Chemistry for Nanosciences and QSAR, Biology-Chemistry Department, Faculty of Chemistry, Biology, Geography at West University of Timișoara, Pestalozzi Street No. 16, Timișoara, RO–300115, Romania*

[2]*Laboratory of Renewable Energies-Photovoltaics, R&D National Institute for Electrochemistry and Condensed Matter, Dr. A. Paunescu Podeanu Str. No. 144, RO–300569 Timișoara, Romania, Tel.: +40-256-592-638, Fax: +40-256-592-620, E-mail: mv_putz@yahoo.com, mihai.putz@e-uvt.ro*

31.1 DEFINITION

Taft parameters are constants that describe the steric effect of a substituent over the molecule's reactivity. They describe the way a particular substituent in a molecule will influence the molecule's reactions through polar (inductive, field, and resonance) and proper steric effects (Anslyn et al., 2006).

31.2 HISTORICAL ORIGINS

Taft parameters were introduced by Robert W. Taft in 1952, within Taft equation (Taft, 1952). In fact, Taft equation is a modification of a previous equation, introduced by Louis Plack Hammett in 1937 which is describing a linear free-energy relationship between reaction rates and equilibrium constants of those reactions (IUPAC, 2006).

31.3 NANO-CHEMICAL IMPLICATIONS

The initial Hammett equation in organic chemistry describes a linear free-energy relationship (that relates the logarithm of a reaction rate constant or equilibrium constant for one series of reactions with the logarithm of the rate or equilibrium constant for a relates series of reactions) between the reaction rates and equilibrium constants for many reactions that are involving benzoic acid analogs, with meta- and para-substituents to each other, with just two parameters: a substituent constant and a reaction constant (Abraham et al., 2002; Hammett, 1937).

Hammett's idea was that for any two reactions with two benzoic acid derivatives only differing in the substituent's type, the change in the free energy of activation is proportional to the change in Gibbs free energy.

The basic equation is:

$$\frac{\log K}{K_0} = \sigma\rho \tag{1}$$

where K is the equilibrium constant for a given equilibrium reaction with a substituent $-R$ in the molecule and K_0 is the equilibrium constant when $-R$ is a Hydrogen atom. σ is the substituent $-R$ constant which is depending only on the specific substituent and ρ is the reaction constant which is depending only on the type of reaction and not on the substituent.

The equation also can be expressed as a logarithm of ratio of the reaction rates:

$$\frac{\log k}{k_0} = \sigma\rho \tag{2}$$

where k_0 is the reference reaction rate of the unsubstituted reactant and k the reaction rate of the $-R$ substituted reactant (Keenan et al., 2008).

The Taft equation is also a linear free energy relationship (LFER) that relates the logarithm of a substituted reaction rate constant with field, inductive, resonance, and steric effects of a substituent over various reactions of the molecule. It is used in organic chemistry for studying reaction mechanisms and in QSAR studies of organic compounds. Taft equation is written as follows:

$$\log\left(\frac{k_s}{k_{CH_3}}\right) = \sigma^*\rho^* + \delta E_s \tag{3}$$

where

- $\log\left(k_s / k_{CH_3}\right)$ is the ratio of the rate of the reaction of the substituted molecule compared to the reference reaction;
- σ^* is the polar substituent constant that reflects the polar (field and inductive) effects of the substituent s;

- E_s is the steric constant;
- ρ^* is the sensitivity factor of the substituent for the studied reaction to polar effects;
- δ is the sensitivity factor of the substituent for the studied reaction to steric effects.

Polar substituent constant σ^* describes the manner in which a particular substituent is influencing a chemical reaction through polar (inductive, field, and resonance) effects. σ^* was determined by studies of the hydrolysis of methyl esters (R-COO-Me) both in acid and base catalysis. Sir Christopher Kelk Ingold suggested in 1930 he use of ester hydrolysis rates to study polar effects (Ingold, 1930) (Figure 31.1).

FIGURE 31.1   Hydrolysis reaction of the methyl esters.

To measure σ^* and E_s the assumption that polar effects would only influence the base-catalyzed reaction was made because of the difference in charge buildup in the rate-determining steps. The basic pathway implies a neutral reactant which is transforming into a negatively charged intermediate. So polar effects would only influence the reaction rate of the base catalyzed reaction when a new electric charge is formed (partial negative). Either of the acid or the base hydrolysis of esters is leading to a tetrahedral intermediate. But this assumption was made because in the base-catalyzed mechanism the reactant goes from a neutral species to negatively charged intermediate in the rate determining (slow) step, while in the acid catalyzed mechanism a positively charged reactant goes to a positively charged inter-mediate (Anslyn et al., 2006) (Figure 31.2).

Because the tetrahedral intermediates have very similar structures (they differ simply by two protons), Taft further proposed that under identical conditions, steric effects are influencing both acid and base pathways equally and that is why the ratio of the transformation rates would not be influenced. σ^* was defined as:

$$\sigma^* = \left(\frac{1}{2.48\rho^*} \right) \left[\log\left(\frac{k_s}{k_{CH_3}} \right)_B - \log\left(\frac{k_s}{k_{CH_3}} \right)_A \right] \tag{4}$$

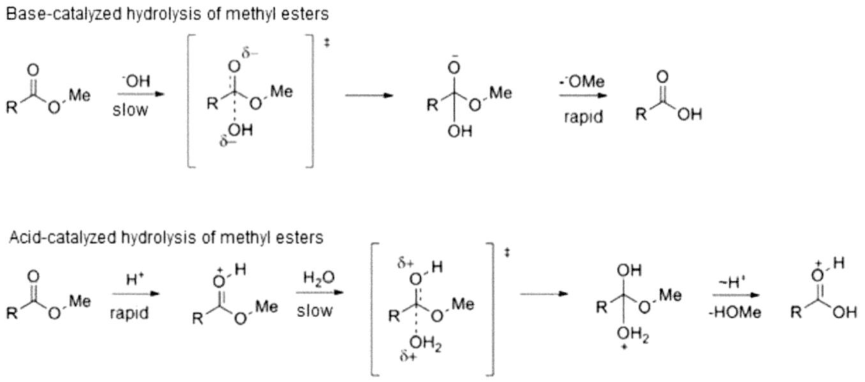

Base-catalyzed hydrolysis of methyl esters

Acid-catalyzed hydrolysis of methyl esters

FIGURE 31.2 Acidic and basic hydrolysis of methyl esters.

where

- $\log\left(k_s / k_{CH_3}\right)_B$ is the ratio of the rate of the base catalyzed transformation compared to the reference reaction (hydrolysis of methyl acetate);

- $\log\left(k_s / k_{CH_3}\right)_A$ is the ratio between the rate of the acid-catalyzed reaction compared to the reference reaction (hydrolysis of methyl acetate);

- ρ^* is a reaction constant that describes the sensitivity of the reaction series (the hydrolysis of the methyl esters)

So, for the hydrolysis of methyl acetate (reference reaction–R = CH$_3$) ρ^* is 1 and σ^* is 0. The factor of 1/2.48 is included just to make σ^* similar in magnitude to the σ values from the previous Hammett equation, where σ measures either the electron-withdrawing or electron-donating capability of a substituent (the functional electron-moiety) (Kar, 2007).

Figure 31.2 is, in fact, representing a mean of the values of Hammett reaction constants ρ for base catalyzed hydrolysis of meta- and para- substituted ethyl and methyl benzoates in different solvents and at different temperatures (Isaacs, 1987).

If σ^* was determined from the base-catalyzed reaction, when an intermediate with a different electric charge from the reactant is formed, E_s is calculated from the acid-catalyzed reaction, because this one doesn't include polar effects (the reactant and the intermediate have same sign electric charges). So E_s is defined as:

$$E_s = \frac{1}{\delta} \log\left(\frac{k_s}{k_{CH_3}}\right) \tag{5}$$

where

- k_s is the rate of the studied reaction;
- k_{CH_3} is the rate of the reference reaction (methyl acetate hydrolysis);
- δ is a reaction constant that describes the susceptibility to steric effects.

So, for the hydrolysis of methyl acetate (reference reaction–R = CH_3), δ is 1 and E_s^* is 0.

In fact, E_s is a measure of the bulkiness of the group it represents and of its effects on the closeness of contact between the drug and receptor site when talking about drug-receptor interactions (Table 31.1).

TABLE 31.1 Constants Used in the Taft Equation (Taft Parameters for Different Simple Substituents) (Taft, 1956)

Substituent	E_s	σ^*
–H	1.24	0.49
–CH_3	0	0
–CH_2CH_3	–0.07	–0.10
–$CH(CH_3)_2$	–0.47	–0.19
–$C(CH_3)_3$	–1.54	–0.30
–$CH_2(C_6H_5)$	–0.38	0.22
–(C_6H_5)	–2.55	0.60

From the values in the table above we can see that the hydrolysis rate is higher for hydrogen than methyl and it is slower as the substituent increases in size. Still, the steric influence of a group is strongly dependant on the context (the whole molecule, the chemical reaction) – as it can be seen from the phenyl and benzyl values of E_s and σ^*.

So, like in Figure 31.3, the size of –R affects the rate of reaction by blocking nucleophilic attack by water. k, the rate constant for ester hydrolysis decreases as the size of –R increases. The bulkier the substituent is, the more negative the constant E_s values are (Todeschini et al., 2000) (Figure 31.3).

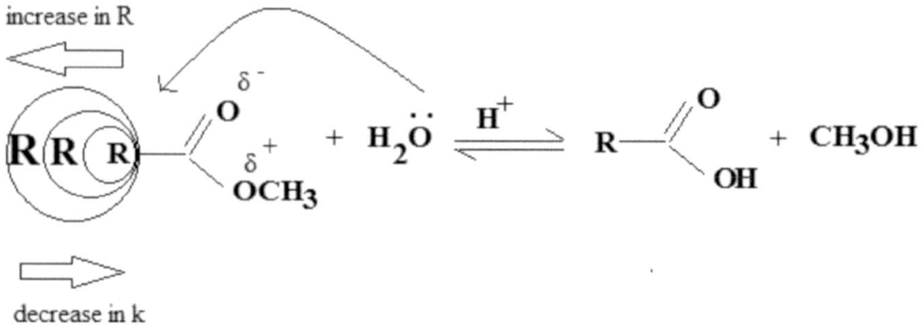

FIGURE 31.3   Dependence of the rate of esteric hydrolysis on the size of substituent –R.

31.4 SENSITIVITY FACTORS

31.4.1 *POLAR SENSITIVITY FACTOR, ρ^**

The polar sensitivity factor ρ^* for Taft plots will describe the susceptibility of a reaction series to polar effects. When the steric effects of substituents do not significantly influence the reaction rate the Taft equation simplifies to a form of the Hammett equation:

$$\log\left(\frac{k_s}{k_{CH_3}}\right) = \sigma^* \rho^* \qquad (6)$$

The polar sensitivity factor ρ^* can be obtained by plotting the ratio of the measured reaction rates (k_s) compared to the reference reaction (k_{CH_3}) versus the (σ^*) values for the substituents. This plot will give a straight line with a slope equal to ρ^*.

If $\rho^* > 1$, a negative charge in the transition state is formed and the process is accelerated by electron withdrawing groups. If $1 > \rho^* > 0$, negative charge is formed and the reaction is mildly sensitive to polar effects. If $\rho^* = 1$, the reaction is not influenced by polar effects. If $0 > \rho^* > -1$, a positive charge is formed and the reaction is mildly sensitive to polar effects. If $-1 > \rho^*$, the reaction is accumulating positive charge and it is accelerated by electron donating groups (Pavellich et al., 1957).

31.4.2 *STERIC SENSITIVITY FACTOR, δ*

The steric sensitivity factor δ for a new reaction series will describe the level in which the reaction rate is influenced by steric effects of the substituent.

When a reaction series is not significantly influenced by polar effects of the substituent, the Taft equation can be expressed as:

$$\log\left(\frac{k_s}{k_{CH_3}}\right) = \delta E_s \qquad (7)$$

A plot of the ratio of the rates versus the E_s value for the substituent will give a straight line with a slope equal to δ. The magnitude of δ will show to what proportion a reaction is influenced by steric effects of a substituent: a very steep slope will correspond to high steric sensitivity, while a shallow slope will correspond to little or to no sensitivity at all.

Since E_s values are large and negative for large substituents, it can be said that if δ is positive, increasing steric bulk (the larger the substituent is) decreases the reaction rate and steric effects are greater in the transition state – the intermediate phase. If δ is negative, increasing steric bulk increases the reaction rate, and steric effects are alleviated in the transition state. (Pawellich et al., 1957)

31.5 TRANS-DISCIPLINARY CONNECTIONS

31.5.1 TAFT PLOTS

Taft equations are used in pharmaceutical industry/chemistry for QSAR studies of molecules with biological activity and for drug design. Recently, Taft plots were used for studying the aminolysis of β-lactam drugs after binding them to a polymer in vitro which resembled to the binding of the penicillins to serum albumin in vivo. In the binding process, a covalent bond is formed between the molecule and the albumin and lysine residues are therefore exposed, residues which are thought to be a cause for penicillin allergies. In this study, there were found no correlations between the polar and steric effects of the substituent –R on the β-lactam ring and the rate of the hydrolysis (Arcelli et al., 2008) (Figure 31.4).

Cobaloximes are some compounds that have a similar structure to cobalamins (Vitamins B$_{12}$). Because they have a smaller molecule but a similar 3D orientation they are used for in vitro studies of biological activity as co-enzymes. It is well known that for cobalamins, the substituent –R is very influent in their in vivo co-enzyme activity, because the Co-R bond varies with its nature and also the spatial conformation of the entire molecule can be changed–the substituent can extract Co from its plane formed with the four nitrogen atoms in different distances. Of course, electron donating and withdrawing moiety of the substituent are very important for the distance of the R-Co bond, and they can be correlated to σ^*, but also the rate of the extraction of the Co atom from the plane can be correlated with steric effects of the –R group and also with Taft's parameter E_s (Randaccio et al., 1994) (Figure 31.5).

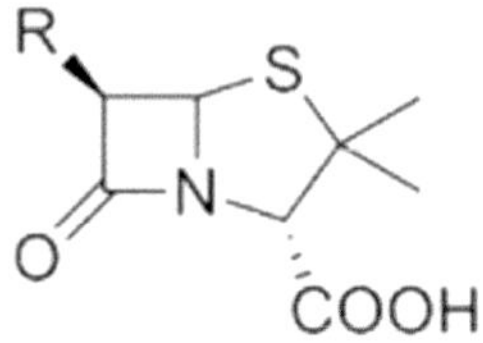

FIGURE 31.4   β-Lactam ring.

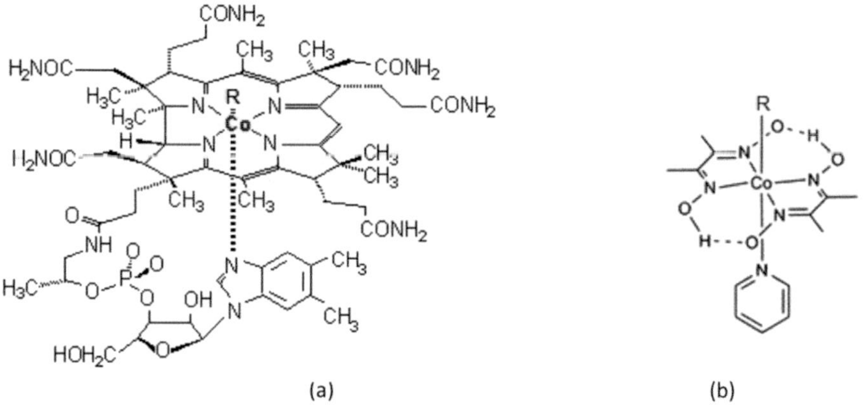

(a)   (b)

FIGURE 31.5   (a) Cobalamin ring, (b) Cobaloxime ring.

31.6   OPEN ISSUES

E_s succeeded well in reproducing the steric effect of substituent but not perfectly. So that is the reason that a variety of steric substituent constants have been proposed by many investigators in order to better describe the steric effect. Some of them are, in fact, modifications of E_s. The most used parameter is E_s' and E_s^c (Hirota et al., 2005).

E_s' is similar to E_s but it is defined on the basis of more unified reactions and over a wider range of substituents. It is defined using the acid-catalyzed esterification of carboxylic acids, at 40°C, as a reference reaction. It is also known as *Dubois steric constant.* (Todeschini et al., 2000; Hirota et al., 2005)

The corrected steric constant E_s^c contains an additional term in order to correct the hyperconjugation effect of α–hydrogen atoms (n_H–number of α–H atoms) (Hirota et al., 2005)

$$E_s^c = E_s - 0.306(3 - n_H) \qquad (8)$$

E_s^c is also known as the *Hancock steric constant*.

A rescaled set of the Taft steric constant was defined for the series of analog esters of the substituted formic acids: XCOOR, as:

$$E_s' = E_s - 1.24 \qquad (9)$$

In Eq. (9), 1.24 corresponds to the E_s value of the formic acid or ester (Todeschini et al., 2000).

In order to account for both C-C and C-H hyperconjugation effects a different correction for Taft steric constant was proposed, obtaining the *Palm steric constant* (Todeschini et al., 2000):

$$E_s^0 = E_s - 0.33(H_\alpha - 3) + 0.13C_\alpha \qquad (10)$$

with H_α – the number of α Hydrogen atoms, N_C – the number of α Carbon atoms in the substituent (Todeschini et al., 2000).

The *Taft-Kutter-Hansch steric constants* (TKH E_s) represent a combined set of original Taft steric constants and those extended by Kutter and Hansch, by using a correlation with the average of the minimum and maximum van der Waals radii as defined by the equation (Todeschini et al., 2000):

$$\hat{E}_s = 3.484 - 1.839\overline{R}_{vdW} \qquad (11)$$

The Fujita steric constant E_s^F was defined to evaluate the global steric effect for branched alkyl substituents of the type $CR_1R_2R_3$, by the following correlation equation (Todeschini et al., 2000):

$$E_s^F = -2.104 + 3.429E_s^C(R_1) + 1.978E_s^C(R_2) + 0.649E_s^C(R_2) \qquad (12)$$

KEYWORDS

- **polar substituent constant**
- **steric substituent constant**
- **Taft equation**
- **Taft plot**

REFERENCES AND FURTHER READING

Abraham, M. H., Ibrahim, A., Zissimos, A. M., Zhao, Y. H., Comer, J., & Reynolds, D. P., (2002). Application of hydrogen bonding calculations in property based drug design. *Drug Discovery Today,* 1056–1063.

Anslyn, E. V. D., & Dennis, A., (2006). *Modern Physical Organic Chemistry.* University Science Books: Sausalito, CA.

Arcelli, A., Porzi, G., Rinaldi, S., & Sandri, M., (2008). *J. Phys. Org. Chem., 21,* 163.

Hammettt, L. P., (1937). The effect of structure upon the reactions of organic compounds. Benzene derivatives. *J. Am. Chem. Soc., 59*(1), p. 96.

Hirota, M., Sakakibara, K., Yuzuri, T., & Kuroda, S., (2005). Re-examination of steric substituent constants by molecular mechanics. *Int. J. Mol. Sci., 6,* 18–29.

Ingold, C. K., (1930). The Mechanismand constitutional factors controlling the hydrolysisof carboxylic esters. I. The constitutional significance of hydrolytic stability maxima. *J. Chem. Soc., 1032,* 1930.

Issacs, N. S., (1987). *Physical Organic Chemistry* (pp. 154, 297). ELBS/Longman.

IUPAC, (1997). *Online Corrected Version: (2006) Compendium of Chemical Terminology* (2nd edn.). (the "Gold Book"): Hammett equation (Hammett relation).

Kar, A., (2007). *Medicinal Chemistry.* New Age International (P) Limited, Publishers, New Delhi.

Keenan, S. L., Peterson, K. P., Peterson, K., & Jacobson, K., (2008). Determination of Hammett equation ρ constant for the hydrolysis of p-nitrophenyl benzoate esters. *J. Chem. Educ., 85*(4), p. 558.

Pavelich, W. A., & Taft, R. W., (1957). *J. Am. Chem. Soc., 79,* 4935.

Randaccio, L., Geremia, S., Zangrando, E., & Ebert, C., (1994). *Inorganic Chemistry, 33*(21), 4641–4650.

Taft, R. W., (1952). Linear Free Energy Relationships from Rates of Esterification and Hydrolysis of Aliphatic and Ortho-substituted Benzoate Esters. *J. Am. Chem. Soc., 74*(11), 2729–2732; idem. Polar and Steric Substituent Constants for Aliphatic and o-Benzoate Groups from Rates of Esterification and Hydrolysis of Esters *74*(12) 3120–3128.

Taft, R. W., (1953). Linear Steric Energy Relationships. *J. Am. Chem. Soc., 75*(18), 4538–4539.

Taft, R. W., (1956). In: Newman, M. S., (ed.), *Steric Effects in Organic Chemistry.* John Wiley & Sons, New York.

Todeschini, R., & Consonni, V., (2000). Handbook of Molecular Descriptors, Wiley-VCH Verlag, Weinheim.

CHAPTER 32

Sterimol Verloop Parameters

BOGDAN BUMBĂCILĂ[1] and MIHAI V. PUTZ[1,2]

[1]Laboratory of Computational and Structural Physical Chemistry for Nanosciences and QSAR, Biology-Chemistry Department, Faculty of Chemistry, Biology, Geography at West University of Timișoara, Pestalozzi Street No.16, Timișoara, RO–300115, Romania

[2]Laboratory of Renewable Energies-Photovoltaics, R&D National Institute for Electrochemistry and Condensed Matter, Dr. A. Paunescu Podeanu Str. No. 144, RO–300569 Timișoara, Romania, Tel.: +40-256-592-638; Fax: +40-256-592-620, E-mail: mv_putz@yahoo.com or mihai.putz@e-uvt.ro

32.1 DEFINITION

Verloop parameters are indices calculated with the Sterimol Software which express dimensional characteristics for one substituent. The calculus is made using the Corey-Pauling-Koltun atomic models (Verloop et al., 1976, 1977, 1983, 1987).

32.2 HISTORICAL ORIGINS

Sterimol parameters were developed in the 1970s, by Arie Verloop and his coworkers as a result to the (then) lack of steric describing parameters (Verloop et al., 1976).

32.3 NANO-CHEMICAL IMPLICATIONS

Verloop preferred to create two different parameters which are describing different dimensional properties of the studied substituent: the width and the length. There can be introduced two or more subparameters which are appreciating the width, and this is shown in Figures 32.1 and 32.2.

At first, Verloop described a length parameter (the length of the substituent along the axis of the bond between the group and its molecule of insertion) and four (B_{1-4}) width parameters which are describing the subdimensions of the functional groups along the fix axes. Later, together with his coworkers, he renewed the parameters (second generation Verloop parameters) and included a fifth width parameter (B_5) defined as the largest width orthogonal to L (Krogsgaard-Larsen et al., 2002).

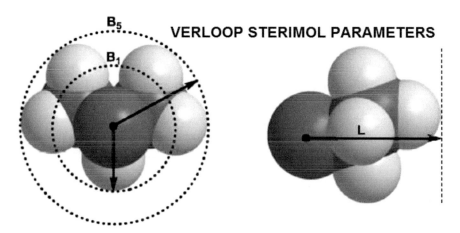

FIGURE 32.1 B_1, B_5 and L Sterimol second generation parameters for an isopropyl group as developed by Verloop and coworkers (Harper, 2013).

B_1 and B_5 are breadth/width subparameters, and L describes the length of the substituent. The width parameters are measured along the substituent's profile when the substituent is viewed from down of the axis of the primary bond. B_1 is describing the minimum width of the substituent orthogonal to the primary bond's axis, and B_5 is describing the maximum width in the same perspective. The length parameter, L, measures the total length of the substituent's profile along the primary bond's axis. As it can be seen in Figure 32.1, B_1 is perpendicular to the primary bond's axis. Thus it is obvious that B_1 values are larger with the increase of the substitution (disubstitution and trisubstitution) at the first carbon center and also B_1 values are smaller for methyl, for example than for ethyl, propyl, etc. (Harper, 2013).

The Sterimol parameters are expressed as length units (Å) so it easy to perceive the influence of a steric effect.

The calculus of these parameters is made using software: STERIMOL QCMP93 program, and it is based on physical classic parameters like van der Waals radii, bond lengths, bond angles (between atoms in the substituent) and hypothetically most likely conformations of the substituent.

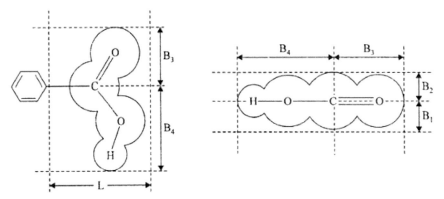

FIGURE 32.2 STERIMOL first generation parameters for benzoic acid: a carboxyl substituent. (Kar, 2007).

Sterimol parameters, in contrast, are not based on a mechanistically discrete reaction.

In the figure above, L describes the length of the substituent along the main axis of the energy-minimized (minimal-overlap) of the substituent, and B_1, B_2, B_3, and B_4 are describing the longitudinal and horizontal radii of two functional groups: carbonyl and hydroxyl.

The ratio between the breadth/width value and the length one or ratios L/B_1 and B_5/B_1 can show us information about the deviation of the shape of a substituent from a sphere (Todeschini, 2009). These ratios can particularly be used for appreciating "key-lock" interactions between a ligand and a receptor.

To determine the STERIMOL parameters, you have to follow the steps:

a. establishing the bond between the basic ring of the molecule and the (variable) substituent;
b. placing the bond into a plane (paper, software);
c. projecting the van der Waals radii-described surface of the substituent onto the same plane;
d. measuring the length from the attachment point to the projected van der Waals surface of the substituent along the bond of attachment, L value is obtained (see Figure 32.3).

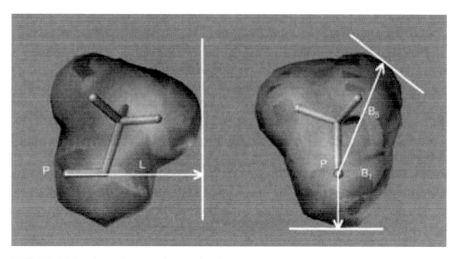

FIGURE 32.3 Second generation sterimol parameters–plane determination from the van der Waals surface using a methoxy-group. (Krogsgaard-Larsen et al., 2002).

If the structure is rotated by 90°, for a view along the bond between the parent molecule and the substituent, the width parameters can be measured (Table 32.1):

a. rotating by 90° the substituent;
b. projecting the van der Waals surface of the substituent onto a plane: paper, screen;
c. measuring the shortest (B1) and the longest (B5) length from the point of attachment (P) to the projected van der Waals surface (Krogsgaard-Larsen et al., 2002).

The second generation of the parameters appeared because of a certain ambiguity of the original formulations of B_2, B_3, and B_4 parameters. Especially for substituents whose first atom is ambivalent (ex. Oxygen, Sulfur), the tangent of minimum distance does not occur at one singular direction, only, like in the example below (Figure 32.4). In cases like this, the value of B_1 itself is uniquely defined, but not its position in the substituent.

In the figure above, it can be seen that the B parameters are projected in a plane, perpendicular to the L-axis. The B_4 parameter is chosen in the direction of the maximum width. The B_1 parameter, instead, can be defined in other directions, too so ambiguities can appear.

As the B_2 and B_3 parameters proved no significant correlations with the biological activity in the equations found in the major QSAR studies,

Verloop proposed that the length parameter (L) should be maintained, also the minimum width parameter B_1. B_2 and B_3 should be omitted, and the B_5 parameter should be introduced to replace the B_4 parameter. By these changes, the problem of the choice of the direction of B_1 is no longer existent, because B_5 has no directional relationship with B_1 as it is illustrated in Figure 32.4. Therefore, the second generation Verloop parameters were defined.

TABLE 32.1 First Generation Verloop Parameters Calculated with STERIMOL Program (Verloop et al., 1976)

Substituent	L	B1	B2	B3	B4
–H	2.06	1.00	1.00	1.00	1.00
–Cl	3.52	1.80	1.80	1.80	1.80
–CH_3	3.00	1.52	2.04	1.90	1.90
n–$(CH_2)_3$-CH_3	6.17	1.52	4.42	1.90	1.90
–OH	2.74	1.35	1.93	1.35	1.35
–NO_2	3.44	1.70	1.70	2.44	2.44
n–$(CH_2)_2$-CH_3	5.05	1.32	3.49	1.90	1.90
–C_6H5	6.28	1.70	1.70	3.11	3.11
–CN	4.23	1.60	1.60	1.60	1.60
–NH_2	2.93	1.50	1.50	1.84	1.84
–SH	3.47	1.70	2.33	1.70	1.70
–F	2.65	1.35	1.35	1.35	1.35
–Br	3.83	1.95	1.95	1.95	1.95
–tBu	4.11	2.59	2.97	2.86	2.86

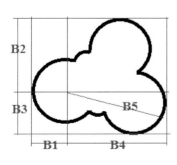
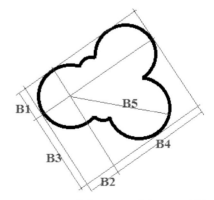

FIGURE 32.4 Different possibilities for the measurement of the B_1 – minimum width parameter of the –OCH_3 substituent (Verloop, 1983).

32.4   TRANS-DISCIPLINARY CONNECTIONS

QSAR studies (C-QSAR program) using Verloop parameters were performed for assessing the interaction between ligands and dopamine receptors. For example, the binding activity of ligands to dopamine D_2 receptor increases with the increase in hydrophobicity up to a certain value of logP and then decreases (Hansch et al., 2005).

A QSAR study of acute toxicity of N-substituted fluoroacetamides in a rat model was conducted over 19 analog N-alkyl and N-cycloalkyl fluoro-acetamides. BILIN program was used for the calculus of the QSAR parameters, also for L and B5 Verloop parameters of the R- (alkyl or cycloalkyl) substituent. All obtained data was correlated with the acute toxicity dose (LD50) (five different doses of N-substituted fluoroacetamides were administrated as aqueous solutions to five groups of 6 rats each; the mortality in each group was recorded after 24 hours). It was discovered that factors that have an influence on the acute toxicity of the studied 19 compounds are lipophilicity of molecules and shape of alkyl and cycloalkyl substituents. So the toxicity in this series is increasing with the lipophilicity of the molecule and with the size of the R-substituent (Juranic et al., 2006).

Another work was trying to develop new compounds for the treatment of obesity. A novel approach in this field uses cannabinoid receptor ligands (antagonists) for blocking the receptor. The first discovered and marketed molecule was rimonabant. Shortly after its release on the pharmaceutical market, it was withdrawn due to psychiatric secondary effects though it had shown success in reducing the weight. The researchers of this study tried to propose new molecules, maybe having a better binding affinity to the CB1 cannabinoid receptor. The binding process was evaluated in vivo, using rat cerebellum tissue (collected after oral administration of the compounds to rats) in comparison with a QSAR parameters assay of the ligand molecules (using ACD/CHEMS-KETCH – version 11.0, Toronto, ON, Canada, ChemDraw ULTRA – version 9.0.1, Cambridge Soft, Cambridge, MA, and Molecular Modeling Pro Plus – Chem SW, Fairfield, CA). It was found that hydrophobicity and steric characteristics of the molecules (particularly of the R-substituent, because the compounds were analogs having a common basic structure and being different through a R-substituent) are very important for the receptor binding process. The bigger the substituent is, the better the binding is fulfilled. Aryl substituents have the best receptor affinity (Song, 2009).

In 2002, a study was conducted to relate the antifungal activity (anti-*Candida albicans* and anti-*Rhodotorula glutinis*) of some imidazole derivatives with their structural properties (Figure 32.5).

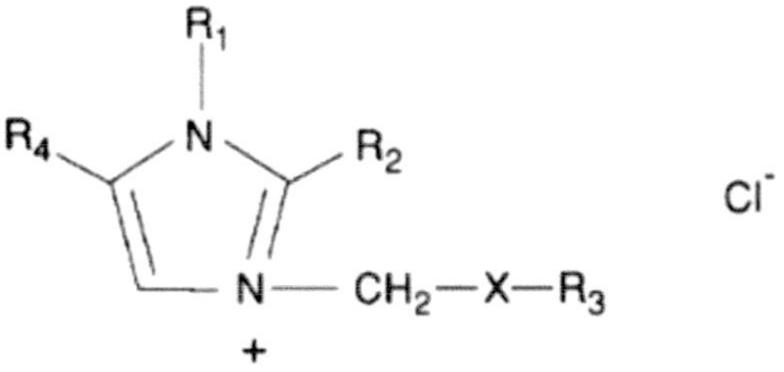

FIGURE 32.5 Basic structure of imidazole derivatives where X can be an Oxygen or Sulphur atom (Wiktorowicz et al., 2002).

Antifungal activity of 265 compounds with different X and R_{1-4} substituents was assessed (expressed as logarithm of the reciprocal of minimal inhibitory concentration) with aqueous solutions tested on these two yeasts. The QSAR studies for calculus of the STERIMOL parameters and other QSAR measures were made using HyperChem 5.01 program (HyperCube, Inc. Waterloo, Canada). The correlations proved that lipophilicity of the whole molecule is very important for the antifungal activity but for the sizes and bulkiness of the substituents, especially of R_3 and R_4, the more increased they are, the less active the molecule is (Wiktorowicz et al., 2002).

Sterimol parameters in QSAR studies are used not only in the pharmaceutical research. For example, photosynthesis inhibiting herbicides were developed using determined steric parameters to assess the inhibitory effect as early as 1969 and the precise mechanism of action (the target-protein receptor inhibited by triazinones) was also established using QSAR assays (Draber, 1996).

Harper, K.C. et al., stated in a paper from 2012 that the use of Sterimol parameters in the evaluation of the steric–enantioselection relationship of the asymmetric catalytic reactions between a substrate and an enzymatic ligand led to strong correlations. Their use also allowed a more detailed explanation of the catalytic mechanism for these asymmetric reactions with chiral specificity for a substrate (Harper et al., 2012).

Iwamura, H. used Verloop STERIMOL parameters to assess the optimal size and shape of the binding site for a series of 38 oximes. Hydrophobicity was well correlated with their potency as sweeteners. The STERIMOL parameters (the length of a substituent and the width in two directions) contributed to the final regression equation, therefore, indicating an optimal size of the substituent and an approximate shape for the hydrophobic binding site (Iwamura, 1980).

Major structure-activity studies were led after the discovery of aspartame. More than 1000 analog molecules have been synthesized and tested.

For example, van der Heijden and his colleagues performed a QSAR study of 33 aspartame analogs sweeteners using STERIMOL. The correlation coefficient between the calculated and the predicted potencies as sweeteners was found to be 0.81. They also found that increased hydrophobicity led to increased potency (Spillane, 2006).

Reill, E. and her colleagues performed a QSAR study on the quinolones. Quinolones are produced by the bacteria *Pseudomonas aeruginosa* and *Stigmatella aurantiaca* and were shown to be inhibitors of respiratory chains of other bacteria. They were taken as 'lead substances' for synthesis and evaluation of biological activity of a series of 2-alkyl- (1), 2-alkyl-3-methyl- (2), 2-methyl-3-alkyl–4(1H)-quinolones (3) and 1-hydroxy-2-methyl-3-alkyl- (4) and 1-hydroxy-2-alkyl-4(1H)-quinolones (5). The length of the alkyl side chain was varied from five to seventeen carbon atoms. Regression equations were calculated on a Power Macintosh 7200r90 using the statistical package Minitab 10 Xtra from Minitab, State College, PA, USA. As physicochemical parameters, the lipophilicity (π) and Verloop's STERIMOL parameter (L) of the alkyl side chain in positions 2 (R2) and 3 (R3) have been used. The quinolones have been grouped in two classes: those bearing a hydrogen atom at the nitrogen one (1–3) and those bearing a hydroxyl group at the nitrogen atom (4–5). Their biological activity was expressed as pI_{50} (inhibitory activity over the cytochrome b/c_1-complex from beef heart and from *Rhdospirillum rubrum*. In complex, I of the mitochondrial respiratory chain 2-alkyl-4(1H)-quinolones (1) showed pronounced activity, which could be substantially increased by an additional methyl substituent in the 3-position. Conversely, the 2-methyl-3-alkyl-4(1H)-quinolones (3), where alkyl and methyl group have changed positions, are completely inactive. This indicates that the optimal position for a hydrophobic substituent is the 2- and not the 3-position. Furthermore, substitution of the nitrogen with a hydroxyl instead of a hydrogen is also detrimental for biological activity. 2-Alkyl- (1) and 2-alkyl-3-methyl-4-(1H)-quinolones (2) reached maximal activity at a length of the alkyl chain of eleven carbon atoms which corresponds to about 14 Å. So the equations indicated that the biological activity of quinolones is only governed by the lipophilicity π and/or the length of the alkyl substituent, or the squares of them, respectively (Reill et al., 1997).

A large number of herbicides that inhibit the photosynthesis (about 30% of the commercialized herbicides have this mechanism of action) are known today. Still, when trying to find a new herbicide, the best start in the QSAR method, because a correlation between the chemical structure and the herbicidal activity is made at first, before synthesizing the compound in the lab. QSAR helped in finding new compounds but also in the elucidation of the photosynthesis mechanism. The investigation of inhibitors led to a detailed understanding of the target protein: D–1 polypeptide subunit of photosystem II (PS II). Quite early, it was found that the Fujita-Ban equation had to be supplemented with steric parameters to describe the relationship between the structure and activity of the photosynthesis inhibitors. The sterimol parameters were found very successful, superior to many other steric descriptors because when they were used in the equations that expressed the inhibitory activity, the correlation coefficient "r" was found to be higher than the one of the equations where logP was used (Draber, 1996).

32.5 OPEN ISSUES

The literature today shows that sterimol steric parameters are already calculated for 1078 different substituents. With a huge variety of substituent parameters and a great ability of the computer programs to use them in the QSAR field, the comparative use of the QSAR/QSPR studies is very helpful for a theoretical understanding the chemical and biological reaction mechanisms. The latest findings in the evaluations of the substituents and their characterizing parameters show that the STERIMOL parameters, especially B_1 is more effective in the correlation of intramolecular steric effects than Taft's classic E_s parameter. Since, the STERIMOL parameters are calculated and not measured as Taft parameters, they can easily be determined for many more substituents (Verloop et al., 1976).

KEYWORDS

- **length parameter**
- **width parameter**

REFERENCES AND FURTHER READING

Draber, W., (1996). STERIMOL and Its Role in Drug Research. *Z. Naturforsch, 51c*, 1–7.
Hansh, C., Verma., R. P., Kurup, A., & Mekapati, S. B., (2005). The role of QSAR in dopamine interactions. *Bioorganic & Medicinal Chemistry Letters, 15*, 2149–2157.
Harper, K. C., (2013). *Multidimensional Free Energy Relationships in Asymmetric Catalysis.* PhD Dissertation, University of Utah.

Harper, K. C., Bess, E. N., & Sigman, M. S., (2012). Multidimensional steric parameters in the analysis of asymmetric catalytic reactions. *Nature Chemistry, 4,* 366–374.

Iwamura, H., (1980). Structure-taste relationship of perillartine and nitro- and cyanoaniline derivatives. *J. Med. Chem., 23*(3), 308–312.

Juranic, I. O., Drakulic, B. J., Petrovic, S. D., Mijin, D. Z., & Stankovic, M. V., (2006). *Chemosphere, 62*, 641–649.

Kar, A., (2007). *Medicinal Chemistry.* New Age International (P) Limited, Publishers, New Delhi.

Krogsgaard-Larsen, P., Liljefors, T., & Madsen, U., (2002). *Textbook of Drug Design and Discovery.* Taylor & Francis, Inc., New York.

Reill, E., Höfle, G., Draber, W., & Oettmeier, W., (1997). Quinolones and their N-oxides as inhibitors of mitochondrial complexes I and III. *Biochimica et Biophysica Acta, 1318,* 291–298.

Song, K. S., Kim, M. J., Seo, H. J., Lee, S. H., Jung, M. E., Kim, S. U., Kim, J., & Lee, J., (2009). Synthesis and structure-activity relationship of novel diarylpyrazole imideanalogues as CB1 cannabinoid receptor ligands. *Bioorganic & Medicinal Chemistry, 17,* 3080–3092.

Spillane, J. W., (2006). *Optimizing Sweet Taste in Foods.* Wood head Publishing Limited, Abington, Cambridge.

Todeschini, R., & Consonni, V., (2009). *Molecular Descriptors for Chemoinformatics.* Wiley-VCH Verlag GmbH & Co. KGaA, Weinheim.

Verloop, A., (1983). Quantitative structure-relationship methods and computer-assisted design. In: Miyamoto, J., (ed.), *The STERIMOL Approach: Further Development of the Method and New Applications* (Vol. 1, p. 339). IUPAC Pesticide Chemistry, Pergamon: Oxford.

Verloop, A., & Ariens, E. J., (1976). *Drug Design* (Vol. III, p. 33). Academic Press: New York.

Verloop, A., & Buisman, J. A., (1977). *Biological Activity and Chemical Structure* (p. 63). Elsevier: Amsterdam.

Verloop, A., & Hadzi, B. J. B., (1987). *QSAR in Drug Desing and Toxicology* (p. 97). Elsevier: Amsterdam.

Wiktorowicz, W., Markuszewski, M., Krysinski, J., & Kaliszan, R., (2002). Quantitative Structure-Activity Relationship Study of a Series of Imidazole Derivatives as Potential new Antifugal Drugs. *Acta Poloniae Pharmaceutica - Drug Research, 59*(4), 295–306.

CHAPTER 33

Superaugmented Eccentric Connectivity Indices and Their Significance in Drug Discovery Process

ROHIT DUTT,[1] HARISH DUREJA,[2] and A. K. MADAN[3]

[1]School of Medical and Allied Sciences, GD Goenka University, Gurugram–122103, India

[2]Department of Pharmaceutical Sciences, M.D. University, Rohtak–124001, India

[3*]Faculty of Pharmaceutical Sciences, Pt. B.D. Sharma University of Health Sciences, Rohtak–124001, India, E-mail: madan_ak@yahoo.com

ABSTRACT

The development of models for the prediction of physicochemical or biological parameters through the application of classification or correlation modeling techniques has acquired utmost importance during the past few decades. The use of graph invariants in (quantitative) structure-activity/ property relationship [(Q)SAR/QSPR] has served as a valuable source for nesting mechanistic information from chemical structure and property relationship studies. Such *in silico* intervention in the initial stages of the drug discovery process is now routinely used as a tool to prioritize experiments with an aim to improve the drug attrition rate. The timely prediction of this failure will naturally save considerable cost, valuable time, human efforts and minimize animal sacrifice. The availability of diverse types of topological indices (TIs) imparts meaningful information about the physicochemical framework would play a certain role in understanding the relationships between chemical structures and experimental outcomes. Various *super-augmented eccentric connectivity indices,* as well as their applications, have been briefly reviewed in the present chapter.

33.1 INTRODUCTION

In the modern drug discovery era, *in silico-aided drug design* (IADD) covers all stages in the drug discovery pipeline, from target identification to lead discovery, from lead optimization to preclinical or clinical trials (Tang et al., 2006). Unlike the classical trial and error method of drug discovery, IADD begins with specific knowledge of chemical responses in the body or target organism and tailoring combinations to fit a treatment profile (Rao and Srinivas, 2011). It *"involves all computer-assisted techniques used to discover, design, and optimize compounds with desired structure and prop-erties"* (van de Waterbeemd et al., 1998). Random screening of continuously increasing database of compounds (both natural and synthetic) is extremely tedious, expensive, and time-consuming process (Shen et al., 2003). Indeed, recent promising advancement in virtual screening through the use of computational approaches has facilitated virtual "synthesis" of billions of chemical compounds and to analyze their properties. Further, the compound which possesses desired activity against the biological target(s) of interest is selected (Xu and Agrafiotis, 2002). European policy for the evaluation of chemicals (REACH: Registration, Evaluation, and Authorization of Chemicals) and USFDA has been a strong advocate of using *in silico* methods in the early phases of drug discovery pipeline as possible alternative to animal testing and this could lead to a reduction in the cost of the drug development process by up to 50 % (FDA, 2004; McGee, 2005; Dutt and Madan, 2012). These factors prompted the use of *in silico* approaches and accordingly have gained immense popularity in directing drug design and discovery (Reddy et al., 2007).

Historically, the origination of the modern (quantitative) structure-activity/property relationship [(Q)SAR/QSPR] formalism is attributed to the pioneer works of Hansch and Fujita (1964) and Free and Wilson (1964) in the 1960s. Since then, the concept of IADD has evolved very quickly, especially in the recent decade with an unprecedented development of structural biology and machine learning approaches (Tang et al., 2006). There can be both qualitative SARs and quantitative SARs [(Q)SARs], depending on the means used to describe the molecular structure and on the nature of the derived relationship. The main assumption in QSPR/(Q)SARs studies is that all properties of a chemical substance (physical, chemical, biological, and pharmacokinetic) can be computed from its molecular structure (encoded in a numerical form) with the aid of various molecular descriptors (MDs) (Ivanciuc, 2000). In this way, MDs are used to extract the structural information for development of a model which can serve as the bridge between the molecular structure and physicochemical/biological/pharmacokinetic activities of the molecules (Hong et al., 2008). An initial major step in building the (Q)SAR/QSPR models is to find one or more MDs that represent variations in the structural property of the molecules by a number (Natarajan, 2011). Graph theoretical or topological indices (TIs) are one of the oldest and most widely used MDs in (Q)SAR (Moghani and Ashrafi, 2006). The emergence of chemical graph theory represents one of the pioneering footsteps in the depiction of molecular structures in terms of quantitative descriptors (Roy, 2015). Chemical graph theory is the branch of mathematical chemistry which applies graph theory to mathematical modeling of chemical phenomenon (Rouvray, 1991). The use of graph theory has been stated to have been used as early as 1736 when Leonard Euler solved the famous Konigsberg bridge problem (Pogliani, 2003). By using this theory, a molecular structure of compounds can be represented as molecular graph, where vertices represent atoms and edges represent bonds in the molecule (Trinajstic, 1983). The pattern of connectedness of atoms in a molecule, called molecular topology, has become a rich source of developing TIs for (Q)SAR/QSPR modeling. In this way, TIs can be characterized as "*mathematical quantification of molecular architecture*" (Natarajan and Nirdosh, 2003). Beside swift computation, TIs have the upper edge of being true structural invariant, which indicate that their values are independent of three dimensional (3D) conformations (Mahmoudi et al., 2006). This conformational independence is especially important in the study of flexible molecules and in the case where the proper confirmation is not well defined (Stanton, 2008). Because of simplicity of topological structural representation, TIs are preferred to more complicated geometric, electrostatic, and quantum chemical descriptors, especially in those cases where their use significantly reduces the computation time (Pompe and Randic, 2006). The topostructural and topochemical are two-dimensional (2D) descriptors which are collectively known as TIs. Topostructural indices depict information, particularly on the adjacency and connectedness of atoms within a molecule whereas topochemical versions consider both molecular topology and the chemical nature of atoms (Basak et al., 1997).

Topostructural indices are derived from various matrices such as distance matrix and/or adjacency matrix, representing a molecular graph. When a distance matrix or adjacency matrix is weighted according to the heteroatom(s) present in a molecule, then the resulting matrix may be termed as a chemical distance or chemical adjacency matrix respectively. Descriptors derived from such matrices are known as topochemical indices (Dureja et al., 2008). The interest in developing highly discriminating graph theoretical descriptors devoid of any degeneracy has revived in recent years. This can be attributable to new applications of TIs in main research areas of drug development which include lead discovery, lead optimization, virtual screening, drug design, combinatorial library design, structure-pharmaco-kinetics, structure-toxicity relationships and so forth (Estrada and Uriate, 2001; Estrada and Molina, 2001; Bonchev and Buck, 2007; Natarajan, 2011).

Various *superaugmented eccentric connectivity indices,* as well as their applications in drug discovery process, have been briefly reviewed in present chapter.

33.2 SUPERAUGMENTED ECCENTRIC CONNECTIVITY INDICES

Dureja and Madan (2007) reported first set of three *superaugmented eccentric connectivity indices* ($^{SA}\xi^c$) derived through distance matrix *(D)* and augmentative adjacency matrix *(A^a)*. The augmentative adjacency matrix (A^a) is obtained by modifying adjacency matrix. In augmentative adjacency matrix, the non-zero row elements representing degrees of vertices (adjacent to corresponding vertex *i* in a molecular graph *G*) upon multiplication yield a product termed as augmented adjacency (Dureja and Madan, 2007).

33.3 SUPERAUGMENTED ECCENTRIC CONNECTIVITY INDEX 1–3

Superaugmented eccentric connectivity index 1–3, (denoted by: $^{SA}\xi^c_1$, $^{SA}\xi^c_2$, and $^{SA}\xi^c_3$) are defined as the summation of the quotients of the product of adjacent vertex degrees and eccentricity (raised to the power 2, 3 or 4 for $^{SA}\xi^c_1$, $^{SA}\xi^c_2$ and $^{SA}\xi^c_3$ respectively) of the concerned vertex, for all vertices in the hydrogen suppressed molecular graph (Dureja and Madan, 2007). The equations of these descriptors are depicted in Table 33.1.

33.4 SUPERAUGMENTED ECCENTRIC CONNECTIVITY TOPOCHEMICAL INDEX 1–3

The topochemical version of $^{SA}\xi^c_1$, $^{SA}\xi^c_2$, and $^{SA}\xi^c_3$ (denoted by $^{SAc}\xi^c_1$, $^{SAc}\xi^c_2$ and $^{SAc}\xi^c_3$) reported by Dureja et al., (2008) can be easily calculated from the chemical distance matrix (D_c) and augmentative chemical adjacency matrix (A_c^a). The chemical distance and chemical adjacency matrices are weighted variant of distance and adjacency matrix respectively accounting the presence and relative position of heteroatoms. When the distance or adjacency matrix is weighted corresponding to the heteroatom(s) like N, O, Cl, etc., in a molecule, the matrix may be termed as chemical distance or chemical adjacency matrix, respectively (Bajaj et al., 2004a).

In chemical distance matrix, the non-zero row elements represents chemical length of path that contain least number of edges between vertex i and j in the molecular graph G. Similarly, the modified or chemical degree of a vertex can be obtained from the adjacency matrix by substituting row element corresponding to heteroatom, with relative atomic weight with respect to carbon atom (Bajaj et al., 2004a). In augmentative chemical adjacency matrix, the non-zero row elements representing chemical degrees of vertices (adjacent to corresponding vertex i in a molecular graph G) upon multiplication yield a product termed as chemical augmented adjacency (Dureja and Madan, 2007; Dureja et al., 2008). Relevant equations of these descriptors are depicted in Table 33.1.

33.5 SUPERAUGMENTED ECCENTRIC CONNECTIVITY INDEX-4−7

Dutt and Madan (2010) reported *superaugmented eccentric connectivity indices* (denoted by ${}^{SA}\xi^c_4$, ${}^{SA}\xi^c_5$, ${}^{SA}\xi^c_6$, and ${}^{SA}\xi^c_7$) along with their topochemical versions (denoted by ${}^{SAc}\xi^c_4$, ${}^{SAc}\xi^c_5$, ${}^{SAc}\xi^c_6$, and ${}^{SAc}\xi^c_7$) for the purpose of (Q) SAR/QSPR modeling. *Superaugmented eccentric connectivity indices* (${}^{SA}\xi^c_4$, ${}^{SA}\xi^c_5$, ${}^{SA}\xi^c_6$, and ${}^{SA}\xi^c_7$) can be computed from the aforementioned distance matrix (D) and augmented adjacency matrix (A$^\alpha$). The equations of these descriptors are given in Table 33.1.

33.6 SUPERAUGMENTED ECCENTRIC CONNECTIVITY TOPOCHEMICAL INDEX-4−7

The topochemical version of ${}^{SA}\xi^c_4$, ${}^{SA}\xi^c_5$, ${}^{SA}\xi^c_6$ and ${}^{SA}\xi^c_7$ (denoted by: ${}^{SAc}\xi^c_4$, ${}^{SAc}\xi^c_5$, ${}^{SAc}\xi^c_6$ and ${}^{SAc}\xi^c_7$ respectively) can be similarly calculated from chemical distance matrix (D$_c$) and augmentative chemical adjacency matrix (A$_c^\alpha$) (Table 33.1). All aforementioned topostructural versions (${}^{SA}\xi^c_{1-7}$) mentioned above were reported to have improved performance in terms of degeneracy and discriminating power for all possible structures containing only five vertices (Dureja and Madan, 2007; Dureja et al., 2008; Dutt and Madan, 2010). The values of degeneracy and discriminating power of all superaugmented eccentric connectivity indices are enlisted Table 33.2.

TABLE 33.1 Definitions of Various Superaugmented Eccentric Connectivity Indices (Dureja et al., (2007) and Dutt and Madan (2010))

Name of Descriptor	Topostructural	Definition	Topochemical	Definition
Superaugmented eccentric connectivity index–1, $^{SA}\xi^c_1$	$^{SA}\xi^c_1 = \sum_{i=1}^{n}\left(\dfrac{M_i}{E_i^2}\right)$	Summation of the quotients of the product of adjacent vertex degrees and squared eccentricity of the concerned vertex, for all vertices in the hydrogen suppressed molecular graph.	$^{SA_c}\xi^c_1 = \sum_{i=1}^{n}\left(\dfrac{M_{ic}}{E_{ic}^2}\right)$	Summation of the quotients of the product of adjacent vertex chemical degrees and squared chemical eccentricity of the concerned vertex, for all vertices in the hydrogen suppressed molecular graph.
Superaugmented eccentric connectivity index–2, $^{SA}\xi^c_2$	$^{SA}\xi^c_2 = \sum_{i=1}^{n}\left(\dfrac{M_i}{E_i^3}\right)$	Summation of the quotients of the product of adjacent vertex degrees and cubic eccentricity of the concerned vertex, for all vertices in the hydrogen suppressed molecular graph.	$^{SA_c}\xi^c_2 = \sum_{i=1}^{n}\left(\dfrac{M_{ic}}{E_{ic}^3}\right)$	Summation of the quotients of the product of adjacent vertex chemical degrees and cubic chemical eccentricity of the concerned vertex, for all vertices in the hydrogen suppressed molecular graph.
Superaugmented eccentric connectivity index–3, $^{SA}\xi^c_3$	$^{SA}\xi^c_3 = \sum_{i=1}^{n}\left(\dfrac{M_i}{E_i^4}\right)$	Summation of the quotients of the product of adjacent vertex degrees and fourth power of eccentricity of the concerned vertex, for all vertices in the hydrogen suppressed molecular graph.	$^{SA_c}\xi^c_3 = \sum_{i=1}^{n}\left(\dfrac{M_{ic}}{E_{ic}^4}\right)$	Summation of the quotients of the product of adjacent vertex chemical degrees and fourth power of chemical eccentricity of the concerned vertex, for all vertices in the hydrogen suppressed molecular graph.
Superaugmented eccentric connectivity index–4, $^{SA}\xi^c_4$	$^{SA}\xi^c_4 = \sum_{i=1}^{n}\left(\dfrac{M_i^2}{E_i}\right)$	Summation of the quotients of squared product of adjacent vertex degrees and eccentricity of the concerned vertex, for all vertices in a hydrogen suppressed molecular graph.	$^{SA_c}\xi^c_4 = \sum_{i=1}^{n}\left(\dfrac{M_{ic}^2}{E_{ic}}\right)$	Summation of the quotients of squared product of adjacent vertex chemical degrees and chemical eccentricity of the concerned vertex, for all vertices in a hydrogen suppressed molecular graph.
Superaugmented eccentric connectivity index–5, $^{SA}\xi^c_5$	$^{SA}\xi^c_5 = \sum_{i=1}^{n}\left(\dfrac{M_i^2}{E_i^2}\right)$	Summation of the quotients of squared product of adjacent vertex degrees and squared eccentricity of the concerned vertex, for all vertices in a hydrogen suppressed molecular graph.	$^{SA_c}\xi^c_5 = \sum_{i=1}^{n}\left(\dfrac{M_{ic}^2}{E_{ic}^2}\right)$	Summation of the quotients of squared product of adjacent vertex chemical degrees and squared chemical eccentricity of the concerned vertex, for all vertices in a hydrogen suppressed molecular graph.
Superaugmented eccentric connectivity index–6, $^{SA}\xi^c_6$	$^{SA}\xi^c_6 = \sum_{i=1}^{n}\left(\dfrac{M_i^2}{E_i^3}\right)$	Summation of the quotients of squared product of adjacent vertex degrees and cubic eccentricity of the concerned vertex, for all vertices in a hydrogen suppressed molecular graph.	$^{SA_c}\xi^c_6 = \sum_{i=1}^{n}\left(\dfrac{M_{ic}^2}{E_{ic}^3}\right)$	Summation of the quotients of squared product of adjacent vertex chemical degrees and cubic power of chemical eccentricity of the concerned vertex, for all vertices in a hydrogen suppressed molecular graph.

TABLE 33.1 *(Continued)*

Name of Descriptor	Topostructural	Definition	Topochemical	Definition
Superaugmented eccentric connectivity index−7, $^{SA}\xi^c_7$	$$^{SA}\xi_7 = \sum_{i=1}^{n}\left(\frac{M_i^2}{E_i^4}\right)$$	Summation of the quotients of squared product of adjacent vertex degrees and fourth power of eccentricity of the concerned vertex, for all vertices in a hydrogen suppressed molecular graph.	$$^{SAc}\xi_7 = \sum_{i=1}^{n}\left(\frac{M_{ic}^2}{E_{ic}^4}\right)$$	Summation of the quotients of squared product of adjacent vertex chemical degrees and fourth power of chemical eccentricity of the concerned vertex, for all vertices in a hydrogen suppressed molecular graph.

M_i is the product of degrees of all vertices (v_j), adjacent to vertex i, E_i is the eccentricity, and n is the number of vertices in graph G. M_{ic} is the product of chemical degrees of all vertices (v_j) adjacent to vertex i, E_{ic} is the chemical eccentricity, and n is the number of vertices in graph G.

TABLE 33.2 Comparison of Discriminating Power and Degeneracy of Reported Superaugmented Eccentric Connectivity Indices (Dureja et al., (2007) and Dutt and Madan (2010)).

	$SA\xi^c_1$	$SA\xi^c_2$	$SA\xi^c_3$	SAc_4	$SA\xi^c_5$	$SA\xi^c_6$	$SA\xi^c_7$
For Three Vertices							
Minimum value	2	1.5	1.25	5	3	2	1.5
Maximum Value	12	12	48	48	48	48	48
Ratio	1:6	1:8	1:9.6	1:9.6	1:16	1:24	1:32
Degeneracy	0/2	0/2	0/2	0/2	0/2	0/2	0/2
For Four Vertices							
Minimum value	1.44	0.65	0.3	6.67	2.89	1.30	0.60
Maximum Value	108	108	108	2916	2916	2916	2916
Ratio	1:74.77	1:166.6	1:360.7	1:166.6	1:360.7	1:360.7	1:360.7
Degeneracy	0/6	0/6	0/6	0/6	0/6	0/6	0/6
For Five Vertices							
Minimum value	1.69	0.71	0.32	12.67	5.39	2.42	1.08
Maximum Value	1280	1280	1280	327680	327680	327680	327680
Ratio	1:755.4	1:1801	1:4063	1:25869	1:60807	1:135332	1:303407
Degeneracy	1/21	0/21	0/21	0/21	0/21	0/21	1/21

Degeneracy = Number of compounds having same values/ total number of compounds with the same number of vertices.

33.7 APPLICATION OF SUPERAUGMENTED ECCENTRIC CONNECTIVITY INDICES

The superaugmented eccentric connectivity indices have been reportedly utilized for development of models for prediction of diverse biological activities. Dureja et al., (2008) found significant correlation of $^{SAc}\xi^c_1$, $^{SAc}\xi^c_2$, and $^{SAc}\xi^c_3$ with anti-human immunodeficiency virus (HIV)-1 activity of a data set comprising differently substituted 6-arylbenzonitriles (Chan et al., 2001). Resulting models exhibited a prediction accuracy of >80%. Average IC_{50} (µM) values of 6-arylbenzonitriles of correctly predicted analogs in various ranges of topochemical models have been depicted in Figure 33.1 (Dureja et al., 2008).

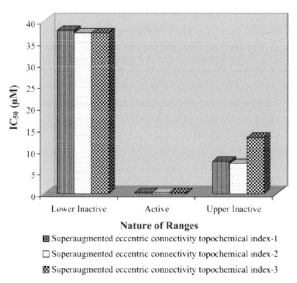

FIGURE 33.1 Average IC_{50} (µM) values of correctly predicted analogs of 6-arylbenzonitriles in various ranges of topochemical models (Reproduced with permission from Dureja et al., 2008).

Similarly, $^{SAc}\xi^c_4$, $^{SAc}\xi^c_5$, $^{SAc}\xi^c_6$, and $^{SAc}\xi^c_7$ were successfully utilized by Dutt et al., (2012) for development of models for predicting inhibitory activity of isatins (Hyatt et al., 2007) against two important mammalian carboxylesterases, i.e., human intestinal hiCE and human liver CEs (hCE1) with overall accuracy of prediction of 85–94% for hCE1 and 81–92% for hiCE activity. The active ranges in these models indicated exceptionally high potency in terms of average Ki value with regard to hiCE and hCE1 inhibitory activities (Figures 33.2 and 33.3).

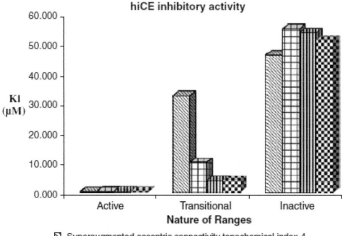

FIGURE 33.2　Average Ki (μM) values of correctly predicted analogs of isatins in various ranges of topochemical models for prediction of hiCE inhibitory activity (Reproduced with permission from Dutt et al., (2012)).

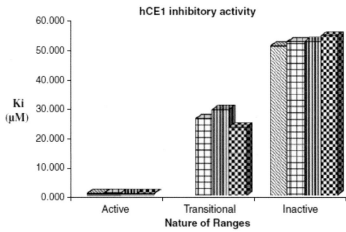

FIGURE 33.3　Average Ki (μM) values of correctly predicted analogs of isatins in various ranges of topochemical models for prediction of hCE1 inhibitory activity [Reproduced with permission from (Dutt et al., (2012)).

33.8 CONCLUSION

The superaugmented eccentric connectivity indices are simple topology based MDs which can be easily calculated by means of mathematical matrices. The sensitivity towards branching and exceptionally high discriminating power coupled with extremely low degeneracy render superaugmented eccentric connectivity indices promising tools in drug design. These indices are helpful in the characterization of structures, similarity/dissimilarity studies, lead identification, lead optimization, combinatorial library design and quantitative structure-activity/property/toxicity studies for the prediction of various biological, physicochemical, pharmacokinetic, and toxicological properties of molecules for facilitating drug design.

KEYWORDS

- **adjacency**
- **decision tree**
- **eccentricity**
- **moving average analysis**
- **topological indices**

REFERENCES

Basak, S. C., Grunwald, G. D., & Niemi, G. J., (1997). Use of graph-theoretic and geometrical molecular descriptors in structure-activity relationships. In: Balaban, A. T., (ed.), *From Chemical Topology to Three Dimensional Molecular Geometry* (pp. 73–116). Plenum Press: New York, ISBN 978–0–306–45462–2.

Bonchev, D., & Buck, G. A., (2007). From molecular to biological structure and back. *J. Chem. Inf. Model., 47*, 909–917.

Chan, J. H., Hong, J. S., Hunter, III, R. N., Orr, G. F., Cowan, J. R., Sherman, D. B., Sparks, S. M., et al., (2001). 2-Amino–6-arylsulfonylbenzonitriles as non-nucleoside reverse transcriptase inhibitors of HIV-1. *J. Med. Chem., 44*, 1866–1882.

Dureja, H., & Madan, A. K., (2007). Superaugmented eccentric connectivity indices: New-generation highly discriminating topological descriptors for QSAR/QSPR modeling. *Med. Chem. Res., 16*, 331–341.

Dureja, H., Gupta, S., & Madan, A. K., (2008). Predicting anti-HIV-1 activity of 6-arylbenzonitriles: Computational approach using superaugmented eccentric connectivity topochemical indices. *J. Mol. Graph. Model., 26*, 1020–1029.

Dutt, R., & Madan, A. K., (2010). Improved superaugmented eccentric connectivity indices for QSAR/QSPR modeling part 1: Development and evaluation. *Med. Chem. Res.*, *19*, 431–447.

Dutt, R., & Madan, A. K., (2012). Classification models for anticancer activity. *Curr. Top. Med. Chem.*, *12*, 2705–2726.

Dutt, R., Singh, M., & Madan, A. K., (2012). Improved superaugmented eccentric connectivity indices Part II: Application in development of models for prediction of hiCE and hCE1 inhibitory activities of isatins. *Med. Chem. Res.*, *21*, 1226–1236.

Estrada, E., & Molina, E., (2001). Novel local (fragment–based) topological molecular descriptors for QSPR/QSAR and molecular design. *J. Mol. Graph. Model.*, *20*, 54–64.

Estrada, E., & Uriarte, E., (2001). Recent advances on the role of topological indices in drug discovery research. *Curr. Med. Chem.*, *8*, 1573–1588.

FDA, (2004). *Challenge and Opportunity on the Critical Path to New Medical Products.* Food and Drug Administration, U.S Department of Health and Human Services. Available from: http://www.fda.gov (accessed on: 27/08/15).

Free, S. M. Jr., & Wilson, J. W., (1964). A mathematical contribution to structure-activity studies. *J. Med. Chem.*, *7*, 395–399.

Hansch, C., & Fujita, T., (1964). ε-σ-π Analysis: A method for the correlation of biological activity and chemical structure. *J. Am. Chem. Soc.*, *86*, 1616–1626.

Hong, H., Xie, Q., Ge, W., Qian, F., Fang, H., Shi, L., Su, Z., Perkins, R., & Tong, W., (2008). Mold 2, molecular descriptors from 2D structure for cheminformatics and toxicoinformatics. *J. Chem. Inf. Model.*, *48*, 1337–1344.

Hyatt, J. L., Moak, T., Jason, H. M., Tsurkan, L., Edwards, C. C., Wierdl, M., Danks, M. K., Wadkins, R. M., & Potter, P. M., (2007). Selective inhibition of carboxylesterases by isatins, indole–2,3-diones. *J. Med. Chem.*, *50*, 1876–1885.

Ivanciuc, O., (2000). QSAR comparative study of Wiener descriptors for weighted molecular graphs. *J. Chem. Inf. Comput. Sci.*, *40*, 1412–1422.

Mahmoudi, N., De Julian-Oritiz, J. V., Ciceron, L., Gálvez, J., Mazier, D., Danis, M., Derouin, F., & García-Domenech, R., (2006). Identification of new anti-malarial drugs by linear discriminant analysis and topological virtual screening. *J. Antimicrob. Chemother.*, *57*, 489–497.

McGee, P., (2005). Modeling success with *in silico* tools. *Drug Discov. Today*, *8*, 23–28.

Moghani, G. A., & Ashrafi, A. R., (2006). On the PI index of some nanotubes. *J. Phy. Conference Series*, *29*, 159–162.

Natarajan, R., (2011). New topological indices with very high discriminatory power. *SAR QSAR Environ. Res.*, *22*, 1–20.

Natarajan, R., & Nirdosh, I., (2003). Application of topochemical, topostructural, physico-chemical, and geometrical parameters to model the flotation efficiencies of N-arylhydroxamic acid. *Int. J. Miner. Process*, *71*, 113–129.

Pogliani, L., (2003). Graph-theoretical concepts and physiochemical data. *Data Sci. J.*, *2*, 1–11.

Pompe, M., & Randic, M., (2006). Anti connectivity: A challenge for structure-property-activity studies. *J. Chem. Inf. Model.*, *46*, 2–8.

Rao, V. S., & Srinivas, K., (2011). Modern drug discovery process: An *in silico* approach. *J. Bioinform. Seq. Anal.*, *2*, 89–94.

Reddy, A. S., Pati, S. P., Kumar, P. P., Pradeep, H. N., & Sastry, G. N., (2007). Virtual screening in drug discovery-a computational perspective. *Curr. Protein Pept. Sci.*, *8*, 329–351.

Rourvray, D. H., (1991). The origins of chemical graph theory. In: Bonchev, D., & Rouvray, D. H., (eds.), *Chemical Graph Theory: Introduction and Fundamentals* (pp. 1–34). Abacus Press: New York. ISBN 0856264547.

Roy, K., (2015). Topological QSAR. In: Roy, K., Kar, S., & Das, R. N., (eds.), *Understanding the Basics of QSAR for Applications in Pharmaceutical Sciences and Risk Assessment* (pp. 103–149). Academic Press: London, UK. ISBN 9780128015056.

Shen, J., Xu, X., Cheng, F., Liu, H., Luo, X., Shen, J., Chen, K., Zhao, W., Shen, X., & Jiang, H., (2003). Virtual screening on natural products for discovering active compounds and target information. *Curr. Med. Chem.*, *10*, 2327–2342.

Stanton, D. T., (2008). On the importance of topological descriptors in understanding structure-property relationships. *J. Comput. Aided Mol. Des.*, *22*, 441–460.

Tang, Y., Zhu, W., Chen, K., & Jiang, H., (2006). New technologies in computer-aided drug design: Toward target identification and new chemical entity discovery. *Drug Discov. Today, Technol.*, *3*, 307–313.

Trinajstic, N., (1983). *Chemical Graph Theory* (pp. 1–50). CRC Press: Boca Raton, Florida.

Van De Waterbeemd, H., Carter, R. E., Grassy, G., Kubinyi, H., Martin, Y. C., Tute, M. S., & Willett, P., (1998). Glossary of the terms used in computational drug design. *Ann. Rep. Med. Chem.*, *33*, 397.

Xu, H., & Agrafiotis, D. K., (2002). Retrospect and prospect of virtual screening in drug discovery. *Curr. Top. Med. Chem.*, *2*, 1305–1320.

CHAPTER 34

van der Waals Molecular Volume

BOGDAN BUMBĂCILĂ[1] and MIHAI V. PUTZ[1,2]

[1]*Laboratory of Computational and Structural Physical Chemistry for Nanosciences and QSAR, Biology-Chemistry Department, Faculty of Chemistry, Biology, Geography at West University of Timișoara, Pestalozzi Street No.16, Timișoara, RO–300115, Romania*

[2]*Laboratory of Renewable Energies-Photovoltaics, R&D National Institute for Electrochemistry and Condensed Matter, Dr. A. Paunescu Podeanu Str. No. 144, RO–300569 Timișoara, Romania,*
Tel.: +40-256-592-638, Fax: +40-256-592-620,
E-mail: mv_putz@yahoo.com or mihai.putz@e-uvt.ro

34.1 DEFINITION

The van der Waals molecular volume, V_w, is a molecular descriptor directly related to the van der Waals surface. It is the volume "occupied" in space by an individual molecule or the volume described by the enclosure of the van der Waals surface (Bondi, 1964).

34.2 HISTORICAL ORIGINS

The van der Waals surface of a molecule is named after Johannes Diderik van der Waals (1837–1923), a Dutch physicist who proved that atoms and molecules have defined volumes and interact because of their attractive forces.

The pioneers of the molecular models based on the van der Waals radii, surfaces, and volumes were Robert Corey, Linus Pauling, and Walter Koltun. These molecular models, also named space-filling models or calotte models are represented as spheres with radii proportional to the atoms radii and center-to-center distances proportional to the distances between nuclei of the atoms. Different chemical elements are usually represented by spheres of different colors. (Corey et al., 1953)

34.3 NANO-CHEMICAL IMPLICATIONS

It is well known that two atoms which are not covalently bound cannot approach each other closer than a certain minimal distance, depending upon the type of atoms involved.

The van der Waals radius between two molecules is shown in Figure 34.1.

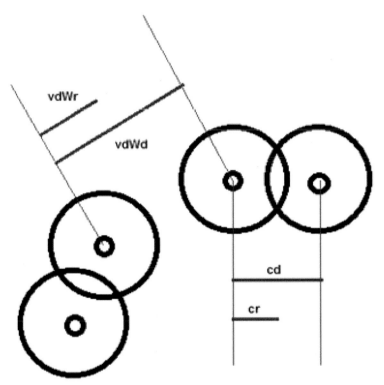

FIGURE 34.1   Intra- and inter- molecular distances.

In Figure 34.1, vdWd is the van der Waals diameter, vdWr is r_W or r_v – the van der Waals radius (half of the vdWd), c_d is the covalent distance, and c_r the covalent radius (half of the c_d).

The van der Waals radius is measured as half of the internuclear distance of two atoms non-bonded or belonging to different molecules, of the same chemical element when these two atoms are situated at the closest possible distance. Intermolecular forces (dispersion, dipole-dipole) and the way of "packing" of the molecules in the solid state are important for establishing the r_v (Rowland et al., 1996).

Moreover, r_v is determined by several methods: from the van der Waals equation, from intra-crystal space measurements, electrical, and optical characteristics of crystalline solids, etc. The methods used to determine the van der Waals radius offer similar but not identical values (expressed in Å or pm). Usually, the r_v values in the tables are, in fact, mean values. (Bondi, 1964) An example of such a table is given in Table 34.1.

TABLE 34.1   Atomic Van Der Waals Radii for Different Chemical Elements (Housecroft et al., 2012)

No.	Atom	r_v (pm)
1	H	120
2	B	208
3	C	185
4	N	154
5	O	140
6	F	135
7	Cl	180
8	Br	195
9	I	215
10	He	99

The van der Waals volume, V_w, also named the van der Waals atomic/molecular volume is the volume represented by an atom or a molecule. It can be calculated from the van der Waals radius of the species (for molecules, using distances and angles between atoms). For a singular atom, having a spherical volume, the van der Waals volume can be calculated with the following formula:

$$V_w = \frac{4}{3}\pi r_w^3 \tag{1}$$

The van der Waals volume of a molecule is always smaller than the sum of the van der Waals atomic volumes of the atoms in the molecule because the atomic volumes are partially overlapped when forming chemical bonds (Chang, 2005).

As it was said in the definition, the van der Waals volume is also the volume described by the enclosure of the van der Waals surface of a molecule. (Bondi, 1964) (Figure 34.2).

FIGURE 34.2   Molecular van der Waals volume model of the ammonia molecule. Overlapping of the atomic van der Waals volumes.

This surface/area is an abstract molecular representation, mathematically computed. As the volume for an atom is expressed in equation MvdWV-1, the surface of a single atom can be assessed with the following formula:

$$A_w = \frac{4}{3}\pi r_w^2 \tag{2}$$

Usually, a molecule can be perceived as a well-defined volume into a Cartesian tridimensional space. The molecular van der Waals surface can be defined as the external surface of the molecular van der Waals volume (the molecule's "well-defined volume") obtained by the overlapping of all the van der Waals atomic volumes (generally, considered as spheres) corresponding to the atoms that form the molecule. All the points included by this envelope have to satisfy at least one of the following conditions:

$$\left(X_i - x\right)^2 + \left(Y_i - y\right)^2 + \left(Z_i - z\right)^2 \leq \left(r_i^w\right)^2, i = \overline{1,m} \tag{3}$$

where x, y, z are points included by the envelope, X_i, Y_i, Z_i are the Cartesian coordinates of atom i and m is the number of atoms included in the molecule (Vlaia, et al., 2009).

The molecular van der Waals volume is computed using numerical integration and today, and certain software tools are completing this job.

The exact shape of the van der Waals surface can be quite difficult to be described, especially for macromolecules as receptors. Usually, their surface may contain gaps, clefts, pockets, hard to be penetrated by the ligand and sometimes by the solvent that surrounds the ligand and the receptor (usually aqueous solutions). Because of those spatial "deformities," the surface of the receptor is not overall accessible for the ligand or for the solvent. That is why, today, the concepts of "contact surface" and "solvent accessible surface" are introduced. The area where the ligand is touching the van der Waals surface of the receptor is called "*contact surface*." The area described by the centers of the ligand molecules touching the contact surface is called "*solvent accessible surface*." This surface is obtained by rolling a solvent molecule (usually the water molecule) with its own van der Waals volume approximated to a sphere, on the van der Waals surface of the receptor (Richards, 1977; Shrake et al., 1973). The small patches described by the imperfect contacts between the van der Waals surface of the spherical ligand molecule and the contact surface of the receptor are called "re-entrant surfaces." In 1983, Connolly described an efficient computer algorithm for sensitively calculating these areas (Connolly, 1983).

A summarization of the above said is presented in Figure 34.3.

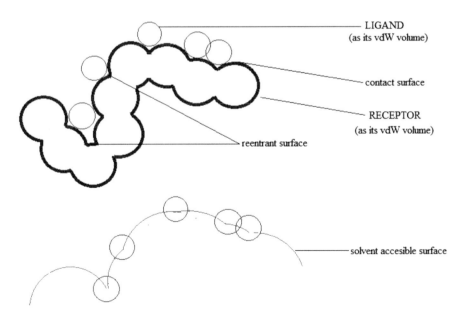

FIGURE 34.3   Described surfaces in the ligand-biomolecule interactions.

In Figure 34.3, the third atom from the left, in the receptor has no accessible surface area. These atoms are considered to be interior atoms, not part of the surface of the molecule. It is accepted that the contact surface together with the re-entrant surface are totalizing the molecular van der Waals surface.

The accessible surface area depends on the size of the ligand, as it can be seen in Figure 34.4. When the radius of the ligand increases, the number of interior atoms of the receptor increases and the accessible surface is much smoother. Thus, the smaller the ligand molecular volume is, the more of the features of the molecular van der Waals surface of the receptor are expressed.

FIGURE 34.4 Different solvent accessible areas of the same macromolecule depending on the size of the ligand.

The accessible surface areas of some individual amino acid residues are given in Table 34.2 (Uttamkumar et al., 2002).

TABLE 34.2 Values for Accessible Surface Areas of Some Amino Acid Residues Expressed in $Å^2$ in a Gly-X-Gly Tripeptide (Uttamkumar et al., 2002)

Ala	113	Gln	183	Leu	180	Ser	122
Arg	241	Glu	189	Lys	211	Thr	146
Asn	158	Gly	85	Met	204	Trp	259
Asp	151	His	194	Phe	218	Tyr	229
Cys	140	Ile	182	Pro	143	Val	160

The X-ray crystallography-determined structures of proteins are indicating that the total accessible surface area of a protein is represented by approximately two-thirds power of its total molecular weight. The surface area (A_S) of a small monomeric protein can be related to its molecular weight (M_w) by the relationship (Lee et al., 1971):

$$A_S = 6.3 M_w^{0.73} \qquad (4)$$

For oligomeric proteins, the relationship becomes (Lee et al., 1971):

$$A_S = 5.3 M_w^{0.76} \qquad (5)$$

These equations are showing that an oligomeric protein has a larger accessible surface area than a monomeric protein with the same molecular weight. The equations are accurate for the molecular range up to 35,000 (Lee et al., 1971).

For an unfolded protein, the total accessible surface area A_T is directly proportional to its molecular weight by the following relationship:

$$A_T = 1.45 \, M_W \qquad (6)$$

So it can be brought in discussion a potential surface area buried by the protein's folding, A_B, calculated with the following equations (Lee et al., 1971):

$$A_B = A_T - A_S = 1.45 M_W - 6.3 M_w^{0.73} \qquad (7)$$

This relationship indicates that 55–75% of the surface area of the unfolded protein is in fact buried in the folding process (the 3D conformation of the protein).

This Lee-Richards methodology is also applicable for the assessment of the accessible surface area in the case of nucleic acids too (Klenin et al., 2011).

34.4 TRANS-DISCIPLINARY CONNECTIONS

This characteristic of a molecule is sometimes used in QSAR studies because it can be correlated with the molar refractivity (a measure of the polarizability) of a mole of substance. The molecular volume can also be a measure of the molecular similarity so it can be used in studies of ligand-receptor interaction (http://www.ccl.net/cca/documents/molecular-modeling/node5.html).

Today, using computer molecular modeling programs, V_w can be easily displayed even for macromolecules like proteins. These models are precious for drug design or solvent interaction studies, because these big irregular volumes contain indentations, gaps, pockets or coils where the interaction with a ligand molecule or with a solvent molecule sometimes occurs. In order to design a molecule to fit the interaction site by the key-lock principle, these volumes/surfaces have to be well understood. Concepts like contact surface, re-entrant surface, and solvent accessible surface are today introduced to assess these interactions (Lee et al., 1971).

The understanding of molecular surface is also important for understanding the mechanism of action of drugs. Sometimes, the complementary principle of binding is not sufficient. There are cases when the walls of the receptor are rigid, and an efficient binding is prevented by a small volume which apparently couldn't make difference. So the volumes of the cavity and of the ligand molecule have to be complementary, but also their surfaces must match. For example, both of them must have the same character – hydrophobic or hydrophilic or intermolecular forces must settle when the two molecules approach. The novel molecular modeling programs are now "painting" the surfaces of the molecules in order to better express the appropriate interaction sites (Chang, 2005).

One study proved that in 9,9-bis (4-fluorophenyl)-3,5-dihydroxy-8-(substituted)-6,8-nonadienoic acids analogs, a class of HMG-CoA reductase inhibitors, the HMG-CoA reductase inhibitory potency is related to the shape and size of both the binding site and C8-substituent of the compound. This was demonstrated using a receptor mapping, and the van der Waals molecular volume was used to assess the shape of the ligand (Motoc et al., 1990).

Hahn proposed a 3D QSAR model which is generating a virtual receptor model, using standard ligand molecules. The virtual receptor is, in fact, a molecular surface formed by points and these points have certain mathematical characteristics that are used when computing interaction energy between a molecule and virtual receptor model. The receptor surface model is program generated to enclose the analog ligand molecules, after the QSAR indices of these molecules are evaluated. The formulas used to generate these models are using van der Waals parameters like van der Waals radii and volumes (Hahn et al., 1995).

34.5 OPEN ISSUES

Chen, C.R. et al., developed Protein Volume, a software to compute geometric volumes of proteins. These volumes are very important for finding the best ligand to a protein or in order to understand conformation changes of the macromolecule when solvated. Protein volume is generating the molecular surface of a protein and uses an algorithm to calculate the individual components of the molecular surface volume, van der Waals and intramolecular void volumes. Protein Volume is available as a free web-server at http://gmlab.bio.rpi.edu (Chen et al., 2015).

One QSAR study on anti-HIV-1 activity of 4-oxo-1,4-dihydroquinoline and 4-oxo-4*H*-pyrido[1,2-*a*] pyrimidine derivatives developed in order to explore the relationship between the structure of synthesized compounds and their anti-HIV-1 activities proved that in the series of the randomly chosen 25 compounds, the van der Waals molecular volume is one of the most important factors to influence the anti-HIV-1 activity (Hajimahdi et al., 2015).

The recent studies in this field are heading to nucleic acids. In order to understand processes like carcinogenesis, gene mutation, DNA, and RNA packing, certain drug mechanisms (which are targeting the RNA transcription, the volumes of these macromolecules, similar in a way with protein structures, have to be computed and novel computer programs are generating today 3D colored models.

Molecular shape and molecular size are two characteristics that are implied in almost all chemical processes. QSAR/QSPR models often use as descriptors total molecular surface area or total molecular volume. Computer software, nowadays, can divide the total molecular surface area (or volume) into atomic contributions and can specify some characteristics for these contributions (e.g., polar/non-polar contributions, accessible/non-accessible surfaces) so the molecular model can be significantly well assessed.

One of the most popular programs to compute the solvent accessible surface areas (SASAs) of proteins and nucleic acids at atomic or residual (multiatomic) level is POPS (parameter optimized surface). Atomic and residual area parameters have been optimized after an accurate all-atom method. Residual areas are modeled as single spheres centered at the alpha-carbon atoms for amino acids and at the phosphorus atoms for nucleotides. The input takes into consideration aspects like the number and type of atoms in the molecule, the van der Waals radius of the solvent and of the potential ligand and some empirical parameters depending on the vicinal atoms, the binding (covalent or non-covalent), etc. (Cavallo et al., 2003).

Investigations of the molecular volume and surfaces are the first and maybe the most important step in the process of understanding the molecular basis of ligand-receptor interactions.

KEYWORDS

- van der Waals radius
- van der Waals surface area

REFERENCES AND FURTHER READING

Bondi, A., (1964). Van der Waals volumes and radii. *J. Phys. Chem., 68*(3), 441–451.

Cavallo, L., Kleinjung, J., & Fraternali, F., (2003). POPS: A fast algorithm for solvent accessible surface areas at atomic and residue level. *Nucleic Acids Research, 31*, 3364–3366.

Chang, R., (2005). *Physical Chemistry for the Biosciences* (1st edn.). University Science Books: Sausalito, CA.

Chen, C. R., & Makhatadze, G. I., (2015). Protein volume: Calculating molecular Van Der Waals and void volumes in proteins. *BMC Bioinformatics, 16*, 101.

Connolly, M. I., (1983). Analytical molecular surface calculation. *Journal of Applied Crystallography, 16*(5), 548–558.

Corey, R. B., & Pauling, L., (1953). Molecular models of amino acids, peptides, and proteins. *Review of Scientific Instruments, 8*(24), 621–627.

Hahn, M., (1995). Receptor surface models. 1. Definition and construction. *J. Med. Chem., 38*, 80–90.

Hahn, M., & Rogers, D., (1995). Receptor surface models. 2. Application to quantitative structure-activity relationships studies. *J. Med. Chem., 38*, 2091–2102.

Hajimahdi, Z., Ranjbar, A., Abolfazl, A., Suratgar, A., & Zarghia, A., (2015). QSAR study on anti-HIV-1 activity of 4-Oxo–1,4-dihydroquinoline and 4-Oxo–4*H*-pyrido[1,2-*a*] pyrimidine derivatives using SW-MLR, artificial neural network and filtering methods. *Iranian Journal of Pharmaceutical Research, Winter, 14*(Suppl.), 69–75.

Housecroft, C., & Sharpe, A. G., (2012). *Inorganic Chemistry*. Pearson Education Ltd.

Klenin, K., Tristram, F., Strunk, T., & Wenzel, W., (2011). Derivatives of molecular surface area and volume: Simple and exact analytical formulas. *Journal of Computational Chemistry, 32*(12), 2647–2653.

Lee, B., & Richards, F. M., (1971). The interpretation of protein structures: Estimation of static accessibility. *Journal Molecular Biology, 55*(3), 379–400.

Motoc, I., Harte, W. E., Sit, S. Y., Balasubramanian, N., & Wright, J. J., (1990). 3-hydroxy–3-methylglutaryl-coenzyme a reductase: Three-dimensional structure-activity relationships and inhibitor design. *Mathematical and Computer Modeling, 14*, 517–521.

Oxtoby, D. W., Gillis, H. P., & Campion, A., (2008). *Principles of Modern Chemistry* (6th edn.). Thomson Brooks/Cole.

Richards, F. M., (1977). Areas, volumes, packing, and protein structure. *Annual Review of Biophysics and Bioengineering, 6*, 151–176.

Rowland, R., & Scott, T. R., (1996). Intermolecular non-bonded contact distances in organic crystal structures: Comparison with distances expected from Van Der Waals radii. *J. Phys. Chem., 100*(18), 7384–7391.

Shrake, A., & Rupley, J. A., (1973). Environment and exposure to solvent of protein atoms. *Lysozyme and Insulin, Journal of Molecular Biology, 79*(2), 351–371.

Uttamkumar, S., Ranjit, P. B., & Pinak, C., (2002). Quantifying the accessible surface area of protein residues in their local environment. *Protein Engineering, Design & Selection, 15*(8), 659–667.

Vlaia, V., Olariu, T., Vlaia, L., Butur, M., Ciubotariu, C., Medeleanu, M., & Ciubotariu, D., (2009). Quantitative structure-activity relationship (QSAR) v. Analysis of the toxicity of aliphatic esters by means of molecular compressibility descriptors. *Farmacia, 57*(5), 549. http://www.ccl.net/cca/documents/molecular-modeling/node5.html.

Index

E

O